The Center and
Focus Problem

Monographs and Research Notes in Mathematics

Series Editors:
John A. Burns, Thomas J. Tucker, Miklos Bona, Michael Ruzhansky

About the Series

This series is designed to capture new developments and summarize what is known over the entire field of mathematics, both pure and applied. It will include a broad range of monographs and research notes on current and developing topics that will appeal to academics, graduate students and practitioners. Interdisciplinary books appealing not only to the mathematical community but also to engineers, physicists and computer scientists are encouraged.

This series will maintain the highest editorial standards, publishing well-developed monographs and research notes on new topics that are final, but not yet refined into a formal monograph. The notes are meant to be a rapid means of publication for current material where the style of exposition reflects a developing topic.

Applications of Homogenization Theory to the Study of Mineralized Tissue
Robert P. Gilbert, Ana Vasilic, Sandra Klinge,
Alex Panchenko, Klaus Hackl

Semigroups of Bounded Operators and Second-Order Elliptic and Parabolic Partial Differential Equations
Luca Lorenzi, Abdelaziz Rhandi

Markov Random Flights
Alexander D. Kolesnik

Level-Crossing Problems and Inverse Gaussian Distributions
Closed-Form Results and Approximations
Vsevolod K. Malinovskii

The Center and Focus Problem
Algebraic Solutions and Hypotheses
M.N. Popa & V.V. Pricop

Abstract Calculus
A Categorical Approach
Francisco Javier Garcia-Pacheco

For more information about this series please visit: https://www.crcpress.com/ Chapman–HallCRC-Monographs-and-Research-Notes-in-Mathematics/ book-series/CRCMONRESNOT

The Center and Focus Problem

Algebraic Solutions and Hypotheses

M.N. Popa & V.V. Pricop

CRC Press
Taylor & Francis Group
Boca Raton London New York

CRC Press is an imprint of the
Taylor & Francis Group, an **informa** business

A CHAPMAN & HALL BOOK

First edition published 2022
by CRC Press
2 Park Square, Milton Park, Abingdon, Oxon, OX14 4RN

and by CRC Press
6000 Broken Sound Parkway NW, Suite 300, Boca Raton, FL 33487-2742

© 2022 M.N. Popa & V.V. Pricop

CRC Press is an imprint of Informa UK Limited

British Library Cataloguing-in-Publication Data
A catalogue record for this book is available from the British Library

ISBN: 978-1-032-01725-9 (hbk)
ISBN: 978-1-032-04410-1 (pbk)
ISBN: 978-1-003-19307-4 (ebk)

DOI: 10.1201/9781003193074

Typeset in Palatino
by codeMantra

Contents

Contents

Authors

Popa Mihail Nicolae was born on May 15, 1948, in Vălcineţ village, Călăraşi district, today Republic of Moldova. He graduated from the State University of Chisinau (today, the State University of Moldova) in 1971. Since 1975 he has been working at the Institute of Mathematics with Computing Center of the Academy of Sciences of Moldova (ASM) (today, Vladimir Andrunachievici Institute of Mathematics and Computer Science (IMCS)) in the Laboratory of Differential Equations. In the period 1980–1999, he worked as scientific secretary, 1999–2005 – deputy director, 2005–2010 – director of IMCS. From April 1, 2010, to until now, he is a senior scientific researcher at IMCS. In 1974–1979 he was a PhD student at ASM, and defended his doctoral thesis on "Affine classification of differential systems with quadratic nonlinearities" in 1979, at Gorky University (now Nizhny Novgorod, Russia). The doctor habilitation thesis entitled "Invariant processes in differential systems and their applications in qualitative theory" was defended in 1992 at the Institute of Mathematics of the National Academy of Ukraine, Kiev. Since 1996 he has been a tenured professor at the State University of Tiraspol (based in Chisinau), and in 2001 he was a visiting professor for a semester at the University of Limoges (France). He has trained 10 doctors of science.

Mihail Popa's scientific interests are related to the invariant processes in the qualitative theory of differential equations, Lie algebras and commutative graded algebras, generating functions and Hilbert series, orbit theory, Lyapunov stability theory.

Pricop Victor Vasile was born on December 1, 1981, in Antoneşti village, Ştefan-Vodă district, today the Republic of Moldova. He graduated from the State University of Moldova in 2004. Since 2005 he has been working at the "Ion Creangă" State Pedagogical University from Chişinău. Since 2008 he has been working at Vladimir Andrunachievici Institute of Mathematics and Computer Science (IMCS) in the Laboratory of Differential Equations, currently scientific researcher. In 2009–2012 he was a PhD student at IMCS, and defended his doctoral thesis on "Combinatorial and asymptotic approaches based on graduated algebras and Hilbert series, applied to differential systems" in 2014, at IMCS. Since 2017 he has been a professor at the State Institute of International Relations of Moldova.

Victor Pricop's scientific interests are related to Lie algebras and graded algebras of invariants and comitants, generating functions and Hilbert series, applications of algebras to polynomial differential systems.

Introduction

One of the old problems of the qualitative theory of differential equations is the center and focus problem. It appears, for example, when the characteristic equation of system of differential equations

$$\frac{dx}{dt} = X(x,y), \quad \frac{dy}{dt} = Y(x,y) \tag{1}$$

has a singular point with purely imaginary eigenvalues ($\lambda_{1,2} = \pm i$).

This problem was formulated by French scientist H. Poincaré (1854–1912) more than 130 years ago. It was shown that if a differential system cannot be solved explicitly, then it is possible to study the behavior of its solutions (integral curves) without knowledge of these solutions. In this way, the qualitative theory of differential equations was initiated. One of the most important problems of this theory is the study of behavior of integral curves (trajectories) around singular points, i.e. such points for which $X(x,y) = Y(x,y) = 0$. In this connection, Poincaré proposed the following classification of nondegenerate singular points: saddle, node, center and focus.

As it was noted above, the presence of center or focus at a singular point of differential system (1) is ensured by purely imaginary eigenvalues of the characteristic equation. Under this condition, in the case of a center, the singular point is surrounded by closed trajectories, and in the case of a focus, it is surrounded by spirals. The center and focus problem is to determine the condition under which a singular point is a center. In general, the center problem is algebraically unsolvable [2,19]. It should be noted that a large number of works in scientific centers of France, Russia, Belarus, China, Great Britain, Spain, Poland, Slovenia, Canada, the USA, etc. are dedicated to the center and focus problem and published in the world literature.

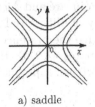
a) saddle

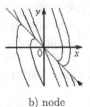
b) node

FIGURE 1
Singular points of the first type.

DOI: 10.1201/9781003193074-1

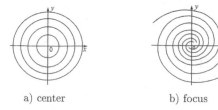

a) center   b) focus

FIGURE 2
Singular points of the second type.

In the Republic of Moldova, the first to deal with the center and focus problem for differential systems with polynomial nonlinearities was academician C. S. Sibirsky (1928–1990). His first work *On conditions for the presence of a center and a focus* (Kishinev Gos. Univ. Uch. Zap. 11, (1954), p. 115–117) caused interest to this problem in our country as well.

His PhD thesis was focused on some aspects of the center and focus problem and was defended in 1955 at the Kazan University (Russia). At different stages, the disciples of the academician C. S. Sibirsky (N. I. Vulpe, A. S. Şubă, Iu. F. Calin, V. A. Baltag, D. V. Cozma and other) examined various issues of this problem and obtained important results.

Later on, we will consider the case when the functions $X(x, y)$ and $Y(x, y)$ of differential system (1) are polynomials. The center and focus problem is algebraically solvable, if the right-hand sides of this system are nonzero. We write the considered system in the form

$$\frac{dx}{dt} = \sum_{i=0}^{\ell} P_{m_i}(x, y), \quad \frac{dy}{dt} = \sum_{i=0}^{\ell} Q_{m_i}(x, y) \ (\ell < \infty), \tag{2}$$

where P_{m_i} and Q_{m_i} are homogeneous polynomials of degree $m_i \geq 1$ in x, y, and $m_0 = 1$. The set $\{1, m_i\}_{i=1}^{\ell}$ consists of a finite number ($\ell < \infty$) of distinct natural numbers. The coefficients and variables in polynomials of system (2) take values from the fields of real numbers $\mathbb{R}$. Hereafter, we denote system (2) by $s(m_0, m_1, ..., m_\ell)$.

The fundamental results on the center and focus problem were obtained by A. M. Lyapunov (1857–1918) [20]. Henri Poincaré and Aleksandr Lyapunov laid the foundations of the qualitative theory of differential systems.

Using the methods proposed in [20,26], with the presence of singular point of the second group at the origin of coordinates $O(0, 0)$, the condition of a center is that the infinite sequence of the following polynomials is equal to zero

$$L_1, L_2, ..., L_k, ..., \tag{3}$$

whose variables are parameters of differential system (2), called focus quantities, Lyapunov's constants or Poincaré-Lyapunov's constants.

If at least one of quantities (3) is not zero, then coordinates origin $O(0,0)$ for system (2) is a focus. These conditions are necessary and sufficient.

From Hilbert's Theorem on the finiteness of basis of polynomial ideals, it follows that *the essential center conditions*, which imply vanishing of an infinite sequence of polynomials (3), consist of a finite number of polynomials, the rest ones are the consequences of them.

Considering this result, the center and focus problem can be formulated in the following way: *what finite number ω of polynomials from* (3) (*essential center conditions*)

$$L_{n_1}, L_{n_2}, ..., L_{n_\omega} \ (n_i \in \{1, 2, ..., k, ...\}; \ i = \overline{1, \omega}, \ \omega < \infty) \tag{4}$$

is necessary for their equality to zero annuls all polynomials from (3)?

Hence, the center and focus problem consists of two parts. *The first part* relates to finding the number ω that determines the upper bound of the number of focus quantities which constitute the essential center conditions. *The second part* consists in finding the set $\Omega = \{n_1, n_2, ..., n_\omega\}$ of indices n_i ($i = \overline{1, \omega}$), corresponding to the focus quantities, which constitute the essential center conditions.

The generalized center and focus problem is to determine the upper bound of the number λ of algebraically independent focus quantities from $\Pi = \{L_i : i \in \Omega\}$.

There is an opinion that if the center and focus problem is solved negatively for system $s(1, m_1, ..., m_\ell)$, having at the origin a singular point of center or focus type, then the solution of the generalized center and focus problem can be considered as the final solution of this problem.

The problem of determining essential center conditions (4) with number ω is a rather complicated problem and it is completely solved only for systems $s(1, 2)$ and $s(1, 3)$, for which we have $\omega = 3$ and 5, respectively (see, e.g., [5], [23], [45]).

Until now, it is not known the number ω for a system $s(1, 2, 3)$, which seems to be not a complicated system.

There exists a hypothesis formulated by Professor H. Żołądek, mostly based on intuition, that for system $s(1, 2, 3)$ the number ω is ≤ 13. Till now this hypothesis has not been disproved. But in [15], it was proved that the 12 focus quantities is not enough for solving the center and focus problem in the complex plane for system $s(1, 2, 3)$.

We note that initially some methods to solve the center and focus problem were proposed by Poincaré and Lyapunov, who allowed obtaining solutions for systems $s(1, 2)$ and $s(1, 3)$ and other special cases. However, the specified way in solving the center and focus problem for system $s(1, 2, 3)$ is connected with cumbersome computations with application of supercomputers. These difficulties are also insurmountable for other more complicated systems $s(1, m_1, ..., m_\ell)$.

Therefore, as a basis, it was taken the generalized center and focus problem, which was formulated above for any systems of the form $s(1, m_1, ..., m_\ell)$. This

allowed avoiding the calculation of the focus quantities (4) for given systems
and replacing this process by investigating some Lie algebras of operators and
Sibirsky graded algebras of comitants for the considered systems. To estimate
the maximal number of algebraically independent focus quantities for system
$s(1, m_1, ..., m_\ell)$, these algebras were used. As a result, a finite upper bound
for the number of algebraically independent focus quantities was obtained,
which are involved in solving the center and focus problem for any system
$s(1, m_1, ..., m_\ell)$ from (1.1)–(1.2); this was announced for the first time at an
international conference [31]. Results on the solution of the generalized center
and focus problem were also presented at the international conferences on
differential equations and algebra [30,32].

In addition, for the Lyapunov system $s\mathcal{L}(1, m_1, m_2, ..., m_\ell)$ from
(7.9)–(7.10), it was found the upper bound of the number of functionally
independent focus quantities, which are involved in solving of the center and
focus problem for these systems.

Chapter 1 (1.1–1.7) is devoted to the construction of Lie algebra of
operators of representation of a centro-affine group in the space of coefficients and variables of differential systems with polynomial nonlinearities of
the form (2).

Chapter 2 (2.1–2.6) is dedicated to the investigation of differential equations for centro-affine invariants and comitants of system (2) and to the study
of their algebraic bases.

Chapter 3 (3.1–3.5) is devoted to the study of generating functions and
Hilbert series for Sibirsky algebras of comitants and invariants of polynomial
differential systems of the form (2).

Chapter 4 (4.1–4.8) is dedicated to the construction of Hilbert series for
Sibirsky algebras of different differential systems of the form (2) and to the
computations of Krull dimension for these algebras.

Chapter 5 (5.1–5.5) contains a brief summary of the concepts related to the
new formulation of the center and focus problem for systems of the form (2).

Chapter 6 (6.1–6.8) describes the examples of differential systems for which
the upper bound of the number of algebraically independent focus quantities,
which are involved in solving of the center and focus problem, is determined.
These results are generalized for any system $s(1, m_1, ..., m_\ell)$.

Chapter 7 (7.1–7.6) is devoted to obtaining the upper bound of the number
of functionally independent focus quantities, which are involved in solving the
center and focus problem for the Lyapunov system $s\mathcal{L}(1, m_1, m_2, ..., m_\ell)$. This
estimation is compared with the results established in Chapter 6. It is worth
noting that the reader's first acquaintance with the results of Chapter 7 will
help him in better understanding the results of the first six chapters.

The main result of this book can be concisely formulated as follows: Let
$$N = 2 \sum_{i=0}^{\ell} (m_i + 1)$$ *be the maximal possible number of nonzero coefficients of*
system (2), *where* $m_0 = 1$. *Then, the number of algebraically independent focus*
quantities from (3) *does not exceed* $N - 1$, *which is the Krull dimension of*

Sibirsky algebra of comitants for system (2). *It is also shown that this number can be reduced to $N - 3$, which is the Krull dimension of Sibirsky algebra of invariants for the mentioned system. It is assumed that the number of essential focus quantities ω from* (4) *does not exceed $N - 1$, and it can be improved up to $N - 3$; their construction will begin with the first algebraically independent nonzero focus quantities obtained consecutively up to the mentioned estimations.*

In this book, modern methods of algebra have found wide application. There were precisely these methods by which academician V. A. Andrunachievici (1917–1997) inspired the first author.

The authors are extremely grateful to Professor N. I. Vulpe for useful discussions on the published articles [29], the main results of which are included in this monograph, and for advertising of the obtained results in many scientific centers of other countries.

The authors are deeply grateful to the participants of the seminar of the Institute of Mathematics and Computer Science of ASM and the Tiraspol State University (Chisinau) "Differential Equations and Algebras".

The authors are sincerely grateful to the reviewers, Professors A. S. Şubă and D. V. Cozma, for their critical comments and valuable advices in the elaboration of the present work. All disadvantages are on the authors' conscience.

Special thanks to Academician M. M. Ciobanu and journalist T. Rotaru for a popular presentation of the initial results of this book published in the article [9] to a wide range of readers.

1

Lie Algebra of Operators of Centro-Affine Group Representation in the Coefficient Space of Polynomial Differential Systems

1.1 Two-Dimensional Polynomial Differential Systems

Consider a two-dimensional autonomous polynomial system of differential equations

$$\dot{x} = \sum_{i=0}^{\ell} P_{m_i}(x,y) \equiv P(x,y),$$

$$\dot{y} = \sum_{i=0}^{\ell} Q_{m_i}(x,y) \equiv Q(x,y) \ (\ell < \infty), \tag{1.1}$$

where $\dot{x} = \frac{dx}{dt}$, $\dot{y} = \frac{dy}{dt}$. We denote by $\Gamma = \{m_i\}_{i=0}^{\ell}$ some finite set of distinct nonnegative integers and homogeneities P_{m_i} and Q_{m_i} of degree m_i with respect to the phase variables x and y (i.e. $P_{m_i}(\alpha x, \alpha y) = \alpha^{m_i} P_{m_i}(x,y)$, $Q_{m_i}(\alpha x, \alpha y) = \alpha^{m_i} Q_{m_i}(x,y)$ $\alpha \in \mathbb{R}$) on the right side of system (1.1) under $\binom{m_i}{k} = \frac{m_i!}{(m_i-k)!k!}$ given by equalities

$$P_{m_i}(x,y) = \sum_{k=0}^{m_i} \binom{m_i}{k} \overset{i1}{a_k} x^{m_i-k} y^k,$$

$$Q_{m_i}(x,y) = \sum_{k=0}^{m_i} \binom{m_i}{k} \overset{i2}{a_k} x^{m_i-k} y^k, \ (m_i \in \Gamma, \ i = \overline{0,\ell}). \tag{1.2}$$

Note that all variables and coefficients of system (1.1)–(1.2) take values from the field of real numbers $\mathbb{R}$.

For the convenience, for the systems of the form (1.1)–(1.2), in some cases, we will use the notation $s(m_0, m_1, ..., m_\ell)$ or $s(\Gamma)$, where $\Gamma = \{m_i\}_{i=0}^{\ell}$, in which you can immediately see what degree of homogeneity is contained in the right parts of these systems.

We give a simplified writing of some systems of the form (1.1)–(1.2), which will be needed in the future.

DOI: 10.1201/9781003193074-2

1.1.1 Affine System

If in (1.1)–(1.2), we take $\Gamma = \{0,1\}$, then we obtain the differential system $s(0,1)$

$$\dot{x} = a + cx + dy, \quad \dot{y} = b + ex + fy, \tag{1.3}$$

where

$$a = \overset{01}{a}_0,\ b = \overset{02}{a}_0,\ c = \overset{11}{a}_0,\ d = \overset{11}{a}_1,\ e = \overset{12}{a}_0,\ f = \overset{12}{a}_1. \tag{1.4}$$

We note that in system (1.3) a and b are called free members, $cx + dy$ and $ex + fy$ – linear parts.

1.1.2 System with Quadratic Nonlinearities

If in (1.1)–(1.2), we take $\Gamma = \{1,2\}$, then we will obtain the differential system $s(1,2)$, which, in the simplified notation, accepted in many papers will have the form

$$\begin{aligned} \dot{x} &= cx + dy + gx^2 + 2hxy + ky^2, \\ \dot{y} &= ex + fy + lx^2 + 2mxy + ny^2, \end{aligned} \tag{1.5}$$

where

$$c = \overset{01}{a}_0,\ d = \overset{01}{a}_1,\ e = \overset{02}{a}_0,\ f = \overset{02}{a}_1,\ g = \overset{11}{a}_0,\ h = \overset{11}{a}_1,\ k = \overset{11}{a}_2,$$
$$l = \overset{12}{a}_0,\ m = \overset{12}{a}_1,\ n = \overset{12}{a}_2. \tag{1.6}$$

We note that in the absence of linear part in system (1.5), we obtain the differential system $s(2)$, which is called

1.1.3 Quadratic System

Which can be written as

$$\begin{aligned} \dot{x} &= gx^2 + 2hxy + ky^2, \\ \dot{y} &= lx^2 + 2mxy + ny^2. \end{aligned} \tag{1.7}$$

1.1.4 System with Cubic Nonlinearities

If in (1.1)–(1.2), we take $\Gamma = \{1,3\}$, then we will obtain the differential system $s(1,3)$, which, in the simplified notation, accepted in many papers and will have the form

$$\begin{aligned} \dot{x} &= cx + dy + px^3 + 3qx^2y + 3rxy^2 + sy^3, \\ \dot{y} &= ex + fy + tx^3 + 3ux^2y + 3vxy^2 + wy^3, \end{aligned} \tag{1.8}$$

where

$$c = \overset{01}{a}_0,\ d = \overset{01}{a}_1,\ e = \overset{02}{a}_0,\ f = \overset{02}{a}_1,\ p = \overset{11}{a}_0,\ q = \overset{11}{a}_1,\ r = \overset{11}{a}_2,$$
$$s = \overset{11}{a}_3,\ t = \overset{12}{a}_0,\ u = \overset{12}{a}_1,\ v = \overset{12}{a}_2,\ w = \overset{12}{a}_3. \tag{1.9}$$

We note that in the absence of linear part in system (1.8), we obtain the differential system $s(3)$, which is called

1.1.5 Cubic System

Which can be written as

$$\dot{x} = px^3 + 3qx^2y + 3rxy^2 + sy^3,$$
$$\dot{y} = tx^3 + 3ux^2y + 3vxy^2 + wy^3. \tag{1.10}$$

In the future, other systems of the form (1) will be considered with various $\Gamma = \{m_i\}_{i=0}^{\ell}$, other than the abovementioned ones.

1.2 One-Parameter Linear Groups of Transformations of the Phase Plane of System (1.1)–(1.2)

Consider the transformations included in the one-parameter family $\{T_\alpha\}$:

$$\overline{x} = f^1(x, y, \alpha), \ \overline{y} = f^2(x, y, \alpha), \tag{1.11}$$

where α – a real parameter continuously changing in a certain range of $\mathbb{R}$. To each value of the parameter α, there corresponds some transformation of family. Transformation (1.11) of the phase plane $E^2(x, y)$ means that the point (x, y) is transferred into new position $(\overline{x}, \overline{y})$ in the same plane $E^2(x, y)$.

Definition 1.1. *Let's say that the family $G_1 = \{T_\alpha\}$, consisting of functions (1.11), continuously dependent on parameter α, forms a one-parameter transformation group G_1, if*

(1) $T_\alpha T_\beta = T_\gamma$, *where $T_\gamma \in \{T_\alpha\}$ and $\gamma = \varphi(\alpha, \beta)$ are considered differentiable for the enough number of times;*

(2) $T_{\alpha_0} = I$ *(or $T_0 = I$) (existence of a unity);*

(3) $T_\alpha^{-1} = T_{\alpha^{-1}}$ *(existence of the inverse element);*

(4) $T_\alpha(T_\beta T_\gamma) = (T_\alpha T_\beta)T_\gamma$ *(associativity of multiplication in a group).*

Note that α^{-1} denotes the value of the parameter corresponding to the inverse transformation, and the condition (2) means the existence of a unique value of parameter α that guarantees the identity transformation in a group.

Example 1.1. Consider the change of variables

$$\overline{x} = \mu x, \ \overline{y} = y(\mu \in \mathbb{R}\backslash\{0\}). \tag{1.12}$$

Take now two particular values μ and μ' of parameter from $\mathbb{R}\backslash\{0\}$ and successively apply substitution (1.12) and substitution $T_{\mu'}$

$$\overline{\overline{x}} = \mu'\overline{x}, \ \overline{\overline{y}} = \overline{y}(\mu' \in \mathbb{R}\backslash\{0\}). \tag{1.13}$$

Substituting (1.12) in these equalities, we obtain

$$\overline{\overline{x}} = \mu\mu'x, \overline{\overline{y}} = y(\mu, \mu' \in \mathbb{R}\backslash\{0\}).$$

This shows that the result of applying two consecutive changes of variables (1.12) and (1.13) is identical to the result of applying the third transformation of this family with the value of the parameter $\mu'' = \mu\mu'$. It is symbolically written as $T_\mu T_{\mu'} = T_{\mu''}$, and they say that substitution (1.12) determines *a group property*. The existence of a unity in (1.12) is determined by the value of the parameter $\mu_0 = 1$. Backward substitution for (1.12) has the form $x = \mu^{-1}\overline{x}$, $y = \overline{y}$, i.e. $T_\mu^{-1} = T_{\mu^{-1}}$ and $\mu^{-1} = 1/\mu$. In this case, it is said that replacement (1.12) *forms a transformation*. To prove the associativity of transformation (1.12), another transformation $T_{\mu''}$ is taken besides (1.13):

$$\overline{\overline{\overline{x}}} = \mu''\overline{x}, \overline{\overline{\overline{y}}} = \overline{\overline{y}} \ (\mu'' \in \mathbb{R}\backslash\{0\})$$

and the property (4) of Definition 1.1 is directly checked.

Therefore, the family $\{T_\mu\}$, given by transformation (1.12), forms a group which we denote by $M(2,\mathbb{R})$.

Example 1.2. Consider the replacing of variables defined by a family $\{T_z\}$:

$$\overline{x} = x + zy, \overline{y} = y \ (z \in \mathbb{R}). \tag{1.14}$$

Similar to the previous Example 1.1, it can be shown that all the conditions of Definition 1.1 are performed, and family of substitutions (1.14) forms a group of continuous transformations from the parameter z, which we define through $Z_+(2,\mathbb{R})$.

It should be noted that in the case of transformation (1.14), the inverse transformation T_z^{-1} will consist of

$$x = \overline{x} - z\overline{y}, y = \overline{y}. \tag{1.15}$$

In this case, the value of the parameter corresponding to the inverse transformation will be $z^{-1} = -z$.

In the future, we need two more one-parameter transformation groups, given by the following examples:

Example 1.3. The group of transformation $Z_-(2,\mathbb{R})$, defined by the family of substitution $\{T_h\}$:

$$\overline{x} = x, \overline{y} = hx + y \ (h \in \mathbb{R}). \tag{1.16}$$

Example 1.4. The group of transformation $L(2,\mathbb{R})$, defined by the family of substitution $\{T_\lambda\}$:

$$\overline{x} = x, \overline{y} = \lambda y \ (\lambda \in \mathbb{R}\backslash\{0\}). \tag{1.17}$$

Remark 1.1. *The groups* $M(2,\mathbb{R})$, $Z_+(2,\mathbb{R})$, $Z_-(2,\mathbb{R})$, $L(2,\mathbb{R})$ *will be called linear one-parameter groups.*

1.3   Centro-Affine and Unimodular Transformation Groups of the Phase Plane of System (1.1)–(1.2)

In this section, we give two linear groups that continuously depend on more than one parameter, the number of which is finite. If the number of parameters is r and there is no way to reduce it, the group is called r–parametric. The definition of r–parametric group coincides with Definition 2.1; only in this case, under the parameter α we will understand a certain vector with r–coordinates.

Example 1.5. $GL(2, \mathbb{R})$ – group of all centro-affine transformations of the phase plane $E^2(x, y)$:

$$\overline{x} = \alpha x + \beta y, \; \overline{y} = \gamma x + \delta y,$$

$$\left(\Delta = det \begin{pmatrix} \alpha & \beta \\ \gamma & \delta \end{pmatrix} \neq 0 \right), \tag{1.18}$$

where $\overline{x}$ and $\overline{y}$ are new variables, and $\alpha, \beta, \gamma, \delta \in \mathbb{R}$ get continuously changing values.

Check the execution of conditions of Definition 1.1 from family (1.18). If we consider the product of successive transformations (1.18) and

$$\overline{\overline{x}} = \alpha' \overline{x} + \beta' \overline{y}, \; \overline{\overline{y}} = \gamma' \overline{x} + \delta' \overline{y},$$

$$\left(\Delta' = det \begin{pmatrix} \alpha' & \beta' \\ \gamma' & \delta' \end{pmatrix} \neq 0 \right), \tag{1.19}$$

given by equalities

$$\overline{\overline{x}} = \alpha'(\alpha x + \beta y) + \beta'(\gamma x + \delta y) = \alpha'' x + \beta'' y,$$
$$\overline{\overline{y}} = \gamma'(\alpha x + \beta y) + \delta'(\gamma x + \delta y) = \gamma'' x + \delta'' y, \tag{1.20}$$

then for their matrices, we have

$$\begin{pmatrix} \alpha'' & \beta'' \\ \gamma'' & \delta'' \end{pmatrix} = \begin{pmatrix} \alpha\alpha' + \beta'\gamma & \alpha'\beta + \beta'\delta \\ \alpha\gamma' + \gamma\delta' & \beta\gamma' + \delta\delta' \end{pmatrix}, \tag{1.21}$$

or which is the same

$$\begin{pmatrix} \alpha'' & \beta'' \\ \gamma'' & \delta'' \end{pmatrix} = \begin{pmatrix} \alpha' & \beta' \\ \gamma' & \delta' \end{pmatrix} \begin{pmatrix} \alpha & \beta \\ \gamma & \delta \end{pmatrix}. \tag{1.22}$$

If we denote by Δ'' the determinant of matrix (1.21) for transformation (1.20), then from (1.22), we have

$$\Delta'' = \Delta'\Delta, \tag{1.23}$$

and this determinant, according to (1.18) and (1.19), is nonzero. Therefore, transformation (1.20) is also a centro-affine transformation, i.e., the first condition of Definition 1.1 is satisfied.

A transformation with the following parameters is taken as a unity of family (1.18) (identity centro-affine transformation):

$$\alpha = 1,\ \beta = 0,\ \gamma = 0,\ \delta = 1. \qquad (1.24)$$

With the help of (1.18), one can show the existence of the inverse element

$$x = \frac{\delta}{\Delta}\overline{x} + \frac{\beta'}{\Delta}\overline{y},\ y = \frac{\gamma'}{\Delta}\overline{x} + \frac{\alpha}{\Delta}\overline{y}, \qquad (1.25)$$

where

$$\beta' = -\beta,\ \gamma' = -\gamma. \qquad (1.26)$$

Similar to the first condition, one can check the associativity in the group.

We note that the parameters α, β, γ, δ in (1.18) are independent since $\Delta \neq 0$. Therefore, the number of parameters in this group cannot be reduced. This shows that the group $GL(2,\mathbb{R})$ is a four-parameter group.

Example 1.6. $SL(2,\mathbb{R})$ – a group of all unimodular transformations of the phase plane $E^2(x,y)$:

$$\overline{x} = \alpha x + \beta y,\ \overline{y} = \gamma x + \delta y,$$

$$\left(\Delta = det \begin{pmatrix} \alpha & \beta \\ \gamma & \delta \end{pmatrix} = 1 \right), \qquad (1.27)$$

where $\overline{x}$ and $\overline{y}$ are new variables, and α, β, γ, $\delta \in \mathbb{R}$ get continuously changing values.

We note that since $\Delta = 1$, then the number of parameters can be reduced by one. Therefore, the unimodular group is a three-parameter group.

Since all unimodular transformations are contained in the group of centro-affine transformations, we will say that they form a subgroup in the group of all centro-affine transformations.

Remark 1.2. *Each centro-affine transformation (1.18) can be considered as a composite of two transformations:*

$$\overline{\overline{x}} = |\Delta|^{\frac{1}{2}} x,\ \overline{\overline{y}} = |\Delta|^{\frac{1}{2}} y \qquad (1.28)$$

and

$$\overline{x} = \frac{\alpha}{|\Delta|^{\frac{1}{2}}}\overline{\overline{x}} + \frac{\beta}{|\Delta|^{\frac{1}{2}}}\overline{\overline{y}},\ \overline{y} = \frac{\gamma}{|\Delta|^{\frac{1}{2}}}\overline{\overline{x}} + \frac{\delta}{|\Delta|^{\frac{1}{2}}}\overline{\overline{y}}, \qquad (1.29)$$

where the second of them is unimodular.

For proof of Remark 1.2, it is enough to calculate the product of successive transformations (1.28) and (1.29).

We will consider transformations (1.18) or (1.27) to be given, if transformation matrices are given by $q = \begin{pmatrix} \alpha & \beta \\ \gamma & \delta \end{pmatrix}$, and write $q \in GL(2,\mathbb{R})$, or $q \in SL(2,\mathbb{R})$.

Following [1], we can show that there takes place

Theorem 1.1. *Transformation* (1.18), *belonging to the four-parameter group* $GL(2,\mathbb{R})$, *can be represented as a product of transformations* (1.12), (1.14), (1.16) *and* (1.17), *belonging, respectively, to one-parameter groups* $M(2,\mathbb{R})$, $Z_+(2,\mathbb{R})$, $Z_-(2,\mathbb{R})$, $L(2,\mathbb{R})$.

Proof. Denote matrices corresponding to transformations (1.12), (1.14), (1.16) and (1.17), by

$$q_1 = \begin{pmatrix} \mu & 0 \\ 0 & 1 \end{pmatrix}, \ q_2 = \begin{pmatrix} 1 & z \\ 0 & 1 \end{pmatrix},$$

$$q_3 = \begin{pmatrix} 1 & 0 \\ h & 1 \end{pmatrix}, \ q_4 = \begin{pmatrix} 1 & 0 \\ 0 & \lambda \end{pmatrix}, \tag{1.30}$$

where μ, $\lambda \in \mathbb{R}\backslash\{0\}$. It is obvious that q_1, q_2, q_3, $q_4 \in GL(2,\mathbb{R})$. We note that performing transformations with matrices (1.30) in the following order $q_4((q_1 q_2)q_3)$ gives a transformation with matrix

$$q = \begin{pmatrix} \mu + \mu z h & \mu z \\ \lambda h & \lambda \end{pmatrix}. \tag{1.31}$$

In order that transformations (1.18) with the matrix $q = \begin{pmatrix} \alpha & \beta \\ \gamma & \delta \end{pmatrix} \in GL(2,\mathbb{R})$ will be presented as (1.31), it is enough to take

$$\mu = \frac{\Delta}{\delta}, \ z = \frac{\beta \delta}{\Delta}, \ h = \frac{\gamma}{\delta}, \ \lambda = \delta \, (\delta \neq 0).$$

If $\delta = 0$, then $\Delta = \beta\gamma \neq 0$, and for this transformation, there can be written a matrix

$$q = \begin{pmatrix} \alpha & \beta \\ \frac{\Delta}{\beta} & 0 \end{pmatrix},$$

which is equal to the product

$$q_3 \cdot \begin{pmatrix} \alpha & \beta \\ -\alpha h + \frac{\Delta}{\beta} & -\beta h \end{pmatrix}.$$

According to the previous arguments, we find that the second factor in this product can be represented as $q_4((q_1 q_2)q_3)$, because in equality

$$\begin{pmatrix} \alpha & \beta \\ -\alpha h + \frac{\Delta}{\beta} & -\beta h \end{pmatrix} = \begin{pmatrix} \alpha_1 & \beta_1 \\ \gamma_1 & \delta_1 \end{pmatrix}$$

the parameter $\delta_1 = -\beta h$ is not zero, as in the case of arbitrary $h \, \beta = 0$, and it entails $\Delta = \beta\gamma = 0$. Then, Theorem 1.1 is proven.

1.4   Lie Operators of One-Parameter Linear Groups and Their Representations in the Coefficient Space of System (1.1)–(1.2)

Suppose that transformations (1.11) form a one-parameter linear group. It is clear that after transformation with elements of this group in system (1.1)–(1.2), this system does not change its form and can be written as follows:

$$
\dot{\bar{x}} = \sum_{i=0}^{\ell} \sum_{k=0}^{m_i} \binom{m_i}{k} \overset{i}{b}_k^1 \bar{x}^{m_i-k} \, \bar{y}^k,
$$

$$
\dot{\bar{y}} = \sum_{i=0}^{\ell} \sum_{k=0}^{m_i} \binom{m_i}{k} \overset{i}{b}_k^2 \bar{x}^{m_i-k} \, \bar{y}^k.
$$

(1.32)

Remark 1.3. *We note that the coefficients $\overset{i}{b}_k^j$ ($j = 1, 2$) of system (1.32) are linear functions of coefficients of system (1.1)–(1.2) with coefficients depending on the parameter α. The last statement can be written in the form*

$$
\overset{i}{b}_k^j = \overset{i}{g}_k^j(A, \alpha) \, (i = \overline{0, \ell}; \; j = 1, 2; \; k = \overline{0, m_i}),
$$

(1.33)

where by A the set of coefficients of the right parts of system (1.1)–(1.2) is denoted.

    We note that equalities (1.33) define some group of linear transformations of the space of coefficients $E^N(A)$ of system (1.1)–(1.2), homomorphic with group (1.11) or, as they say, relations (1.33) define some linear representation of group (1.11) in the space $E^N(A)$. Similarly, a linear representation of the r–parametric linear group (1.11) is defined with $\alpha = (\alpha^1, \alpha^2, ..., \alpha^r)$ on the space $E^N(A)$.

    By N we denote the number of coefficients of the right parts of system (1.1)–(1.2), which is defined by the equality

$$
N = 2 \left(\sum_{i=0}^{\ell} m_i + \ell + 1 \right).
$$

(1.34)

Suppose that in group (1.11), identity transformation is provided by $\alpha = 0$ (in the paper [24] it is proved that in any one-parameter group, you can take a new parameter with this condition, i.e., if $T_{\alpha_0} = I$ and $\alpha_0 \neq 0$, then with parameter redefinition, we can achieve $T_0 = I$).

    We decompose functions (1.11) and (1.33) in a Taylor series by parameter α in neighborhood of $\alpha = 0$. By condition $T_0 = I$, we have $f^1(x, y, 0) = x$, $f^2(x, y, 0) = y$ and $\overset{ij}{g}_k(A, 0) = \overset{ij}{a}_k$ ($i = \overline{0, \ell}; j = 1, 2; k = \overline{0, m_i}$). Therefore, by denoting

$$\xi^j(x,y) = \left.\frac{\partial f^j(x,y,\alpha)}{\partial \alpha}\right|_{\alpha=0}, \; \overset{ij}{\eta_k}(A) = \left.\frac{\partial \overset{ij}{g_k}(A,\alpha)}{\partial \alpha}\right|_{\alpha=0} \tag{1.35}$$

$$(i = \overline{0,\ell}; \; j = 1,2; \; k = \overline{0,m_i}),$$

we write transformations (1.11) and (1.33) in the form

$$\overline{x} = x + \xi^1(x,y)\alpha + o(\alpha), \; \overline{y} = y + \xi^2(x,y)\alpha + o(\alpha),$$

$$\overset{ij}{b_k} = \overset{ij}{a_k} + \overset{ij}{\eta_k}(A)\alpha + o(\alpha) \; (i = \overline{0,\ell}; \; j = 1,2; \; k = \overline{0,m_i}). \tag{1.36}$$

If we denote by B, a set of the coefficients of the right parts of systems (1.32), then, in this case, they say that groups (1.11) and (1.33) are determined by their tangent vector field $(\xi,\eta) = \left(\xi^1, \xi^2, \overset{ij}{\eta_k}\right)$ $(i = \overline{0,\ell}; \; j = 1,2; \; k = \overline{0,m_i})$, since by the formulas (1.35), the tangent vector is set at the point (x,y,A) to a curve described by the points $(\overline{x}, \overline{y}, B)$ with group transformation (1.11) and (1.33).

One-parameter groups (1.11) and (1.33) are fully recovered if the coordinates of the vector (ξ,η) are known. This process is carried out using the Lie equations with the initial condition [34]:

$$\frac{d\overline{x}}{d\alpha} = \xi^1(\overline{x}, \overline{y}), \; \overline{x}|_{\alpha=0} = x,$$

$$\frac{d\overline{y}}{d\alpha} = \xi^2(\overline{x}, \overline{y}), \; \overline{y}|_{\alpha=0} = y, \tag{1.37}$$

$$\frac{dB}{d\alpha} = \eta(B), \; B|_{\alpha=0} = A.$$

For any one-parameter group (1.11) and its representation (1.33), Lie equations (1.37) are written in a unique way and vice versa.

The tangent vector field (space) $(\xi,\eta) = \left(\xi^1, \xi^2, \overset{ij}{\eta_k}\right)$ $(i = \overline{0,\ell}; \; j = 1,2;$ $k = \overline{0,m_i})$ is also written using the first-order differential operator:

$$X = \xi^1(x,y)\frac{\partial}{\partial x} + \xi^2(x,y)\frac{\partial}{\partial y} + D, \tag{1.38}$$

where

$$D = \sum_{j=1}^{2}\sum_{i=0}^{\ell}\sum_{k=0}^{m_i} \overset{ij}{\eta_k}(A)\frac{\partial}{\partial \overset{ij}{a_k}}. \tag{1.39}$$

The functions $\xi^j(x,y)$ and $\overset{ij}{\eta_k}(A)$ are called coordinates of operators (1.38) and (1.39), and operator (1.39) is called operator of group representation (1.11) in the space of the coefficients $E^N(A)$ of system (1.1)–(1.2).

The operator X (D) is called the infinitesimal Lie operator or simply Lie operator of a group of transformations (1.11) in the space $E^{N+2}(x,y,A)$ $(E^N(A))$, where N is from (1.34).

1.5 Operators of Representation of the Linear Groups (1.12), (1.14), (1.16) and (1.17) in the Space of Variables and Coefficients of System (1.1)–(1.2)

Theorem 1.2. *The operator of representation of the group $M(2, \mathbb{R})$ from (1.12) in the space $E^{N+2}(x, y, A)$ of system (1.1)–(1.2) has the form*

$$X_1 = x\frac{\partial}{\partial x} + D_1, \tag{1.40}$$

where

$$D_1 = \sum_{i=0}^{\ell}\sum_{k=0}^{m_i}\left[(-m_i + k + 1)\overset{i1}{a_k}\frac{\partial}{\partial \overset{i1}{a_k}} + (-m_i + k)\overset{i2}{a_k}\frac{\partial}{\partial \overset{i2}{a_k}}\right]. \tag{1.41}$$

Proof. With transformation (1.12) in system (1.1)–(1.2), the coefficients of system (1.32) can be written as

$$\overset{i1}{b_k} = \mu^{-m_i+k+1}\overset{i1}{a_k}, \; \overset{i2}{b_k} = \mu^{-m_i+k}\overset{i2}{a_k} \; (i = \overline{0, \ell}, \; k = \overline{0, m_i}). \tag{1.42}$$

Note that for transformations (1.12) and (1.42) identity transformation is obtained with $\mu_0 = 1$. If we take $|\mu| = exp(\overline{\mu})$ ($\overline{\mu} \in \mathbb{R}$), then from (1.12) and (1.42) we have the following family of transformations $\{T_\mu\}$:

$$\overline{x} = xexp(\overline{\mu}), \; \overline{y} = y, \; \overset{i1}{b_k} = \overset{i1}{a_k}exp(-m_i + k + 1)\overline{\mu},$$
$$\overset{i2}{b_k} = \overset{i2}{a_k}exp(-m_i + k)\overline{\mu} \; (i = \overline{0, \ell}, \; k = \overline{0, m_i}), \; \overline{\mu} \in \mathbb{R}, \tag{1.43}$$

that satisfy the condition $T_0 = I$, and herewith $\overline{\mu}^{-1} = -\overline{\mu}$. Since in (1.43), $\overline{\mu} = 0$ provides identity transformation in the space $E^{N+2}(x, y, A)$, then taking into account (1.35) from (1.43), we have

$$\xi^1 = x, \; \xi^2 = 0, \; \overset{i1}{\eta_k} = (-m_i + k + 1)\overset{i1}{a_k}, \; \overset{i2}{\eta_k} = (-m_i + k)\overset{i2}{a_k}.$$

Substituting these equalities into (1.38)–(1.39) we find (1.40)–(1.41). Theorem 1.2 is proved.

Consequence 1.1. *The operator of representation of the group $M(2, \mathbb{R})$ in the space of coefficients $E^N(A)$ of system (1.1)–(1.2) has the form (1.41).*

Theorem 1.3. *The operator of representation of the group $Z_+(2, \mathbb{R})$ from (1.14) in the space $E^{N+2}(x, y, A)$ of system (1.1)–(1.2) has the form*

$$X_2 = y\frac{\partial}{\partial x} + D_2, \tag{1.44}$$

where

$$D_2 = \sum_{i=0}^{\ell} \sum_{k=0}^{m_i} \left[\left(\overset{i}{a}_k^2 - k \overset{i}{a}_{k-1}^1 \right) \frac{\partial}{\partial \overset{i}{a}_k^1} - k \overset{i}{a}_{k-1}^2 \frac{\partial}{\partial \overset{i}{a}_k^2} \right]. \tag{1.45}$$

Proof. If we make transformation (1.14) in system (1.1)–(1.2), then the coefficients of the resulting system (1.32) can be written as

$$\overset{i}{b}_k^1 = \overset{i}{a}_k^1 + (\overset{i}{a}_k^2 - k \overset{i}{a}_{k-1}^1)z + o(z), \quad \overset{i}{b}_k^2 = \overset{i}{a}_k^2 - k \overset{i}{a}_{k-1}^2 z + o(z) \tag{1.46}$$
$$(i = \overline{0, \ell}, \ k = \overline{0, m_i}),$$

where $o(z)$ is a polynomial containing z in the degree of at least two in all members.

Using (1.14), (1.35) and (1.46), the coordinates of the tangent vector (ξ, η) of the group $Z_+(2, \mathbb{R})$ are built, which have the form

$$\xi^1 = y, \ \xi^2 = 0, \ \overset{i}{\eta}_k^1 = \overset{i}{a}_k^2 - k \overset{i}{a}_{k-1}^1, \ \overset{i}{\eta}_k^2 = -k \overset{i}{a}_{k-1}^2.$$

Substituting these equalities into (1.38)–(1.39), we find (1.44)–(1.45). Theorem 1.3 is proved.

Consequence 1.2. *The operator of representation of the group $Z_+(2, \mathbb{R})$ in the space of coefficients $E^N(A)$ of system (1.1)–(1.2) has the form (1.45).*

Theorem 1.4. *The operator of representation of the group $Z_-(2, \mathbb{R})$ from (1.16) in the space $E^{N+2}(x, y, A)$ of system (1.1)–(1.2) has the form*

$$X_3 = x \frac{\partial}{\partial y} + D_3, \tag{1.47}$$

where

$$D_3 = \sum_{i=0}^{\ell} \sum_{k=0}^{m_i} \left[(-m_i + k) \overset{i}{a}_{k+1}^1 \frac{\partial}{\partial \overset{i}{a}_k^1} + (\overset{i}{a}_k^1 + (-m_i + k) \overset{i}{a}_{k+1}^2) \frac{\partial}{\overset{i}{a}_k^2} \right]. \tag{1.48}$$

Proof. If we make a transformation (1.16) in system (1.1)–(1.2), then the coefficients of the resulting system (1.32) can be written as

$$\overset{i}{b}_k^1 = \overset{i}{a}_k^1 - (m_i - k) \overset{i}{a}_{k+1}^1 h + o(h),$$
$$\overset{i}{b}_k^2 = \overset{i}{a}_k^2 + [\overset{i}{a}_k^1 - (m_i - k) \overset{i}{a}_{k+1}^2] h + o(h) \ (i = \overline{0, \ell}, \ k = \overline{0, m_i}), \tag{1.49}$$

where $o(h)$ is a polynomial containing h in the degree of at least two in all members.

Using (1.16), (1.35) and (1.49), the coordinates of the tangent vector (ξ, η) of the $Z_-(2, \mathbb{R})$ are built, which have the form

$$\xi^1 = 0, \ \xi^2 = x, \ \overset{i}{\eta}_k^1 = (-m_i + k) \overset{i}{a}_{k+1}^1, \ \overset{i}{\eta}_k^2 = \overset{i}{a}_k^1 - (m_i - k) \overset{i}{a}_{k+1}^2.$$

Substituting these equalities into (1.38)–(1.39), we find (1.47)–(1.48). Theorem 1.4 is proved.

Consequence 1.3. *The operator of representation of the group $Z_-(2,\mathbb{R})$ in the space of coefficients $E^N(A)$ of system (1.1)–(1.2) has the form (1.48).*

Theorem 1.5. *The operator of representation of the group $L(2,\mathbb{R})$ from (1.17) in the space $E^{N+2}(x,y,A)$ of system (1.1)–(1.2) has the form*

$$X_4 = y\frac{\partial}{\partial y} + D_4, \tag{1.50}$$

where

$$D_4 = \sum_{i=0}^{\ell}\sum_{k=0}^{m_i}\left[-k\overset{i1}{a_k}\frac{\partial}{\partial\overset{i1}{a_k}} - (k-1)\overset{i2}{a_k}\frac{\partial}{\partial\overset{i2}{a_k}}\right]. \tag{1.51}$$

Proof. With transformation (1.17) in system (1.1)–(1.2), the coefficients of the resulting system (1.32) can be written as

$$\overset{i1}{b_k} = \lambda^{-k}\overset{i1}{a_k},\ \overset{i2}{b_k} = \lambda^{-k+1}\overset{i2}{a_k}\ (i=\overline{0,\ell};\ k=\overline{0,m_i}). \tag{1.52}$$

We note that for transformations (1.17) and (1.52), the identity transformation is obtained by $\lambda_0 = 1$. If $|\lambda| = exp\overline{\lambda}$, $(\overline{\lambda}\in\mathbb{R})$, then from (1.17) and (1.52) we have the following family of transformations $\{T_{\overline{\lambda}}\}$:

$$\overline{x} = x,\ \overline{y} = y\,exp\overline{\lambda},\ \overset{i1}{b_k} = \overset{i1}{a_k}exp(-k\overline{\lambda}),\ \overset{i2}{b_k} = \overset{i2}{a_k}exp(-k+1)\overline{\lambda}$$
$$(i=\overline{0,\ell};\ k=\overline{0,m_i}),\ \overline{\lambda}\in\mathbb{R}, \tag{1.53}$$

that satisfies the condition $T_0 = I$, and at the same time $\overline{\lambda}^{-1} = -\overline{\lambda}$.

Since in (1.53), $\overline{\lambda} = 0$ provides identity transformation in the space $E^{N+2}(x,y,A)$, then taking into account (1.35), from (1.53) we have

$$\xi^1 = 0,\ \xi^2 = y,\ \overset{i1}{\eta_k} = -k\overset{i1}{a_k},\ \overset{i2}{\eta_k} = (-k+1)\overset{i2}{a_k}.$$

Substituting these equalities into (1.38)–(1.39), we find (1.50)–(1.51). Theorem 1.5 is proved.

Consequence 1.4. *The operator of representation of the group $L(2,\mathbb{R})$ from (1.17) in the space of coefficients $E^N(A)$ of system (1.1)–(1.2) has the form (1.51).*

1.6 Lie Algebra of Operators of Centro-Affine Group Representation in the Space of Variables and Coefficients of System (1.1)–(1.2)

Definition 1.2. *Following L. V. Ovsyannikov* [24], *let's say that a linear space L on the field* $\mathbb{R}$ *is called Lie algebra, if for any two of its elements* u, v *the commutation operation* $[u, v]$ *is defined, giving again an element L* (*commutator of elements u, v*) *and satisfying the following axioms:*

(1) *bilinearity: for any* $u, v, w \in L$ *and* $\alpha, \beta \in R$

$$[\alpha u + \beta v, w] = \alpha[u, w] + \beta[v, w], \quad [u, \alpha v + \beta w] = \alpha[u, v] + \beta[u, w];$$

(2) *antisymmetry: for any* $u, v \in L$

$$[u, v] = -[v, u];$$

(3) *Jacobi identity justice: for any* $u, v, w \in L$

$$[[u, v], w] + [[v, w], u] + [[w, u], v] = 0.$$

The dimension of Lie algebra L is the dimension of its vector space L, and in the case of a finite dimension r, this algebra is denoted by the symbol L_r and is called finite-dimensional.

In Section 1.5, a description of Lie operators of representation of one-parameter groups $M(2, \mathbb{R})$, $Z_+(2, \mathbb{R})$, $Z_-(2, \mathbb{R})$, $L(2, \mathbb{R})$ in the spaces $E^{N+2}(x, y, A)$ and $E^N(A)$ of system (1.1)–(1.2) was given, where N is from (1.34). These operators are the first-order differential operators whose general form can be written as follows:

$$Y(F) = \sum_{i=1}^{N+2} P_i \frac{\partial F}{\partial y_i}, \tag{1.54}$$

and a vector

$$(y_1, y_2, ..., y_{N+2}) = (x, y, \overset{0}{a_0^1}, \overset{0}{a_1^1}, ..., \overset{\ell_2}{a_{m_\ell}^2}) \in E^{N+2}(x, y, A) \tag{1.55}$$

and P_i – are polynomials in $y_1, y_2, ..., y_{N+2}$.

It is evident that

$$Y(F_1 + F_2) = Y(F_1) + Y(F_2),$$
$$Y(F_1 F_2) = F_1 Y(F_2) + F_2 Y(F_1), \quad Y(\alpha) = 0, \tag{1.56}$$

if F_1, F_2 – are functions of $y_1, y_2, ..., y_{N+2}$, and $\alpha \in \mathbb{R}$. A composition $Y_1 Y_2$ of two differential operators of the form (1.54) is again a differential operator, but if Y_1 and Y_2 have order 1, then $Y_1 Y_2$ will have order 2, since it will already include the second derivatives. However, the fact that the commutator

$$[Y_1, Y_2] = Y_1 Y_2 - Y_2 Y_1 \tag{1.57}$$

is again the first-order operator, follows from the fact that if for Y_1, Y_2 relations (1.56) are fulfilled, then they are fulfilled for $[Y_1, Y_2]$ as well.

If we set operators Y_1 and Y_2 in a coordinate notation,

$$Y_1 = \sum_{i=1}^{N+2} P_i \frac{\partial}{\partial y_i}, \quad Y_2 = \sum_{i=1}^{N+2} Q_i \frac{\partial}{\partial y_i}, \tag{1.58}$$

where P_i and Q_i – are functions of $y_1, y_2, ..., y_{N+2}$, then taking into account (1.56) and (1.57), we have

$$[Y_1, Y_2] = \sum_{i=1}^{N+2} R_i \frac{\partial}{\partial y_i}, \quad R_i = \sum_{i=1}^{N+2} \left(P_k \frac{\partial Q_i}{\partial y_k} - Q_k \frac{\partial P_i}{\partial y_k} \right). \tag{1.59}$$

This directly shows that $[Y_1, Y_2]$ is the first-order operator. This can be illustrated by:

Example 1.7. Writing operators (1.40)–(1.41), (1.44)–(1.45), (1.47)–(1.48), (1.50)–(1.51) in the space $E^8(x, y, A)$ of system (1.3), where

$$A = (a, b, c, d, e, f), \tag{1.60}$$

we have

$$X_1 = x \frac{\partial}{\partial x} + D_1, \quad X_2 = y \frac{\partial}{\partial x} + D_2,$$
$$X_3 = x \frac{\partial}{\partial y} + D_3, \quad X_4 = y \frac{\partial}{\partial y} + D_4, \tag{1.61}$$

and

$$D_1 = a \frac{\partial}{\partial a} + d \frac{\partial}{\partial d} - e \frac{\partial}{\partial e},$$
$$D_2 = b \frac{\partial}{\partial a} + e \frac{\partial}{\partial c} + (f - c) \frac{\partial}{\partial d} - e \frac{\partial}{\partial f},$$
$$D_3 = a \frac{\partial}{\partial b} - d \frac{\partial}{\partial c} + (c - f) \frac{\partial}{\partial e} + d \frac{\partial}{\partial f}, \tag{1.62}$$
$$D_4 = b \frac{\partial}{\partial b} - d \frac{\partial}{\partial d} + e \frac{\partial}{\partial e}.$$

Calculating all kinds of commutators of operators (1.61)–(1.62), we have

$$[X_1, X_2] = -X_2, \quad [X_1, X_3] = X_3, \quad [X_1, X_4] = 0,$$
$$[X_2, X_1] = X_2, \quad [X_2, X_3] = X_4 - X_1, \quad [X_2, X_4] = -X_2,$$
$$[X_3, X_1] = -X_3, \quad [X_3, X_2] = X_1 - X_4, \quad [X_3, X_4] = X_3, \tag{1.63}$$
$$[X_4, X_1] = 0, \quad [X_4, X_2] = X_2, \quad [X_4, X_3] = -X_3.$$

However, it should be noted that equality (1.57) has the following advantage:

Lemma 1.1. *Commutator* (1.57), *whose operators are of the form* (1.58), *is invariant with respect to coordinate system* (1.55) *in* $E^{N+2}(x, y, A)$.

The proof of Lemma 1.1 can be found in the paper of L. V. Ovsyannikov [24].

It is easy to check that the set of operators (1.54) form a linear space and satisfy Definition 1.2. Therefore, there takes place

Lemma 1.2. *Linear differential operators of the first-order* (1.54) *form a Lie algebra of operators.*

There is no associativity in this algebra, but it is replaced by a Jacobi identity.

Suppose that dimension of a Lie algebra of linear differential operators is finite and is equal to r, and let us fix in this algebra some basis $Y_1, Y_2, ..., Y_r$, which we denote by L_r. Consider commutators $[Y_\mu, Y_\nu]$ of all possible pairs of these basis operators. Since any operator Y from L_r is decomposed over a basis as:

$$Y = \sum_{\mu=1}^{r} e^\mu Y_\mu, \quad e^\mu = const, \tag{1.64}$$

then the values of all $[Y_\mu, Y_\nu]$ allow one to find uniquely the commutator of any operators from algebra L_r using bilinearity. From here it is clear that r–dimensional vector space L_r with the basis $Y_1, Y_2, ..., Y_r$ forms a Lie algebra if and only if, the commutators of basis operators belong to L_r, i.e.

$$[Y_\mu, Y_\nu] = \sum_{\lambda=1}^{r} C_{\mu\nu}^\lambda Y_\lambda, \quad (\mu, \nu = \overline{1, r}), \tag{1.65}$$

where $C_{\mu\nu}^\lambda$ – real numbers, called *structural constants* of the algebra L_r. A convenient way to work directly with Lie algebra is the following: there is given the basis of operators and the table of their commutators, which is easily constructed using (1.56) and in which the value of $[Y_\mu, Y_\nu]$ is located at the intersection of the μ-th row and the ν-th column. If commutators table consists of all zeros, then the Lie algebra is commutative. From here we have

Consequence 1.5. *The one-dimensional Lie algebra of operators, i.e., Lie algebra consisting of one basis operator, is commutative.*

Remark 1.4. *Using formulas* (1.59), *we can show that equalities* (1.63) *take place for operators* (1.40)–(1.41), (1.44)–(1.45), (1.47)–(1.48), (1.50)–(1.51), *applied to system* (1.1)–(1.2) *for any* $\Gamma = \{m_i\}_{i=0}^\ell$.

From Remark 1.4, we consider the totality of these operators and draw up a Table 1.1 of their commutators, in which the value of $[Y_\mu, Y_\nu]$ is located at the intersection of the μ-th row and the ν-th column.

From here we note that the vector space with the basis (1.40), (1.44), (1.47) and (1.50) forms a 4-dimensional Lie algebra L_4.

	X_1	X_2	X_3	X_4
X_1	0	$-X_2$	X_3	0
X_2	X_2	0	$X_4 - X_1$	$-X_2$
X_3	$-X_3$	X_1-X_4	0	X_3
X_4	0	X_2	$-X_3$	0

Remark 1.5. *If in Table* 1.1, *replace* X_1, X_2, X_3, X_4 *with* D_1, D_2, D_3, D_4 *from* (1.41), (1.45), (1.48), (1.51), *respectively, then, for new operators, this table will remain as it is, and, therefore, the vector space with this basis forms a 4-dimensional Lie algebra with* $\Gamma \neq \{1\}$.

Equalities (1.63) are called *structural equations* of a Lie algebra L_4.

If we now consider that operators (1.40), (1.44), (1.47), (1.50) ((1.41), (1.45), (1.48), (1.51)) are operators of representations of one-parameter groups (1.12), (1.14), (1.16), (1.17) in the space $E^{N+2}(x, y, A)$ ($E^N(A)$), respectively, then according to Theorem 1.1, it should be noted that these one-parameter groups are the groups into which the group $GL(2, \mathbb{R})$ (and its representation in the space $E^{N+2}(x, y, A)$ ($E^N(A)$)) is decomposed. Thus, to representation of the group $GL(2, \mathbb{R})$ in the space $E^{N+2}(x, y, A)$ ($E^N(A)$), the algebra Lie of operators X_1, X_2, X_3, X_4 (D_1, D_2, D_3, D_4) is put into correspondence, defined by Table 1.1 of commutators.

1.7   Comments to Chapter One

In this chapter, the continuous groups of linear transformations for two-dimensional polynomial differential systems (1.1)–(1.2) are considered. It is shown that these transformations preserve the form of the systems. This leads us to the group theory and Lie algebras, without which it is impossible today to imagine modern mathematics and even physics.

Therefore, the admission of these groups and the corresponding Lie algebras of operators to the considered systems implies the preservation of form of differential system (1.1)–(1.2) under the above-mentioned groups of linear transformations.

As shown in [33,34], this fact is verified on the coordinates of the obtained operators X_1, X_2, X_3, X_4 from (1.40)–(1.41), (1.44)–(1.45), (1.47)–(1.48), (1.50)–(1.51), respectively, taking into account that (1.38)–(1.39) satisfy the defining equations

$$\begin{aligned}
\xi_x^1 P + \xi_y^1 Q &= \xi^1 P_x + \xi^2 P_y + D(P), \\
\xi_x^2 P + \xi_y^2 Q &= \xi^1 Q_x + \xi^2 Q_y + D(Q),
\end{aligned} \tag{1.66}$$

of the respective system (1.1)–(1.2), where P and Q are from the considered system.

2

Differential Equations for Centro-Affine Invariants and Comitants of Differential Systems and Their Applications

2.1 Concept of Centro-Affine Comitant and an Invariant of Differential System

Recall that in the future, we will consider transformation (1.18) through the matrix $q = \begin{pmatrix} \alpha & \beta \\ \gamma & \delta \end{pmatrix}$; and its membership in the group $GL(2, \mathbb{R})$ we will write as $q \in GL(2, \mathbb{R})$ and $\Delta = det(q)$.

Denote the set of coefficients of system (1.1)–(1.2) by A, and of (1.32) – by B. From Remark 1.3, it is clear that $B = g(A, q)$, i.e., B is a linear function on A, and it is rational on elements of transformation q.

Consider a few examples.

Example 2.1. Consider a polynomial on the coefficients of system (1.3) and phase variables x, y, which we write in a determinant form

$$k_1(x, y, A) = det \begin{pmatrix} a & x \\ b & y \end{pmatrix}, \tag{2.1}$$

where we denote by A the set of coefficients of the right side of this system. Then, after transformation (1.18) in system (1.3), we obtain the system

$$\dot{\overline{x}} = \overline{a} + \overline{c}\,\overline{x} + \overline{d}\,\overline{y}, \ \dot{\overline{y}} = \overline{b} + \overline{e}\,\overline{x} + \overline{f}\,\overline{y}, \tag{2.2}$$

where

$$\overline{a} = \alpha a + \beta b, \ \overline{b} = \gamma a + \delta b,$$
$$\Delta\overline{c} = \delta(\alpha c + \beta e) - \gamma(\alpha d + \beta f),$$
$$\Delta\overline{d} = -\beta(\alpha c + \beta e) + \alpha(\alpha d + \beta f), \tag{2.3}$$
$$\Delta\overline{e} = \delta(\gamma c + \delta e) - \gamma(\gamma d + \delta f),$$
$$\Delta\overline{f} = -\beta(\gamma c + \delta e) + \alpha(\gamma d + \delta f).$$

Similar expression (2.1) for system (2.2) will have a form

$$k_1(\overline{x}, \overline{y}, B) = det \begin{pmatrix} \overline{a} & \overline{x} \\ \overline{b} & \overline{y} \end{pmatrix}. \tag{2.4}$$

DOI: 10.1201/9781003193074-3

We will search which relation exists between (2.1) and (2.4). To do this, we note that the matrix form (2.4) will be written through (1.18) and (2.3) as

$$\begin{pmatrix} \bar{a} & \bar{x} \\ \bar{b} & \bar{y} \end{pmatrix} = \begin{pmatrix} \alpha a + \beta b & \alpha x + \beta y \\ \gamma a + \delta b & \gamma x + \delta y \end{pmatrix}, \tag{2.5}$$

where the right side will get the form

$$\begin{pmatrix} \alpha a + \beta b & \alpha x + \beta y \\ \gamma a + \delta b & \gamma x + \delta y \end{pmatrix} = \begin{pmatrix} \alpha & \beta \\ \gamma & \delta \end{pmatrix} \begin{pmatrix} a & x \\ b & y \end{pmatrix}.$$

Hence, taking into account (2.5), we have

$$\begin{pmatrix} \bar{a} & \bar{x} \\ \bar{b} & \bar{y} \end{pmatrix} = \begin{pmatrix} \alpha & \beta \\ \gamma & \delta \end{pmatrix} \begin{pmatrix} a & x \\ b & y \end{pmatrix}.$$

Applying a determinant property to this equality, we obtain

$$det \begin{pmatrix} \bar{a} & \bar{x} \\ \bar{b} & \bar{y} \end{pmatrix} = det \begin{pmatrix} \alpha & \beta \\ \gamma & \delta \end{pmatrix} det \begin{pmatrix} a & x \\ b & y \end{pmatrix},$$

from where, taking into account (1.18), (2.1), (2.4), we find the equality

$$k_1(\bar{x}, \bar{y}, B) = \Delta k_1(x, y, A) \tag{2.6}$$

for any set of coefficients A of system (1.3), any x and y and any transformations $q \in GL(2, \mathbb{R})$. Note also that from (2.6), the expression (2.4) is equal to the product of transformation determinant (1.18) on search expression (2.1).

Example 2.2. Consider for system (1.3) matrix of the coefficients

$$F = \begin{pmatrix} c & d \\ e & f \end{pmatrix} \tag{2.7}$$

and write the sum of its elements along the main diagonal

$$i_1(F) = c + f. \tag{2.8}$$

It is usually denoted by trF and called as the trace of the matrix F. Then, for the respective system (2.2), received after transformation (1.18), in system (1.3), the similar expression has the form

$$i_1(\bar{F}) = \bar{c} + \bar{f}. \tag{2.9}$$

Using equalities (2.3), we find

$$i_1(\bar{F}) = i_1(F), \tag{2.10}$$

i.e., search expression (2.8) does not change its value after any transformation (1.18) in system (1.3).

Using (2.3), you can easily check that the sum of the elements of the second diagonal $d + e$ of the matrix F from (2.7) does not have the above property.

If we denote the set of coefficients of system (1.32), received after transformation (1.18) in system (1.1)–(1.2), by B, then Examples 2.1 and 2.2 lead us to the following

Definition 2.1. *We say that the integer rational function*

$$K(x, y, \overset{0}{a}\overset{1}{_0}, \overset{0}{a}\overset{1}{_1}, ..., \overset{0}{a}\overset{1}{_{m_1}}, ..., \overset{\ell_2}{a_0}, \overset{\ell_2}{a_1}, ..., \overset{\ell_2}{a_{m_\ell}}),$$

which we will denote by $K(x, y, A)$, on a set A of coefficients of system (1.1)–(1.2) and phase variables x and y, is called the centro-affine comitant of this system, if there exists such function $\lambda(q)$, that the identity

$$K(\overline{x}, \overline{y}, B) \equiv \lambda(q)K(x, y, A) \tag{2.11}$$

holds for any $q \in GL(2, \mathbb{R})$, any coefficients of system (1.1)–(1.2) and any variables x and y. If the comitant K does not depend on variables x and y, then it is called the centro-affine invariant of system (1.1)–(1.2).

Remark 2.1. *In the monograph [37], it is shown that in (2.11), $\lambda(q) = \Delta^{-g}$, where g – integer number.*

The number g is usually called a weight of the comitant $K(x, y, A)$.

If $g = 0$, then the comitant $K(x, y, A)$ is called absolute, *otherwise –* relative.

In certain cases, in the name "centro-affine comitant", we will omit the word "centro-affine", if this does not lead to misunderstandings.

From Examples 2.1 and 2.2, we obtain

Observation 2.1. *The expression k_1 from (2.1) is a relative comitant with weight $g = -1$ for system (1.3), and i_1 from (2.8) is an absolute invariant of system (1.3).*

Directly from Definition 2.1 follows

Property 2.1. *The product of any two centro-affine comitants (invariants) of system (1.1)–(1.2) is a centro-affine comitant (invariant) with a weight equal to the sum of weights of the factors.*

Property 2.2. *The sum of two centro-affine comitants (invariants) of system (1.1)–(1.2) with the same weights is a centro-affine comitant (invariant) with the same weight.*

The sum of the two centro-affine comitants (invariants) is not always the centro-affine comitant (invariant).

Observation 2.2. *Similar to Examples 2.1 and 2.2, using (2.3), you can easily check that the following expressions are centro-affine invariants and* *comitants of system* (1.3) *with corresponding weights g*

$$i_1 = c + f \ (g = 0),$$
$$i_2 = c^2 + 2de + f^2 \ (g = 0),$$
$$i_3 = -ea^2 + (c - f)ab + db^2, \ g = -1,$$
$$k_1 = -bx + ay \ (g = -1),$$
$$k_2 = -ex^2 + (c - f)xy + dy^2 \ (g = -1),$$
$$k_3 = -(ea + fb)x + (ca + db)y \ (g = -1).$$

(2.12)

It is clear that, taking into account Properties 2.1 and 2.2, from (2.12), one can obtain an infinite number of centro-affine comitants and invariants of system (1.3).

2.2 Centro-Affine Transformations of System (1.1)–(1.2)

Lemma 2.1. *The representation of the group of centro-affine transformations* $GL(2, \mathbb{R})$ *with formulas* (1.18) *in the space of coefficients* $E^N(A)$ *of differential system* (1.1)–(1.2) *is a four-parameter group defined by one of series of expressions*

$$\Delta^{m_i} b^i_k = (-1)^k \frac{(m_i - k)!}{(m_i)!} \left(\alpha \frac{\partial}{\partial \gamma} + \beta \frac{\partial}{\partial \delta} \right)^k [\alpha P_{m_i}(\delta, -\gamma) + \beta Q_{m_i}(\delta, -\gamma)],$$

(2.13)

$$\Delta^{m_i} b^{i2}_k = (-1)^k \frac{(m_i - k)!}{(m_i)!} \left[\gamma \left(\alpha \frac{\partial}{\partial \gamma} + \beta \frac{\partial}{\partial \delta} \right)^k P_{m_i}(\delta, -\gamma) \right.$$
$$\left. + \delta \left(\alpha \frac{\partial}{\partial \gamma} + \beta \frac{\partial}{\partial \delta} \right)^k Q_{m_i}(\delta, -\gamma) \right] \ (i = \overline{0, \ell}; \ k = \overline{0, m_i}),$$

(2.14)

or

$$\Delta^{m_i} b^i_k = (-1)^{m_i - k} \frac{k!}{(m_i)!} \left[\alpha \left(\gamma \frac{\partial}{\partial \alpha} + \delta \frac{\partial}{\partial \beta} \right)^{m_i - k} P_{m_i}(-\beta, \alpha) \right.$$
$$\left. + \beta \left(\gamma \frac{\partial}{\partial \alpha} + \delta \frac{\partial}{\partial \beta} \right)^{m_i - k} Q_{m_i}(-\beta, \alpha) \right],$$

(2.15)

$$\Delta^{m_i} b^{i2}_k = (-1)^{m_i - k} \frac{k!}{(m_i)!} \left(\gamma \frac{\partial}{\partial \alpha} + \delta \frac{\partial}{\partial \beta} \right)^{m_i - k} [\gamma P_{m_i}(-\beta, \alpha)$$
$$+ \delta Q_{m_i}(-\beta, \alpha)] \ (i = \overline{0, \ell}, \ k = \overline{0, m_i}),$$

(2.16)

in which the value of the parameters $\alpha = \delta = 1$, $\beta = \gamma = 0$ *corresponds to identity transformation* (1.24).

Proof. According to Remark 1.3, expressions (2.13), (2.14) or (2.15), (2.16) form a group of transformations of the coefficient space $E^N(A)$ of system (1.1)–(1.2).

With centro-affine transformation (1.18) in system (1.1), taking into account (1.25)–(1.26), we find

$$
\begin{aligned}
\dot{\overline{x}} &= \sum_{m_i \in \Gamma} [\alpha \Delta^{-m_i} P_{m_i}(\delta \overline{x} + \beta' \overline{y}, \gamma' \overline{x} + \alpha \overline{y}) \\
&\quad + \beta \Delta^{-m_i} Q_{m_i}(\delta \overline{x} + \beta' \overline{y}, \gamma' \overline{x} + \alpha \overline{y})], \\
\dot{\overline{y}} &= \sum_{m_i \in \Gamma} [\gamma \Delta^{-m_i} P_{m_i}(\delta \overline{x} + \beta' \overline{y}, \gamma' \overline{x} + \alpha \overline{y}) \\
&\quad + \delta \Delta^{-m_i} Q_{m_i}(\delta \overline{x} + \beta' \overline{y}, \gamma' \overline{x} + \alpha \overline{y})].
\end{aligned}
\tag{2.17}
$$

Taking into account (1.2) for P_{m_i} and Q_{m_i}, we have

$$
P_{m_i}(\delta \overline{x} + \beta' \overline{y}, \gamma' \overline{x} + \alpha \overline{y}) = \sum_{k=0}^{m_i} \binom{m_i}{k} \overset{i}{a_k^1}(\delta \overline{x} + \beta' \overline{y})^{m_i-k}(\gamma' \overline{x} + \alpha \overline{y})^k,
$$

$$
Q_{m_i}(\delta \overline{x} + \beta' \overline{y}, \gamma' \overline{x} + \alpha \overline{y}) = \sum_{k=0}^{m_i} \binom{m_i}{k} \overset{i}{a_k^2}(\delta \overline{x} + \beta' \overline{y})^{m_i-k}(\gamma' \overline{x} + \alpha \overline{y})^k,
$$

or, which is the same

$$
\begin{aligned}
P_{m_i}(\delta \overline{x} + \beta' \overline{y}, \gamma' \overline{x} + \alpha \overline{y}) &= \sum_{k=0}^{m_i} \binom{m_i}{k} \overset{i}{B_k^1} \overline{x}^{m_i-k} \overline{y}^k, \\
Q_{m_i}(\delta \overline{x} + \beta' \overline{y}, \gamma' \overline{x} + \alpha \overline{y}) &= \sum_{k=0}^{m_i} \binom{m_i}{k} \overset{i}{B_k^2} \overline{x}^{m_i-k} \overline{y}^k,
\end{aligned}
\tag{2.18}
$$

where the coefficients $\overset{i}{B_k^1}$ and $\overset{i}{B_k^2}$ are rational functions of α, β', γ', δ and linear functions of $\overset{ij}{a_k}$. Then from (2.17), taking into account (2.18), we find

$$
\begin{aligned}
\dot{\overline{x}} &= \sum_{m_i \in \Gamma} \sum_{k=0}^{m_i} \binom{m_i}{k} \overset{i}{b_k^1} \overline{x}^{m_i-k} \overline{y}^k, \\
\dot{\overline{y}} &= \sum_{m_i \in \Gamma} \sum_{k=0}^{m_i} \binom{m_i}{k} \overset{i}{b_k^2} \overline{x}^{m_i-k} \overline{y}^k,
\end{aligned}
\tag{2.19}
$$

where

$$
\Delta^{m_i} \overset{i}{b_k^1} = \alpha \overset{i}{B_k^1} + \beta \overset{i}{B_k^2}, \quad \Delta^{m_i} \overset{i}{b_k^2} = \gamma \overset{i}{B_k^1} + \delta \overset{i}{B_k^2} \ (i = \overline{0, \ell}).
\tag{2.20}
$$

Denote $\xi = \frac{\bar{y}}{\bar{x}}$, $\eta = \frac{\bar{x}}{\bar{y}}$, and from (2.18) we find

$$P_{m_i}(\delta + \beta'\xi, \gamma' + \alpha\xi) = \sum_{k=0}^{m_i} \binom{m_i}{k} \overset{i}{B}^1_k \xi^k,$$

$$Q_{m_i}(\delta + \beta'\xi, \gamma' + \alpha\xi) = \sum_{k=0}^{m_i} \binom{m_i}{k} \overset{i}{B}^2_k \xi^k \ (i = \overline{0,\ell})$$

(2.21)

and

$$P_{m_i}(\delta\eta + \beta', \gamma'\eta + \alpha) = \sum_{k=0}^{m_i} \binom{m_i}{k} \overset{i}{B}^1_k \eta^{m_i-k},$$

$$Q_{m_i}(\delta\eta + \beta', \gamma'\eta + \alpha) = \sum_{k=0}^{m_i} \binom{m_i}{k} \overset{i}{B}^2_k \eta^{m_i-k} \ (i = \overline{0,\ell}).$$

(2.22)

If we expand in a Taylor series polynomials (2.21) and (2.22) on variables ξ and η, respectively, then we obtain

$$P_{m_i}(\delta + \beta'\xi, \gamma' + \alpha\xi) = \sum_{k=0}^{m_i} \frac{\xi^k}{k!} \left(\alpha\frac{\partial}{\partial\gamma'} + \beta'\frac{\partial}{\partial\delta}\right)^k P_{m_i}(\delta, \gamma'),$$

$$Q_{m_i}(\delta + \beta'\xi, \gamma' + \alpha\xi) = \sum_{k=0}^{m_i} \frac{\xi^k}{k!} \left(\alpha\frac{\partial}{\partial\gamma'} + \beta'\frac{\partial}{\partial\delta}\right)^k Q_{m_i}(\delta, \gamma')$$

$$(i = \overline{0,\ell})$$

(2.23)

and

$$P_{m_i}(\delta\eta + \beta', \gamma'\eta + \alpha) = \sum_{k=0}^{m_i} \frac{\eta^{m_i-k}}{(m_i-k)!} \left(\gamma'\frac{\partial}{\partial\alpha} + \delta\frac{\partial}{\partial\beta'}\right)^{m_i-k} P_{m_i}(\beta', \alpha),$$

$$Q_{m_i}(\delta\eta + \beta', \gamma'\eta + \alpha) = \sum_{k=0}^{m_i} \frac{\eta^{m_i-k}}{(m_i-k)!} \left(\gamma'\frac{\partial}{\partial\alpha} + \delta\frac{\partial}{\partial\beta'}\right)^{m_i-k} Q_{m_i}(\beta', \alpha)$$

$$(i = \overline{0,\ell}).$$

(2.24)

Equating coefficients by the same degrees ξ and η, respectively, in (2.21) and (2.23), as well as in (2.22) and (2.24), we obtain

$$\overset{i}{B}^1_k = \frac{(m_i-k)!}{m_i!} \left(\alpha\frac{\partial}{\partial\gamma'} + \beta'\frac{\partial}{\partial\delta}\right)^k P_{m_i}(\delta, \gamma'),$$

$$\overset{i}{B}^2_k = \frac{(m_i-k)!}{m_i!} \left(\alpha\frac{\partial}{\partial\gamma'} + \beta'\frac{\partial}{\partial\delta}\right)^k Q_{m_i}(\delta, \gamma') \ (i = \overline{0,\ell}),$$

or

$$\overset{i}{B}^1_k = \frac{k!}{m_i!} \left(\gamma'\frac{\partial}{\partial\alpha} + \delta\frac{\partial}{\partial\beta'}\right)^{m_i-k} P_{m_i}(\beta', \alpha),$$

$$\overset{i}{B}^2_k = \frac{k!}{m_i!} \left(\gamma'\frac{\partial}{\partial\alpha} + \delta\frac{\partial}{\partial\beta'}\right)^{m_i-k} Q_{m_i}(\beta', \alpha) \ (i = \overline{0,\ell}).$$

From the last four equalities, taking into account (1.26), we have

$$\overset{i}{B}{}^1_k = (-1)^k \frac{(m_i - k)!}{m_i!} \left(\alpha \frac{\partial}{\partial \gamma} + \beta \frac{\partial}{\partial \delta} \right)^k P_{m_i}(\delta, -\gamma),$$

$$\overset{i}{B}{}^2_k = (-1^k) \frac{(m_i - k)!}{m_i!} \left(\alpha \frac{\partial}{\partial \gamma} + \beta \frac{\partial}{\partial \delta} \right)^k Q_{m_i}(\delta, -\gamma) \, (i = \overline{0, \ell}),$$

(2.25)

or

$$\overset{i}{B}{}^1_k = (-1)^{m_i - k} \frac{k!}{m_i!} \left(\gamma \frac{\partial}{\partial \alpha} + \delta \frac{\partial}{\partial \beta} \right)^{m_i - k} P_{m_i}(-\beta, \alpha),$$

$$\overset{i}{B}{}^2_k = (-1)^{m_i - k} \frac{k!}{m_i!} \left(\gamma \frac{\partial}{\partial \alpha} + \delta \frac{\partial}{\partial \beta} \right)^{m_i - k} Q_{m_i}(-\beta, \alpha) \, (i = \overline{0, \ell}).$$

(2.26)

Taking into account (2.20) and (2.25)–(2.26), we obtain for $\overset{i}{b}{}^1_k$ and $\overset{i}{b}{}^2_k$ expressions (2.13)–(2.14) or (2.15)–(2.16). Lemma 2.1 is proved.

Comparing (1.1)–(1.2) and (2.19), we obtain

Consequence 2.1. *On centro-affine transformation (1.18) system (1.1) with new coefficients and new variables retains its form, and its homogeneities of the right-hand sides with respect to x and y go to the homogeneities of the same degree with respect to $\overline{x}$ and $\overline{y}$.*

From (2.13)–(2.14) and (2.15)–(2.16), we have

Consequence 2.2. *The expression $\Delta^{m_i} \overset{i}{b}{}^1_k$ ($\Delta^{m_i} \overset{i}{b}{}^2_k$) has the form (2.13) or (2.15) ((2.14) or (2.16)) and is a homogeneous function of degree $k + 1$ (k) with respect to the pair (α, β), and of degree $m_i - k$ ($m_i - k + 1$) with respect to the pair (γ, δ).*

Consider expressions for the coefficients of affine differential system (1.3) after centro-affine transformation (1.18). Using Lemma 2.1 and equalities (1.4), we obtain (2.3), where in system (2.2) we have $\overline{a} = \overset{0}{b}{}^1_0$, $\overline{b} = \overset{0}{b}{}^2_0$, $\overline{c} = \overset{1}{b}{}^1_0$, $\overline{d} = \overset{1}{b}{}^1_1$, $\overline{e} = \overset{1}{b}{}^2_0$, $\overline{f} = \overset{1}{b}{}^2_1$.

2.3 Differential Equations for Centro-Affine Invariants and Comitants

Example 2.3. Consider Lie algebra of operators corresponding to the representation of the group $GL(2, \mathbb{R})$ in the space $E^6(x, y, A)$ of system (1.3). According to Example 1.7, this Lie algebra consists of operators (1.61)–(1.62).

For invariants and comitants (2.12) of system (1.3) on a group $GL(2, \mathbb{R})$, using the mentioned operators, we obtain the equalities

$$D_m(i_j) = 0 \, (m = \overline{1,4}, \, j = 1, 2), \, D_1(i_3) = D_4(i_3) = i_3, \, D_2(i_3) = D_3(i_3) = 0$$

and

$$X_1(k_j) = X_4(k_j) = i_3, \, X_2(k_j) = X_3(k_j) = 0 \, (j = \overline{1,3}).$$

The following theorem takes place

Theorem 2.1. *In order that an integer rational function of the coefficients of system (1.1)–(1.2) to be a centro-affine invariant of this system with a weight g, it is necessary and sufficient that it satisfies the equations*

$$D_1(I) = D_4(I) = -gI, \, D_2(I) = D_3(I) = 0, \qquad (2.27)$$

where D_m $(m = \overline{1,4})$ are from (1.41), (1.45), (1.48), (1.51) and form a Lie algebra of operators of representation of the group $GL(2, \mathbb{R})$ in the space of coefficients $E^N(A)$ of system (1.1)–(1.2).

Proof. Necessity. Suppose that $I(A)$ is a centro-affine invariant of system (1.1)–(1.2) with the weight g. Then, according to Definition 2.1 and Remark 2.1, the following identity takes place

$$I(B) = \Delta^{-g} I(A), \qquad (2.28)$$

where the totality of B consists of the coefficients of system (2.19), having the form (2.13)–(2.16).

Note that the determinant of transformation (1.18) satisfies the differential equations

$$\alpha \frac{\partial \Delta}{\partial \alpha} + \beta \frac{\partial \Delta}{\partial \beta} = \Delta, \, \gamma \frac{\partial \Delta}{\partial \alpha} + \delta \frac{\partial \Delta}{\partial \beta} = 0,$$

$$\alpha \frac{\partial \Delta}{\partial \gamma} + \beta \frac{\partial \Delta}{\partial \delta} = 0, \, \gamma \frac{\partial \Delta}{\partial \gamma} + \delta \frac{\partial \Delta}{\partial \delta} = \Delta. \qquad (2.29)$$

If we apply to both sides of equality (2.28), the operators

$$\alpha \frac{\partial}{\partial \alpha} + \beta \frac{\partial}{\partial \beta}, \, \gamma \frac{\partial}{\partial \alpha} + \delta \frac{\partial}{\partial \beta}, \, \alpha \frac{\partial}{\partial \gamma} + \beta \frac{\partial}{\partial \delta}, \, \gamma \frac{\partial}{\partial \gamma} + \delta \frac{\partial}{\partial \delta},$$

then, taking into account equalities (2.29), we obtain

$$\alpha \frac{\partial I(B)}{\partial \alpha} + \beta \frac{\partial I(B)}{\partial \beta} = -gI(B), \, \gamma \frac{\partial I(B)}{\partial \alpha} + \delta \frac{\partial I(B)}{\partial \beta} = 0,$$

$$\alpha \frac{\partial I(B)}{\partial \gamma} + \beta \frac{\partial I(B)}{\partial \delta} = 0, \, \gamma \frac{\partial I(B)}{\partial \gamma} + \delta \frac{\partial I(B)}{\partial \delta} = -gI(B). \qquad (2.30)$$

This system of differential equations can be represented in another form, considering   the   fact   that,   according   to   equalities

(2.13)–(2.16), $I(B)$ is a complex function of $\alpha, \beta, \gamma, \delta$ included in $\overset{0}{b^1_0}, \overset{0}{b^1_1}, ...,$ $\overset{\ell}{b^1_{m_1}}, ..., \overset{0}{b^2_0}, \overset{\ell}{b^2_1}, ..., \overset{\ell}{b^2_{m_\ell}}$. Then, from (2.30), we obtain the following system of differential equations:

$$\sum_{i=0}^{\ell}\sum_{k=0}^{m_i}\left[\left(\alpha\frac{\partial \overset{i}{b^1_k}}{\partial\alpha}+\beta\frac{\partial \overset{i}{b^1_k}}{\partial\beta}\right)\frac{\partial I(B)}{\partial \overset{i}{b^1_k}}+\left(\alpha\frac{\partial \overset{i}{b^2_k}}{\partial\alpha}+\beta\frac{\partial \overset{i}{b^2_k}}{\partial\beta}\right)\frac{\partial I(B)}{\partial \overset{i}{b^2_k}}\right]=-gI(B),$$

$$\sum_{i=0}^{\ell}\sum_{k=0}^{m_i}\left[\left(\gamma\frac{\partial \overset{i}{b^1_k}}{\partial\alpha}+\delta\frac{\partial \overset{i}{b^1_k}}{\partial\beta}\right)\frac{\partial I(B)}{\partial \overset{i}{b^1_k}}+\left(\gamma\frac{\partial \overset{i}{b^2_k}}{\partial\alpha}+\delta\frac{\partial \overset{i}{b^2_k}}{\partial\beta}\right)\frac{\partial I(B)}{\partial \overset{i}{b^2_k}}\right]=0,$$

$$\sum_{i=0}^{\ell}\sum_{k=0}^{m_i}\left[\left(\alpha\frac{\partial \overset{i}{b^1_k}}{\partial\gamma}+\beta\frac{\partial \overset{i}{b^1_k}}{\partial\delta}\right)\frac{\partial I(B)}{\partial \overset{i}{b^1_k}}+\left(\alpha\frac{\partial \overset{i}{b^2_k}}{\partial\gamma}+\beta\frac{\partial \overset{i}{b^2_k}}{\partial\delta}\right)\frac{\partial I(B)}{\partial \overset{i}{b^2_k}}\right]=0,$$

$$\sum_{i=0}^{\ell}\sum_{k=0}^{m_i}\left[\left(\gamma\frac{\partial \overset{i}{b^1_k}}{\partial\gamma}+\delta\frac{\partial \overset{i}{b^1_k}}{\partial\delta}\right)\frac{\partial I(B)}{\partial \overset{i}{b^1_k}}+\left(\gamma\frac{\partial \overset{i}{b^2_k}}{\partial\gamma}+\delta\frac{\partial \overset{i}{b^2_k}}{\partial\delta}\right)\frac{\partial I(B)}{\partial \overset{i}{b^2_k}}\right]=-gI(B).$$

$$(2.31)$$

Consider the obtainment of simpler expressions for round brackets from (2.31). In view of equalities (2.29), we have

$$\Delta^{m_i}\left(\alpha\frac{\partial \overset{i}{b^1_k}}{\partial\alpha}+\beta\frac{\partial \overset{i}{b^1_k}}{\partial\beta}\right)=-m_i\overset{i}{b^1_k}\Delta^{m_i}+\alpha\frac{\partial(\Delta^{m_i}\overset{i}{b^1_k})}{\partial\alpha}+\beta\frac{\partial(\Delta^{m_i}\overset{i}{b^1_k})}{\partial\beta}, \quad (2.32)$$

$$\Delta^{m_i}\left(\alpha\frac{\partial \overset{i}{b^2_k}}{\partial\alpha}+\beta\frac{\partial \overset{i}{b^2_k}}{\partial\beta}\right)=-m_i\overset{i}{b^2_k}\Delta^{m_i}+\alpha\frac{\partial(\Delta^{m_i}\overset{i}{b^2_k})}{\partial\alpha}+\beta\frac{\partial(\Delta^{m_i}\overset{i}{b^2_k})}{\partial\beta}, \quad (2.33)$$

$$\Delta^{m_i}\left(\gamma\frac{\partial \overset{i}{b^1_k}}{\partial\alpha}+\delta\frac{\partial \overset{i}{b^1_k}}{\partial\beta}\right)=\gamma\frac{\partial(\Delta^{m_i}\overset{i}{b^1_k})}{\partial\alpha}+\delta\frac{\partial(\Delta^{m_i}\overset{i}{b^1_k})}{\partial\beta}, \quad (2.34)$$

$$\Delta^{m_i}\left(\gamma\frac{\partial \overset{i}{b^2_k}}{\partial\alpha}+\delta\frac{\partial \overset{i}{b^2_k}}{\partial\beta}\right)=\gamma\frac{\partial(\Delta^{m_i}\overset{i}{b^2_k})}{\partial\alpha}+\delta\frac{\partial(\Delta^{m_i}\overset{i}{b^2_k})}{\partial\beta}, \quad (2.35)$$

$$\Delta^{m_i}\left(\alpha\frac{\partial \overset{i}{b^1_k}}{\partial\gamma}+\beta\frac{\partial \overset{i}{b^1_k}}{\partial\delta}\right)=\alpha\frac{\partial(\Delta^{m_i}\overset{i}{b^1_k})}{\partial\gamma}+\beta\frac{\partial(\Delta^{m_i}\overset{i}{b^1_k})}{\partial\delta}, \quad (2.36)$$

$$\Delta^{m_i}\left(\alpha\frac{\partial \overset{i}{b^2_k}}{\partial\gamma}+\beta\frac{\partial \overset{i}{b^2_k}}{\partial\delta}\right)=\alpha\frac{\partial(\Delta^{m_i}\overset{i}{b^2_k})}{\partial\gamma}+\beta\frac{\partial(\Delta^{m_i}\overset{i}{b^2_k})}{\partial\delta}, \quad (2.37)$$

$$\Delta^{m_i}\left(\gamma\frac{\partial \overset{i}{b^1_k}}{\partial\gamma} + \delta\frac{\partial \overset{i}{b^1_k}}{\partial\delta}\right) = -m_i\overset{i}{b^1_k}\Delta^{m_i} + \gamma\frac{\partial(\Delta^{m_i}\overset{i}{b^1_k})}{\partial\gamma} + \delta\frac{\partial(\Delta^{m_i}\overset{i}{b^1_k})}{\partial\delta}, \quad (2.38)$$

$$\Delta^{m_i}\left(\gamma\frac{\partial \overset{i}{b^2_k}}{\partial\gamma} + \delta\frac{\partial \overset{i}{b^2_k}}{\partial\delta}\right) = -m_i\overset{i}{b^2_k}\Delta^{m_i} + \gamma\frac{\partial(\Delta^{m_i}\overset{i}{b^2_k})}{\partial\gamma} + \delta\frac{\partial(\Delta^{m_i}\overset{i}{b^2_k})}{\partial\delta}. \quad (2.39)$$

Considering the degree of homogeneity of the expressions $\Delta^{m_i}\overset{i}{b^1_k}$ and $\Delta^{m_i}\overset{i}{b^2_k}$ in relation to the couples (α, β) and (γ, δ), from Consequence 2.2, by Eulers theorem about homogeneity of the function for equalities (2.32)–(2.33) and (2.38)–(2.39), respectively, we obtain

$$\alpha\frac{\partial(\Delta^{m_i}\overset{i}{b^1_k})}{\partial\alpha} + \beta\frac{\partial(\Delta^{m_i}\overset{i}{b^1_k})}{\partial\beta} = (k+1)\Delta^{m_i}\overset{i}{b^1_k},$$

$$\alpha\frac{\partial(\Delta^{m_i}\overset{i}{b^2_k})}{\partial\alpha} + \beta\frac{\partial(\Delta^{m_i}\overset{i}{b^2_k})}{\partial\beta} = k\Delta^{m_i}\overset{i}{b^2_k},$$

$$\gamma\frac{\partial(\Delta^{m_i}\overset{i}{b^1_k})}{\partial\gamma} + \delta\frac{\partial(\Delta^{m_i}\overset{i}{b^1_k})}{\partial\delta} = (m_i - k)\Delta^{m_i}\overset{i}{b^1_k}, \quad (2.40)$$

$$\gamma\frac{\partial(\Delta^{m_i}\overset{i}{b^2_k})}{\partial\gamma} + \delta\frac{\partial(\Delta^{m_i}\overset{i}{b^2_k})}{\partial\delta} = (m_i - k + 1)\Delta^{m_i}\overset{i}{b^2_k}.$$

Considering (2.15), for the expression from the left side of (2.34), we find

$$(-1)^{m_i-k}\frac{(m_i)!}{k!}\Delta^{m_i}\left(\gamma\frac{\partial \overset{i}{b^1_k}}{\partial\alpha} + \delta\frac{\partial \overset{i}{b^1_k}}{\partial\beta}\right) = \left(\gamma\frac{\partial}{\partial\alpha} + \delta\frac{\partial}{\partial\beta}\right)^{m_i-k}$$

$$\times[\gamma P_{m_i}(-\beta, \alpha) + \delta Q_{m_i}(-\beta, \alpha)] + \alpha\left(\gamma\frac{\partial}{\partial\alpha} + \delta\frac{\partial}{\partial\beta}\right)^{m_i-k+1}P_{m_i}(-\beta, \alpha)$$

$$+\beta\left(\gamma\frac{\partial}{\partial\alpha} + \delta\frac{\partial}{\partial\beta}\right)^{m_i-k+1}Q_{m_i}(-\beta, \alpha).$$

From this equality, using (2.15) and (2.16), for $k - 1$, we have

$$\gamma\frac{\partial \overset{i}{b^1_k}}{\partial\alpha} + \delta\frac{\partial \overset{i}{b^1_k}}{\partial\beta} = \overset{i}{b^2_k} - k\overset{i}{b^1_{k-1}}. \quad (2.41)$$

For the left side of (2.35), considering (2.16), we obtain

$$\Delta^{m_i}\left(\gamma\frac{\partial \overset{i}{b^2_k}}{\partial\alpha} + \delta\frac{\partial \overset{i}{b^2_k}}{\partial\beta}\right) = (-1)^{m_i-k}\frac{k!}{(m_i)!}\left(\gamma\frac{\partial}{\partial\alpha} + \delta\frac{\partial}{\partial\beta}\right)^{m_i-k+1}$$

$$\times[\gamma P_{m_i}(-\beta, \alpha) + \delta Q_{m_i}(-\beta, \alpha)] = -k\Delta^{m_i}\overset{i}{b^2_{k-1}}. \quad (2.42)$$

Using (2.13) for the left side of (2.36), we have

$$\Delta^{m_i}\left(\alpha\frac{\partial \overset{i}{b^1_k}}{\partial\gamma} + \beta\frac{\partial \overset{i}{b^1_k}}{\partial\delta}\right) = (-1)^k\frac{(m_i-k)!}{(m_i)!}\left(\alpha\frac{\partial}{\partial\gamma} + \beta\frac{\partial}{\partial\delta}\right)^{k+1}$$

$$\times[\alpha P_{m_i}(\delta,-\gamma) + \beta Q_{m_i}(\delta,-\gamma)] = -(m_i-k)\Delta^{m_i}\overset{i}{b^1_{k+1}}. \qquad (2.43)$$

Using (2.14) for the left side of (2.37), we have

$$(-1)^k\frac{(m_i)!}{(m_i-k)!}\Delta^{m_i}\left(\alpha\frac{\partial \overset{i}{b^2_k}}{\partial\gamma} + \beta\frac{\partial \overset{i}{b^2_k}}{\partial\delta}\right) = \left(\alpha\frac{\partial}{\partial\gamma} + \beta\frac{\partial}{\partial\delta}\right)^{k}$$

$$\times[\alpha P_{m_i}(\delta,-\gamma) + \beta Q_{m_i}(\delta,-\gamma)] + \gamma\left(\alpha\frac{\partial}{\partial\gamma} + \beta\frac{\partial}{\partial\delta}\right)^{k+1}P_{m_i}(\delta,-\gamma)$$

$$+\delta\left(\alpha\frac{\partial}{\partial\gamma} + \beta\frac{\partial}{\partial\delta}\right)^{k+1}Q_{m_i}(\delta,-\gamma).$$

Hence, considering (2.13) and (2.14) for $k+1$, we have

$$\alpha\frac{\partial \overset{i}{b^2_k}}{\partial\gamma} + \beta\frac{\partial \overset{i}{b^2_k}}{\partial\delta} = \overset{i}{b^1_k} - (m_i-k)\overset{i}{b^2_{k+1}}. \qquad (2.44)$$

Then, using (2.40), (2.42) and (2.43) from (2.32)–(2.33), (2.35)–(2.36) and (2.38)–(2.39) after reduction by Δ^{m_i}, we obtain

$$\alpha\frac{\partial \overset{i}{b^1_k}}{\partial\alpha} + \beta\frac{\partial \overset{i}{b^1_k}}{\partial\beta} = -(m_i-k-1)\overset{i}{b^1_k},$$

$$\alpha\frac{\partial \overset{i}{b^2_k}}{\partial\alpha} + \beta\frac{\partial \overset{i}{b^2_k}}{\partial\beta} = -(m_i-k)\overset{i}{b^2_k},$$

$$\gamma\frac{\partial \overset{i}{b^2_k}}{\partial\alpha} + \delta\frac{\partial \overset{i}{b^2_k}}{\partial\beta} = -k\overset{i}{b^2_{k-1}},$$

$$\alpha\frac{\partial \overset{i}{b^1_k}}{\partial\gamma} + \beta\frac{\partial \overset{i}{b^1_k}}{\partial\delta} = -(m_i-k)\overset{i}{b^1_{k+1}}, \qquad (2.45)$$

$$\gamma\frac{\partial \overset{i}{b^1_k}}{\partial\gamma} + \delta\frac{\partial \overset{i}{b^1_k}}{\partial\delta} = -k\overset{i}{b^1_k},$$

$$\gamma\frac{\partial \overset{i}{b^2_k}}{\partial\gamma} + \delta\frac{\partial \overset{i}{b^2_k}}{\partial\delta} = -(k-1)\overset{i}{b^2_k}.$$

Considering (2.41), (2.44) and (2.45), system (2.31) will be rewritten as

$$\sum_{i=0}^{\ell}\sum_{k=0}^{m_i}\left[(m_i-k-1)\overset{i}{b}_k^1\frac{\partial I(B)}{\partial\overset{i}{b}_k^1}+(m_i-k)\overset{i}{b}_k^2\frac{\partial I(B)}{\partial\overset{i}{b}_k^2}\right]=-gI(B),$$

$$\sum_{i=0}^{\ell}\sum_{k=0}^{m_i}\left[k\overset{i}{b}_{k-1}^1\frac{\partial I(B)}{\partial\overset{i}{b}_k^1}+k\overset{i}{b}_{k-1}^2\frac{\partial I(B)}{\partial\overset{i}{b}_k^2}-\overset{i}{b}_k^2\frac{\partial I(B)}{\partial\overset{i}{b}_k^1}\right]=0,$$

$$\sum_{i=0}^{\ell}\sum_{k=0}^{m_i}\left[(m_i-k)\overset{i}{b}_{k+1}^1\frac{\partial I(B)}{\partial\overset{i}{b}_k^1}+(m_i-k)\overset{i}{b}_{k+1}^2\frac{\partial I(B)}{\partial\overset{i}{b}_k^2}-\overset{i}{b}_k^1\frac{\partial I(B)}{\partial\overset{i}{b}_k^2}\right]=0,$$

$$\sum_{i=0}^{\ell}\sum_{k=0}^{m_i}\left[k\overset{i}{b}_k^1\frac{\partial I(B)}{\partial\overset{i}{b}_k^1}+(k-1)\overset{i}{b}_k^2\frac{\partial I(B)}{\partial\overset{i}{b}_k^2}\right]=-gI(B). \qquad (2.46)$$

This system of equations must exist for any transformation (1.18). In particular, on the identity transformation $\alpha=\delta=1$, $\beta=\gamma=0$, using (1.1)–(1.2) and (1.18), we obtain the equalities $\overset{i}{b}_k^j=\overset{i}{a}_k^j$ ($i=\overline{0,\ell}$, $j=1,2$, $k=\overline{0,m_i}$). Taking this into account, from (2.46) using D_m ($m=\overline{1,4}$), from (1.41), (1.45), (1.48), (1.51) we find the necessity for conditions (2.27).

Sufficiency. Suppose that an integer rational function $I(A)$ satisfies equations (2.27) with operators (1.41), (1.45), (1.48), (1.51) for any $A\in E^N(A)$. Then, taking this into account, equalities (2.46) are valid and, therefore, equalities (2.30) are valid too.

Note that if we formally apply $L=ln|I(B)|$, then from (2.30), we find the equalities

$$\alpha\frac{\partial L}{\partial\alpha}+\beta\frac{\partial L}{\partial\beta}=-g,\ \ \gamma\frac{\partial L}{\partial\alpha}+\delta\frac{\partial L}{\partial\beta}=0,$$

$$\alpha\frac{\partial L}{\partial\gamma}+\beta\frac{\partial L}{\partial\delta}=0,\ \ \gamma\frac{\partial L}{\partial\gamma}+\delta\frac{\partial L}{\partial\delta}=-g. \qquad (2.47)$$

Considering (2.29), the function $L=-gln|\Delta|+ln|C|$, where Δ is from (1.18), and C – arbitrary constant, satisfies these equalities as well. Therefore, from $ln|I(B)|=-gln|\Delta|+ln|C|$ we obtain

$$I(B)=C\Delta^{-g}. \qquad (2.48)$$

This also takes place on identity transformation $\alpha=\delta=1$, $\beta=\gamma=0$ of group (1.18) in system (1.1)–(1.2). Since in this case $\overset{i}{b}_k^j=\overset{i}{a}_k^j$ ($i=\overline{0,\ell}$, $j=1,2$, $k=\overline{0,m_i}$), then $C=I(A)$. Taking into account equality (2.48), we find identity (2.28), which, according to Definition 1.2, confirms the sufficiency of conditions (2.27).

For the fact that the operators D_m ($m = \overline{1,4}$) form a Lie algebra of representation of the group $GL(2, \mathbb{R})$ in the space $E^N(A)$, see Section 1.6. Theorem 2.1 is proved.

Observation 2.3. *The differential operators D_2 and D_3 contained in the equations of system (2.27) coincide with the operators Ω and Θ established in the paper [43].*

Following the similar proof of Theorem 2.1, it can be shown that the following theorem takes place

Theorem 2.2. *In order that integer rational function $K(x, y, A)$ to be a centro-affine comitant of system (1.1)–(1.2) with a weight g, it is necessary and sufficient that it satisfies the equations*

$$X_1(K) = X_4(K) = -gK, \quad X_2(K) = X_3(K) = 0, \qquad (2.49)$$

where X_m ($m = \overline{1,4}$) are from (1.40), (1.44), (1.47), (1.50) and form a Lie algebra of operators of representation of the group $GL(2, \mathbb{R})$ in the space $E^{N+2}(x, y, A)$ of system (1.1)–(1.2).

2.4 Rational Absolute Centro-Affine Invariants and Comitants and Their Applications

Consider the case when in (2.11) $K(x, y, A)$ is a fraction, i.e. its numerator and denominator are polynomials depending on the set A of coefficients of system (1.1)–(1.2) and on phase variables x, y.

So, let

$$K(x, y, A) = \frac{R(x, y, A)}{S(x, y, A)}, \qquad (2.50)$$

where R and S – reciprocal simple integer rational functions on the set A and phase variables x, y.

If $K(x, y, A)$ is a rational absolute centro-affine comitant, then from (2.11) will follow

$$R(\overline{x}, \overline{y}, B)S(x, y, A) = R(x, y, A)S(\overline{x}, \overline{y}, B). \qquad (2.51)$$

In equality (2.51), after substitution, instead of the set B, expressions from (2.13)–(2.14) or (2.15)–(2.16) and $\overline{x}, \overline{y}$ from (1.18), the resulting ratio should turn into an identity with respect to the coefficients of system (1.1)–(1.2) and phase variables x, y for any value of $\alpha, \beta, \gamma, \delta \in \mathbb{R}$, for which $det \begin{pmatrix} \alpha & \beta \\ \gamma & \delta \end{pmatrix} \neq$ 0. Due to equality (2.51), the left side must be divided by $R(x, y, A)$. However, R and S are reciprocal simple. Consequently, $R(\overline{x}, \overline{y}, A)$ considered as a polynomial in coefficients of system (1.1)–(1.2) and phase variables x, y, should be divided by $R(x, y, A)$. Since the degrees of these polynomials are obviously the same, then

$$R(\overline{x}, \overline{y}, B) = \lambda_1(q)R(x, y, A).$$

Similarly, for $S(x, y, A)$ from (2.51) we find

$$S(\overline{x}, \overline{y}, B) = \lambda_2(q)S(x, y, A).$$

According to Remark 2.1, $\lambda_1(q) = \Delta^{-g_1}$ and $\lambda_2(q) = \Delta^{-g_2}$, where g_1 and g_2 – integer numbers.

It is not difficult to check that $g_1 = g_2$. Consequently, R and S are relative centro-affine comitants with equal weights. The same can be said with respect to integer rational absolute centro-affine invariants. From here, we have that the following theorem takes place:

Theorem 2.3. *Every absolute rational centro-affine comitant (invariant) of system* (1.1)–(1.2) *is a quotient from dividing two integer rational centro-affine comitants (invariants) of this system with equal weights and vice versa.*

Observation 2.4. *The idea of proving Theorem 2.3 is taken from paper* [16].

Example 2.4. Consider expressions (2.12) which, according to Remark 2.1, are centro-affine invariants and comitants of system (1.3) with specific weights g.

According to Theorem 2.3, one can easily check that, for example, the relations $\frac{i_1}{i_2}, \frac{k_1}{i_3}, \frac{k_2}{i_3}, \frac{k_3}{i_3}, \frac{k_2}{k_1}, \frac{k_3}{k_1}, \frac{k_3}{k_2}$ are absolute rational centro-affine invariants and comitants of system (1.3).

From Theorems 2.1 and 2.2, it is easy to see that there takes place

Lemma 2.2. *In order that function* (2.50) *to be a rational absolute centro-affine comitant (invariant) of system* (1.1)–(1.2), *it is necessary and sufficient that it satisfies the equations*

$$X_i(K) = 0, \; (D_i(K) = 0) \; (i = \overline{1, 4}), \tag{2.52}$$

where the expressions X_1, X_2, X_3, X_4 (D_1, D_2, D_3, D_4) *from* (1.40), (1.44), (1.47), (1.50) ((1.41), (1.45), (1.48), (1.51)) *form a Lie algebra of operators.*

If you make a matrix M from the coordinates of the operators X_1, X_2, X_3, X_4, and matrix M_1 – the same matrix, only for operators D_1, D_2, D_3, D_4, and denote their common ranks by R and R_1, respectively, then, using Lemma 2.2, from homogeneous linear partial differential equations (2.52), we obtain that there takes place

Lemma 2.3. *Maximal number of functionally independent rational absolute centro-affine comitants (invariants) of differential system* (1.1)–(1.2) *is equal to*

$$2\left(\sum_{i=0}^{\ell} m_i + \ell\right) + 4 - R \quad \left(2\left(\sum_{i=0}^{\ell} m_i + \ell\right) + 2 - R_1\right). \tag{2.53}$$

Note that in order that system (1.1)–(1.2) to possess m rational absolute centro-affine comitants and invariants, it is necessary that it has $m+1$ centro-affine comitants and invariants from Definition 2.1. Taking into account this fact and Lemma 2.3, we obtain that for $R = R_1 = 4$ there takes place

Theorem 2.4. *Maximal number of functionally independent rational absolute centro-affine comitants (invariants) of system (1.1)–(1.2) is equal to*

$$\varrho = 2\left(\sum_{i=0}^{\ell} m_i + \ell\right) + 1 \quad \left(\widetilde{\varrho} = 2\left(\sum_{i=0}^{\ell} m_i + \ell\right) - 1, \ \Gamma \neq \{0\}, \{1\}\right). \quad (2.54)$$

In the formulation of Theorem 2.4, system (1.1)–(1.2) for $\Gamma \neq \{0\}, \{1\}$ for differential invariants is excluded from consideration, since, in the first case, there are no absolute invariants of the group $GL(2, \mathbb{R})$, and in the second case, there are two of them.

It is easy to show that in other cases, the matrix constructed on the coordinates of the operators X_1, X_2, X_3, X_4 (D_1, D_2, D_3, D_4), always has a nonzero fourth-order minor.

Remark 2.2. *Carrying out similar reasoning, as in the theory of invariants of binary forms (see, for example [1]), we can easily see that the indicated numbers in Theorem 2.4 are nothing more than the number of elements in the algebraic basis of comitants (invariants) of differential system (1.1)–(1.2), i.e., if we denote these centro-affine comitants by K_1, K_2,...,K_ϱ (invariants by $I_1, I_2, ..., I_{\widetilde{\varrho}}$), then for any centro-affine comitant K (invariant I) it satisfies the equation*

$$P_0 K^m + P_1 K^{m-1} + ... + P_m = 0 \ (Q_0 I^n + Q_1 I^{n-1} + ... + Q_n = 0), \quad (2.55)$$

where P_{m_i} $(i = \overline{0, m})$ $(Q_i \ (i = \overline{0, n}))$ are polynomials in $K_1, K_2, ..., K_\varrho$ $(I_1, I_2, ..., I_{\widetilde{\varrho}})$. Therefore, in the future, the number ϱ $(\widetilde{\varrho})$ from Theorem 2.4 we will consider to be related to the algebraic basis of centro-affine comitants (invariants) of system (1.1)–(1.2).

Denote by S the coefficient at the highest degree x^δ in the comitant K, which in the paper [43] is called *semi-invariant*. In the same monograph, it is shown that if S is a semi-invariant in the comitant K of system (1.1)–(1.2), then

$$K = S x^\delta - D_3(S) x^{\delta-1} y + \frac{1}{2!} D_3^2(S) x^{\delta-2} y^2 + ... + \frac{(-1)^\delta}{\delta!} D_3^\delta(S) y^\delta, \quad (2.56)$$

where D_3 is taken from (1.48).

Remark 2.3. *Using equality (2.56), we can see that the comitants $K_1, K_2, ..., K_\varrho$ for system (1.1)–(1.2) are algebraically independent if and only if their semi-invariants are algebraically independent.*

38 *The Center and Focus Problem*

Suppose that comitant (2.56) is written as

$$K = (S_0 + \alpha_0)x^\delta + (S_1 + \alpha_1)x^{\delta-1}y + (S_2 + \alpha_2)x^{\delta-2}y^2$$
$$+ \ldots + (S_\delta + \alpha_\delta)y^\delta, \tag{2.57}$$

where S_i $(i = \overline{0,\delta})$ – some known polynomials in coefficients of system (1.1)–(1.2), and α_i $(i = \overline{0,\delta})$ – unknown polynomials in coefficients of the same system.

According to equality (2.56), for comitant (2.57), we obtain a system of partial differential equations

$$D_3(S_0 + \alpha_0) = -(S_1 + \alpha_1), \quad -D_3(S_1 + \alpha_1) = S_2 + \alpha_2,$$
$$D_3(S_2 + \alpha_2) = -(S_3 + \alpha_3), \quad -D_3(S_3 + \alpha_3) = S_4 + \alpha_4,$$

$$\cdots\cdots\cdots\cdots\cdots\cdots\cdots\cdots\cdots\cdots\cdots\cdots\cdots\cdots\cdots\cdots\cdots\cdots\cdots\cdots \tag{2.58}$$

$$(-1)^{k-1}D_3(S_{k-1} + \alpha_{k-1}) = (-1)^k(S_k + \alpha_k) \ (k = \overline{1,\delta}).$$

Since the number of unknowns $\alpha_0, \alpha_1, \ldots, \alpha_\delta$ is greater than the number of equations (2.58) by one, then the following is true:

Lemma 2.4. *System* (2.58) *has an infinite set of solutions that generate infinite number of comitants of the form* (2.57) *for differential system* (1.1)–(1.2).

2.5 Examples of Algebraic Bases of Centro-Affine Comitants and Invariants for Some Differential Systems

Example 2.5. For system (1.3), Lie algebra of operators of representation of a centro-affine group in the space $E^8(x, y, A)$ consists of operators (1.61)–(1.62). Considering this fact and system (1.3), writing the matrix of coordinates of these operators for it, we have

$$M = \begin{pmatrix} x & 0 & a & 0 & 0 & d & -e & 0 \\ y & 0 & b & 0 & e & f-c & 0 & -e \\ 0 & x & 0 & a & -d & 0 & c-f & d \\ 0 & y & 0 & b & 0 & -d & e & 0 \end{pmatrix}. \tag{2.59}$$

It can be easily checked that $R = rank\, M = 4$. Such conclusion was made based on the fact that, for example, for one of the fourth-order minors M, we have

$$det \begin{pmatrix} x & 0 & 0 & d \\ y & 0 & e & f-c \\ 0 & x & -d & 0 \\ 0 & y & 0 & -d \end{pmatrix} = -d\, k_2 \neq 0,$$

where k_2 is from (2.12). Since in this case, for system (1.3), we have $\Gamma = \{0,1\}$, i.e., it is obtained from system (1.1)–(1.2) for $\ell = 1$, $m_0 = 0$, $m_1 = 1$, then from (2.54) we find $\varrho = 5$. Therefore, for system (1.3) an algebraic basis consists of five centro-affine invariants and comitants. We will show that as such, we can take the first five invariants and comitants from (2.12). For this, using all six invariants and comitants from (2.12), we construct the Jacobi matrix

$$
\mathcal{J} = \begin{pmatrix}
\frac{\partial i_1}{\partial x} & \frac{\partial i_1}{\partial y} & \frac{\partial i_1}{\partial a} & \frac{\partial i_1}{\partial b} & \frac{\partial i_1}{\partial c} & \frac{\partial i_1}{\partial d} & \frac{\partial i_1}{\partial e} & \frac{\partial i_1}{\partial f} \\
\frac{\partial i_2}{\partial x} & \frac{\partial i_2}{\partial y} & \frac{\partial i_2}{\partial a} & \frac{\partial i_2}{\partial b} & \frac{\partial i_2}{\partial c} & \frac{\partial i_2}{\partial d} & \frac{\partial i_2}{\partial e} & \frac{\partial i_2}{\partial f} \\
\frac{\partial i_3}{\partial x} & \frac{\partial i_3}{\partial y} & \frac{\partial i_3}{\partial a} & \frac{\partial i_3}{\partial b} & \frac{\partial i_3}{\partial c} & \frac{\partial i_3}{\partial d} & \frac{\partial i_3}{\partial e} & \frac{\partial i_3}{\partial f} \\
\frac{\partial k_1}{\partial x} & \frac{\partial k_1}{\partial y} & \frac{\partial k_1}{\partial a} & \frac{\partial k_1}{\partial b} & \frac{\partial k_1}{\partial c} & \frac{\partial k_1}{\partial d} & \frac{\partial k_1}{\partial e} & \frac{\partial k_1}{\partial f} \\
\frac{\partial k_2}{\partial x} & \frac{\partial k_2}{\partial y} & \frac{\partial k_2}{\partial a} & \frac{\partial k_2}{\partial b} & \frac{\partial k_2}{\partial c} & \frac{\partial k_2}{\partial d} & \frac{\partial k_2}{\partial e} & \frac{\partial k_2}{\partial f} \\
\frac{\partial k_3}{\partial x} & \frac{\partial k_3}{\partial y} & \frac{\partial k_3}{\partial a} & \frac{\partial k_3}{\partial b} & \frac{\partial k_3}{\partial c} & \frac{\partial k_3}{\partial d} & \frac{\partial k_3}{\partial e} & \frac{\partial k_3}{\partial f}
\end{pmatrix}. \tag{2.60}
$$

Calculating all 28 minors of sixth order for this matrix, we find that all of them are equal to zero. Consequently, among the six centro-affine invariants and comitants (2.12) of system (1.3), there is an algebraic dependence that has the form

$$
(i_1 k_1 - k_3)^2 - i_2 k_1^2 - 2i_3 k_2 + k_3^2 = 0. \tag{2.61}
$$

Note that in the theory of invariants of differential systems (see, for example [37]) a relation of the form (2.61) is called *syzygy*.

Further, calculating with (2.12) the fifth-order minor (2.60), built on the lines 1,2,3,4 and 5 and columns 1,2,3,5 and 6, we obtain

$$
\begin{aligned}
\Delta_{12345}^{12356} = &\, 2[(-4a^2e^3 + 4abce^2 - 4abe^2f - b^2c^2e + 2b^2cef - b^2ef^2)x \\
&+ (2a^2ce^2 - 2a^2e^2f - abc^2e + 2abcef + 4abde^2 \\
&- 2abef^2 - 2b^2cde + 2b^2def)y] \not\equiv 0.
\end{aligned}
$$

This allows us to conclude that the expressions i_1, i_2, i_3, k_1 and k_2 constitute an algebraic basis of centro-affine comitants of system (1.3).

Remark 2.4. *Similar to the previous case, it can be shown that any five centro-affine comitants and invariants from (2.12) form an algebraic basis of comitants for system (1.3).*

Remark 2.5. *With the help of operators (1.62) and formulas for $\widetilde{\varrho}$ from (2.54) it is easy to prove that the algebraic basis of invariants of system (1.3) consists of three elements. These may be i_1, i_2, i_3 from (1.65).*

Example 2.6. Consider quadratic system of differential equations (1.7). With notation (1.6) and Lie algebras of operators of representation of centro-affine group in the space $E^6(A)$ of system (1.7) (see Section 1.5), we obtain that it consists of operators

$$D_1 = -g\frac{\partial}{\partial g} + k\frac{\partial}{\partial k} - 2l\frac{\partial}{\partial l} - m\frac{\partial}{\partial m},$$

$$D_2 = l\frac{\partial}{\partial g} + (-g+m)\frac{\partial}{\partial h} + (-2h+n)\frac{\partial}{\partial k} - l\frac{\partial}{\partial m} - 2m\frac{\partial}{\partial n},$$

$$D_3 = -2h\frac{\partial}{\partial g} - k\frac{\partial}{\partial h} + (-2m+g)\frac{\partial}{\partial l} + (-n+h)\frac{\partial}{\partial m} + k\frac{\partial}{\partial n},$$

$$D_4 = -h\frac{\partial}{\partial h} - 2k\frac{\partial}{\partial k} + l\frac{\partial}{\partial l} - n\frac{\partial}{\partial n}.$$

$$(2.62)$$

Composing the matrix M_1, built on coordinates of the vectors of these operators, we obtain

$$M_1 = \begin{pmatrix} -g & 0 & k & -2l & -m & 0 \\ l & -g+m & -2h+n & 0 & -l & -2m \\ -2h & -k & 0 & -2m+g & -n+h & k \\ 0 & -h & -2k & l & 0 & -n \end{pmatrix}. \quad (2.63)$$

It is easy to see that all 15 minors of the fourth order of this matrix are not identical to zero. For example, we give an expression of one of these minors, built on the columns 1, 2, 3, 4 of matrix (2.63), having a form

$$\Delta_{1234} = -2g^3k + 2g^2h^2 - g^2hn + 6g^2km - 4gh^2m - 9ghkl$$
$$-2ghmn + gkln - 4gkm^2 + 8h^3l - 4h^2ln + 8hklm - 3k^2l^2 \not\equiv 0. \quad (2.64)$$

Consequently, for matrix (2.63), the total rank $R_1 = rank M_1 = 4$. Then, taking into account Remark 2.2 and Theorem 2.4 (in this case, we have $\ell = 0$, $m_0 = 2$), we find $\tilde{\varrho} = 3$, i.e., system (1.7) has an algebraic basis of centro-affine invariants consisting of three elements.

We will search this basis among the invariants of this differential system, known from the paper [37] and having the form

$$I_7 = g^3k - g^2h^2 + g^2hn + g^2km - 3gh^2m + 2ghkl + 2ghmn + 2gkln$$
$$-gkm^2 + gmn^2 - h^3l - h^2ln - 4h^2m^2 + 2hklm + hln^2 - 3hm^2n$$
$$+2klmn - km^3 + ln^3 - m^2n^2,$$

$$I_8 = g^3k - g^2h^2 + g^2hn - g^2km - gh^2m + 4ghkl + 3gkm^2 + gmn^2$$
$$-3h^3l + 3h^2ln - 4h^2m^2 - 4hklm - hln^2 - hm^2n + 2k^2l^2 + 4klmn$$
$$-3km^3 + ln^3 - m^2n^2,$$

$$I_9 = g^3k - g^2h^2 + g^2hn + 3g^2km + 2g^2n^2 - 5gh^2m - 4ghmn$$
$$+3gkm^2 + gmn^2 + h^3l + 3h^2ln - 4h^2m^2 + 3hln^2 - 5hm^2n$$
$$+km^3 + ln^3 - m^2n^2,$$
$$I_{15} = g^4kn - g^3h^2n - 2g^3hkm + g^3hn^2 - g^3k^2l - g^3kmn$$
$$+2g^2h^3m + 3g^2h^2kl - 3g^2h^2mn + 3g^2hkln - 3g^2k^2lm - 3g^2km^2n$$
$$-g^2mn^3 - 2gh^4l - gh^3ln + 4gh^3m^2 + 3gh^2klm + 3gh^2ln^2 + 6ghkm^3 \quad (2.65)$$
$$+ghln^3 + 3ghm^2n^2 - 3gk^2lm^2 - 3gklmn^2 + gkm^3n - gln^4 + gm^2n^3$$
$$-4h^4lm + h^3kl^2 - 6h^3lmn + 3h^2kl^2n - 4h^2m^3n + 3hkl^2n^2$$
$$-3hklm^2n + 4hkm^4 + 2hlmn^3 - 2hm^3n^2 - k^2lm^3 + kl^2n^3$$
$$-3klm^2n^2 + 2km^4n.$$

Applying Theorem 2.1 to these expressions, we obtain

$$D_1(I_j) = D_4(I_j) = -2I_j, \quad D_2(I_j) = D_3(I_j) = 0 \quad (j = 7, 8, 9),$$
$$D_1(I_{15}) = D_4(I_{15}) = -3I_{15}, \quad D_2(I_{15}) = D_3(I_{15}) = 0,$$

where D_i $(i = \overline{1,4})$ are from (2.62). This confirms that expressions (2.65) are centro-affine invariants of system (1.7). Composing Jacobi matrix for polynomials (2.65), we find

$$\mathcal{J} = \begin{pmatrix} \frac{\partial I_7}{\partial g} & \frac{\partial I_7}{\partial h} & \frac{\partial I_7}{\partial k} & \frac{\partial I_7}{\partial l} & \frac{\partial I_7}{\partial m} & \frac{\partial I_7}{\partial n} \\ \frac{\partial I_8}{\partial g} & \frac{\partial I_8}{\partial h} & \frac{\partial I_8}{\partial k} & \frac{\partial I_8}{\partial l} & \frac{\partial I_8}{\partial m} & \frac{\partial I_8}{\partial n} \\ \frac{\partial I_9}{\partial g} & \frac{\partial I_9}{\partial h} & \frac{\partial I_9}{\partial k} & \frac{\partial I_9}{\partial l} & \frac{\partial I_9}{\partial m} & \frac{\partial I_9}{\partial n} \\ \frac{\partial I_{15}}{\partial g} & \frac{\partial I_{15}}{\partial h} & \frac{\partial I_{15}}{\partial k} & \frac{\partial I_{15}}{\partial l} & \frac{\partial I_{15}}{\partial m} & \frac{\partial I_{15}}{\partial n} \end{pmatrix}. \quad (2.66)$$

Note that all 15 fourth-order minors of this matrix are zero. Therefore, among centro-affine invariants (2.65), there exists an algebraic relationship which has the form

$$f_1 \equiv (I_8 - I_7)I_9^2 + (I_9 - I_7)I_7^2 - 2I_{15}^2 = 0. \quad (2.67)$$

This syzygy is known from the paper [37]. Next, by calculating a constructed third-order minor, for example, with the first three lines and the first three columns of matrix (2.66), we find that it is nonzero. Consequently, the centro-affine invariants I_7, I_8, I_9 can be taken as elements of the algebraic basis of the centro-affine invariants of system (1.7). It is not difficult to check that any three invariants of (2.65) also form an algebraic basis of the centro-affine invariants of system (1.7), i.e., $\widetilde{\varrho} = 3$.

Example 2.7. Consider the cubic system of differential equations (1.10). Using notation (1.9) and Lie algebra of operators of representation of centro-affine group in the space $E^8(A)$ of system (1.10) (see Section 1.5), we obtain that algebra consists of operators

$$D_1 = -2p\frac{\partial}{\partial p} - q\frac{\partial}{\partial q} + s\frac{\partial}{\partial s} - 3t\frac{\partial}{\partial t} - 2u\frac{\partial}{\partial u} - v\frac{\partial}{\partial v},$$

$$D_2 = t\frac{\partial}{\partial p} + (-p+u)\frac{\partial}{\partial q} + (-2q+v)\frac{\partial}{\partial r} + (-3r+w)\frac{\partial}{\partial s}$$
$$- t\frac{\partial}{\partial u} - 2u\frac{\partial}{\partial v} - 3v\frac{\partial}{\partial w},$$

$$D_3 = -3q\frac{\partial}{\partial p} - 2r\frac{\partial}{\partial q} - s\frac{\partial}{\partial r} + (-3u+p)\frac{\partial}{\partial t} + (-2v+q)\frac{\partial}{\partial u}$$
$$+ (-w+r)\frac{\partial}{\partial v} + s\frac{\partial}{\partial w},$$

$$D_4 = -q\frac{\partial}{\partial q} - 2r\frac{\partial}{\partial r} - 3s\frac{\partial}{\partial s} + t\frac{\partial}{\partial t} - v\frac{\partial}{\partial v} - 2w\frac{\partial}{\partial w}. \tag{2.68}$$

Composing the matrix M_1, built on coordinates of the vectors of these operators, we obtain

$$M_1 = \begin{pmatrix} -2p & -q & 0 & s & -3t & -2u & -v & 0 \\ t & -p+u & -2q+v & -3r+w & 0 & -t & -2u & -3v \\ -3q & -2r & -s & 0 & -3u+p & -2v+q & -w+r & s \\ 0 & -q & -2r & -3s & t & 0 & -v & -2w \end{pmatrix}. \tag{2.69}$$

It is easy to verify that all 70 fourth-order minors of this matrix are not identically equal to zero. For example, we give an expression of one of such minors, built on the columns 1,2,3 and 4 of matrix (2.69), having the form

$$\Delta_{1234} = 6p^2s^2 - 36pqrs + 2pqsw + 24pr^3 - 8pr^2w + 12prsv$$
$$-6ps^2u + 24q^3s - 18q^2r^2 + 6q^2rw - 12q^2sv + 6qrsu + 4qs^2t - 4r^2st. \tag{2.70}$$

Consequently, for matrix (2.69), the total rank $R_1 = rankM_1 = 4$. Then, considering Remark 2.2 and Theorem 2.4 (we have $\ell = 0$, $m_0 = 3$ in it), we find $\widetilde{\varrho} = 5$, i.e., system (1.10) has an algebraic basis of centro-affine invariants consisting of five elements. We will search for this basis among the invariants of this differential system, known from the paper [43] and having a form

$$J_1 = 2pr + 2pw - 2q^2 - 4qv + 2ru + 2uw - 2v^2,$$
$$J_2 = 2pr - 2q^2 + 2qv - 4ru + 2st + 2uw_2 - 2v^2,$$
$$J_3 = -p^2s + 3pqr - pqw + 5prv - 2psu + pvw - 2q^3 - 2q^2v - qru - 5quw$$
$$+ 2qv^2 + r^2t + 2rtw + ruv - su^2 + tw^2 - 3uvw + 2v^3,$$
$$J_4 = -p^2s + 3pqr - pqw + 2prv + psu + pvw - 2q^3 + q^2v - 4qru - 3qst$$
$$- 2quw - qv^2 + 4r^2t - rtw + 4ruv + 3stv - 4su^2 + tw^2 - 3uvw + 2v^3,$$
$$J_5 = -p^2rw + p^2sv + pq^2w - 2pqsu + pr^2u - 2pruw + 2prv^2 - puw^2$$

$$+pv^2w - q^3v + q^2ru + q^2st + 2q^2uw - 2q^2v^2 - qr^2t + 2qstv - 2qsu^2$$
$$+qtw^2 - qv^3 - 2r^2tv + 2r^2u^2 - 2rtvw + ru^2w + ruv^2 + stv^2 - su^2v,$$
$$J_6 = 2p^3w^3 - 18p^2qvw^2 + 6p^2ruw^2 + 12p^2rv^2w - 6p^2stw^2 + 12p^2suvw$$
$$-12p^2sv^3 + 12pq^2uw^2 + 42pq^2v^2w + 12pqrtw^2 - 120pqruvw + 24pqstvw$$
$$-24pqsu^2w + 36pqsuv^2 - 24pr^2tvw + 78pr^2u^2w - 24prstuw$$
$$+24prstv^2 - 36prsu^2v + 6ps^2t^2w - 12ps^2tuv + 12ps^2u^3 - 12q^3tw^2$$
$$-42q^3v^3 + 36q^2rtvw + 126q^2ruv^2 + 24q^2stuw - 78q^2stv^2 - 36qr^2tuw$$
$$-126qr^2u^2v - 12qrst^2w + 120qrstuv - 6qs^2t^2v - 12qs^2tu^2$$
$$+12r^3t^2w + 42r^3u^3 - 12r^2st^2v - 42r^2stu^2 + 18rs^2t^2u - 2s^3t^3.$$

$$(2.71)$$

Applying Theorem 2.1 to these expressions, we obtain

$$D_1(J_j) = D_4(J_j) = -2J_j, \ D_2(J_j) = D_3(J_j) = 0 \ (j = 1, 2),$$
$$D_1(J_j) = D_4(J_j) = -3J_j, \ D_2(J_j) = D_3(J_j) = 0 \ (j = 3, 4),$$
$$D_1(J_5) = D_4(J_5) = -4J_5, \ D_2(J_5) = D_3(J_5) = 0,$$
$$D_1(J_6) = D_4(J_6) = -6J_6, \ D_2(J_6) = D_3(J_6) = 0,$$

where $D_i \ (i = \overline{1,4})$ are from (2.68).

This confirms that expressions (2.71) are centro-affine invariants of system (1.10). Composing the Jacobi matrix for the polynomials (2.71), we have

$$\mathcal{J} = \begin{pmatrix} \frac{\partial J_1}{\partial p} & \frac{\partial J_1}{\partial q} & \frac{\partial J_1}{\partial r} & \frac{\partial J_1}{\partial s} & \frac{\partial J_1}{\partial t} & \frac{\partial J_1}{\partial u} & \frac{\partial J_1}{\partial v} & \frac{\partial J_1}{\partial w} \\ \frac{\partial J_2}{\partial p} & \frac{\partial J_2}{\partial q} & \frac{\partial J_2}{\partial r} & \frac{\partial J_2}{\partial s} & \frac{\partial J_2}{\partial t} & \frac{\partial J_2}{\partial u} & \frac{\partial J_2}{\partial v} & \frac{\partial J_2}{\partial w} \\ \frac{\partial J_3}{\partial p} & \frac{\partial J_3}{\partial q} & \frac{\partial J_3}{\partial r} & \frac{\partial J_3}{\partial s} & \frac{\partial J_3}{\partial t} & \frac{\partial J_3}{\partial u} & \frac{\partial J_3}{\partial v} & \frac{\partial J_3}{\partial w} \\ \frac{\partial J_4}{\partial p} & \frac{\partial J_4}{\partial q} & \frac{\partial J_4}{\partial r} & \frac{\partial J_4}{\partial s} & \frac{\partial J_4}{\partial t} & \frac{\partial J_4}{\partial u} & \frac{\partial J_4}{\partial v} & \frac{\partial J_4}{\partial w} \\ \frac{\partial J_5}{\partial p} & \frac{\partial J_5}{\partial q} & \frac{\partial J_5}{\partial r} & \frac{\partial J_5}{\partial s} & \frac{\partial J_5}{\partial t} & \frac{\partial J_5}{\partial u} & \frac{\partial J_5}{\partial v} & \frac{\partial J_5}{\partial w} \\ \frac{\partial J_6}{\partial p} & \frac{\partial J_6}{\partial q} & \frac{\partial J_6}{\partial r} & \frac{\partial J_6}{\partial s} & \frac{\partial J_6}{\partial t} & \frac{\partial J_6}{\partial u} & \frac{\partial J_6}{\partial v} & \frac{\partial J_6}{\partial w} \end{pmatrix}. \qquad (2.72)$$

Note that all 28 minors of the sixth order of this matrix are zero. Therefore, among centro-affine invariants (2.71), there exists an algebraic relationship (syzygy), which has the form

$$\varphi_1 \equiv 18J_1^6 - 81J_1^5J_2 + 189J_1^4J_2^2 + 108J_1^4J_5 - 279J_1^3J_2^3 - 108J_1^3J_2J_5$$
$$-30J_1^3J_3^2 + 96J_1^3J_3J_4 - 12J_1^3J_4^2 - 144J_1^3J_6 + 243J_1^2J_2^4 - 324J_1^2J_2^2J_5$$
$$+162J_1^2J_2J_3^2 - 432J_1^2J_2J_3J_4 + 108J_1^2J_2J_4^2 + 324J_1^2J_2J_6 - 108J_1J_2^5$$
$$+540J_1J_2^3J_5 - 196J_1J_2^2J_3^2 + 504J_1J_2^2J_3J_4 - 144J_1J_2^2J_4^2 - 432J_1J_2^2J_6$$
$$-648J_1J_2J_5^2 + 288J_1J_3^2J_5 - 144J_1J_3J_4J_5 - 144J_1J_4^2J_5 - 432J_1J_5J_6$$
$$+18J_2^6 - 216J_2^4J_5 + 66J_2^3J_3^2 - 168J_2^3J_3J_4 + 48J_2^3J_4^2 + 144J_2^3J_6 \qquad (2.73)$$
$$+648J_2^2J_5^2 - 720J_2J_3^2J_5 + 1008J_2J_3J_4J_5 - 288J_2J_4^2J_5 - 864J_2J_5J_6$$
$$+128J_3^4 - 416J_3^3J_4 + 480J_3^2J_4^2 + 264J_3^2J_6 - 224J_3J_4^3 - 672J_3J_4J_6$$
$$+32J_4^4 + 192J_4^2J_6 - 2592J_5^3 + 288J_6^2 = 0.$$

Next, by calculating the constructed fifth-order minor, for example, with the first five lines and the first five columns of matrix (2.72), we find that it is nonzero. Consequently, the centro-affine invariants $J_1 - J_5$ can be taken as elements of the algebraic basis of the centro-affine invariants of system (1.10). It is not difficult to check that any five invariants from (2.71) also form an algebraic basis of the centro-affine invariants of system (1.10).

2.6   Comments to Chapter Two

Centro-affine comitants and invariants of differential systems of the form (1.1)–(1.2) have found wide applications in qualitative study of such systems (see, e.g., [34], [37], [43]). However, the existing methods of their construction [37,43] do not allow us to know a priori the number of invariants and comitants in algebraic basis and other bases, the number of which differs from system to system.

In this chapter, general formulas for the number of invariants and comitants included in an algebraic basis of any system of the form (1.1)–(1.2) are given.

3

Generating Functions and Hilbert Series for Sibirsky Graded Algebras of Comitants and Invariants of Differential Systems

3.1 Formulas for Weights of Centro Affine Comitants and Invariants of Given Type

Note that the index i over the coefficients of system (1.1)–(1.2) indicates that a_k^{ij} belong to homogeneities $P_{m_i}(x,y)$ and $Q_{m_i}(x,y)$, and ℓ indicates the number of homogeneities in the right side of system (1.2).

Definition 3.1. *Let's say that centro-affine comitant of system* (1.1)–(1.2) *has type*

$$(d) = (\delta, d_0, d_1, ..., d_\ell), \tag{3.1}$$

if it is a homogeneous polynomial of degree d_i with respect to the coefficients a_k^{ij} of homogeneities of $P_{m_i}(x,y)$ and $Q_{m_i}(x,y)$ and of degree δ with respect to the phase variables x,y. At the same time, the number $d = \sum_{i=0}^{\ell} d_i$ (δ) is called the degree (order) of comitant of type (3.1).

Observation 3.1. *From Definition 2.1 of centro-affine invariant as a comitant, in which the phase variables x and y are absent, it follows that for an invariant of system* (1.1)–(1.2) *of type* (3.1) *in this record $\delta = 0$.*

If we consider, for example, type (3.1) for centro-affine comitants of system (1.3), then it will be written as

$$(d) = (\delta, d_0, d_1), \tag{3.2}$$

since on the right side of this system, there are two homogeneities: of zero degree and the linear one.

Example 3.1. Consider the types of centro-affine invariants of system (1.3), given in (2.12) from (3.2):

(1) invariant i_1 has a type $(0, 0, 1)$;
(2) invariant i_2 has a type $(0, 0, 2)$;

DOI: 10.1201/9781003193074-4

(3) invariant i_3 has a type $(0, 2, 1)$;
(4) comitant k_1 has a type $(1, 1, 0)$;
(5) comitant k_2 has a type $(2, 0, 1)$;
(6) comitant k_3 has a type $(1, 1, 1)$.

Example 3.2. The centro-affine invariants I_7, I_8, I_9 and I_{15} from (2.65) are invariants of type $(0, 4)$ and $(0, 6)$ for system (1.7). If we consider centro-affine invariants (2.71) for system (1.10), then J_1, J_2 are invariants of type $(0, 2)$, J_3, J_4 – of type $(0, 3)$, J_5 – of type $(0, 4)$, J_6 – of type $(0, 6)$.

Lemma 3.1. *If the centro-affine comitant $K(x, y, A)$ of system (1.1)–(1.2) has type (3.1) and weight g, then for $\Gamma = \{m_i\}_{i=0}^{\ell}$, the following equality is true:*

$$2g = \sum_{i=0}^{\ell} d_i(m_i - 1) - \delta. \qquad (3.3)$$

Proof. Let there be given a centro-affine comitant $K(x, y, A)$ in the form

$$K(x, y, A) = \sum C \prod_{i=0}^{\ell} (\overset{i}{a}_0^1)^{\overset{i}{\eta}_0} (\overset{i}{a}_1^1)^{\overset{i}{\eta}_1} ... (\overset{i}{a}_{m_i}^1)^{\overset{i}{\eta}_{m_i}} (\overset{i}{a}_0^2)^{\overset{i}{\xi}_0} (\overset{i}{a}_1^2)^{\overset{i}{\xi}_1} ... (\overset{i}{a}_{m_i}^2)^{\overset{i}{\xi}_{m_i}}$$
$$\times x^{\delta_1} y^{\delta_2}, \qquad (3.4)$$

having type (3.1) and weight g, where C – numerical coefficients. Therefore, from (3.4), according to Definition 3.1, we have

$$\delta_1 + \delta_2 = \delta, \qquad (3.5)$$

$$\overset{i}{\eta}_0 + \overset{i}{\eta}_1 + ... + \overset{i}{\eta}_{m_i} + \overset{i}{\xi}_0 + \overset{i}{\xi}_1 + \overset{i}{\xi}_2 + ... + \overset{i}{\xi}_{m_i} = d_i, \ (i = \overline{0, \ell}). \qquad (3.6)$$

By implementing the following centro-affine transformation in system, (1.1)–(1.2)

$$\bar{x} = \mu^{-1} x, \ \bar{y} = \mu^{-1} y, \ \Delta = \mu^{-2}, \ \mu \in \mathbb{R} \backslash \{0\}, \qquad (3.7)$$

we obtain system (2.19), in which the new coefficients are of the form

$$\overset{i}{b}_k^j = \mu^{m_i - 1} \overset{i}{a}_k^j \ (j = 1, 2, \ i = \overline{0, \ell}, \ k = \overline{0, m_i}). \qquad (3.8)$$

On the other hand, if we use the fact that $K(x, y, A)$ is a centro-affine comitant of system (1.1)–(1.2) of the weight g, then according to Definition 2.1 and Remark 2.1 for transformation (3.7) with the help of (3.4), we have

$$\sum C \prod_{i=0}^{\ell} (\overset{i}{b}_0^1)^{\overset{i}{\eta}_0} (\overset{i}{b}_1^1)^{\overset{i}{\eta}_1} ... (\overset{i}{b}_{m_i}^1)^{\overset{i}{\eta}_{m_i}} (\overset{i}{b}_0^2)^{\overset{i}{\xi}_0} (\overset{i}{b}_1^2)^{\overset{i}{\xi}_1} ... (\overset{i}{b}_{m_i}^2)^{\overset{i}{\xi}_{m_i}} \bar{x}^{\delta_1} \bar{y}^{\delta_2}$$

$$= \mu^{2g} \sum C \prod_{i=0}^{\ell} (\overset{i}{a}_0^1)^{\overset{i}{\eta}_0} (\overset{i}{a}_1^1)^{\overset{i}{\eta}_1} ... (\overset{i}{a}_{m_i}^1)^{\overset{i}{\eta}_{m_i}} (\overset{i}{a}_0^2)^{\overset{i}{\xi}_0} (\overset{i}{a}_1^2)^{\overset{i}{\xi}_1} ... (\overset{i}{a}_{m_i}^2)^{\overset{i}{\xi}_{m_i}} x^{\delta_1} y^{\delta_2}.$$

Substituting (3.7) and (3.8) in the left side of this equality, we find

$$\mu^{\sum_{i=0}^{\ell}(m_i-1)(\overset{i}{\eta}_0+\overset{i}{\eta}_1+...+\overset{i}{\eta}_{m_i}+\overset{i}{\xi}_0+\overset{i}{\xi}_1+...+\overset{i}{\xi}_{m_i})-(\delta_1+\delta_2)} = \mu^{2g},$$

from where, taking into account (3.5) and (3.6), we obtain equality (3.3). Lemma 3.1 is proved.

Consequence 3.1. *Centro-affine comitants or invariants of system* (1.1)–(1.2), *having the same type* (3.1), *have the same weight.*

The proof of Consequence 3.1 follows directly from equality (3.3).

Observation 3.2. *Equality* (3.3) *for centro-affine comitants of system* (1.1)–(1.2) *are known from the paper* [43].

3.2 Initial form of Generating Function for Centro-Affine Comitants of Differential Systems

Lemma 3.2. *The set of centro-affine comitants of system* (1.1)–(1.2) *of the same type* (3.1) *forms a finite-dimensional linear space, i.e. has a finite maximal system of linearly independent comitants (linear basis) of given type through which all others are linearly expressed.*

Proof. Let there exist the centro-affine comitants of system (1.1)–(1.2) of type (3.1). Then, according to Property 2.2 and Consequence 3.1, for any two such comitants K and k of this set and any numbers $\alpha, \beta \in \mathbb{R}$ we have $K + k$, and αK belongs to this set, and there are equalities $\alpha(K + k) = = \alpha K + \alpha k$, $(\alpha + \beta)K = \alpha K + \beta K$, $(\alpha\beta)K = \alpha(\beta K)$, $1 \cdot K = K$. For any comitant K, there is an inverse element $(-K)$.

The sum of comitants is commutative and associative, and the zero element is considered to be a comitant that is identically equal to zero, which can be considered a comitant of any weight and any type. Therefore, this set forms a linear space. From the monographs [37, p. 18] and [43, p. 29] it follows that such space is finite-dimensional. Lemma 3.2 is proved.

Denote the linear space of centro-affine comitants of system (1.1)–(1.2) of type (3.1) by

$$V_{\Gamma}^{(d)}, \tag{3.9}$$

and its dimension

$$dim_{\mathbb{R}}V_{\Gamma}^{(d)}. \tag{3.10}$$

In the monograph [43, pp. 24–26], it is shown that for the spaces of comitants (3.9), there is a famous classic result (see., e.g., [16]).

Theorem 3.1. *The dimension of space (3.9) is determined by the equality*

$$dim_{\mathbb{R}} V_{\Gamma}^{(d)} = N_g - N_{g-1}, \tag{3.11}$$

where N_g (N_{g-1}) is equal to the set of all different systems of nonnegative integers

$$\overset{1}{\eta}_0, \overset{1}{\eta}_1, ..., \overset{1}{\eta}_{m_1}, \overset{1}{\xi}_0, \overset{1}{\xi}_1, ..., \overset{1}{\xi}_{m_1}, ..., \overset{\ell}{\eta}_0, \overset{\ell}{\eta}_1, ..., \overset{\ell}{\eta}_{m_\ell}, \overset{\ell}{\xi}_0, \overset{\ell}{\xi}_1, ..., \overset{\ell}{\xi}_{m_\ell},$$

among which there are also allowed the identical ones, satisfying system (3.6) together with the equation

$$\sum_{i=0}^{\ell} \left[0\overset{i}{\eta}_0 + 1\overset{i}{\eta}_1 + ... + m_i \overset{i}{\eta}_{m_i} + (-1)\overset{i}{\xi}_0 + 0\overset{i}{\xi}_1 + ... + (m_i - 1)\overset{i}{\xi}_{m_i} \right] = g,$$

$$\left(\sum_{i=0}^{\ell} \left[\overset{i}{\eta}_0 + 1\overset{i}{\eta}_1 + ... + m_i \overset{i}{\eta}_{m_i} + (-1)\overset{i}{\xi}_0 + 0\overset{i}{\xi}_1 + ... + (m_i - 1)\overset{i}{\xi}_{m_i} \right] = g - 1 \right),$$

$$\tag{3.12}$$

where g is a weight of a comitant from space (3.9).

Summing equations (3.6) with the first of (3.12), we obtain an equivalent system

$$\overset{i}{\eta}_0 + \overset{i}{\eta}_1 + ... + \overset{i}{\eta}_{m_i} + \overset{i}{\xi}_0 + \overset{i}{\xi}_1 + ... + \overset{i}{\xi}_{m_i} = d_i, \ (i = \overline{0, \ell}),$$

$$\left(\sum_{i=0}^{\ell} \left[\overset{i}{\eta}_0 + 2\overset{i}{\eta}_1 + ... + (m_i + 1)\overset{i}{\eta}_{m_i} + \overset{i}{\xi}_1 + 2\overset{i}{\xi}_2 + ... + m_i \overset{i}{\xi}_{m_i} \right] = g + d \right),$$

$$\tag{3.13}$$

where d is taken from Definition 3.1.

Consider the sum $\sum u^{g+d} z_0^{d_0} z_1^{d_1} ... z_\ell^{d_\ell}$, which is written with the help of (3.13) as

$$\sum u^{\sum_{i=0}^{\ell} \left[\overset{i}{\eta}_0 + 2\overset{i}{\eta}_1 + ... + (m_i+1)\overset{i}{\eta}_{m_i} + \overset{i}{\xi}_1 + 2\overset{i}{\xi}_2 + ... + m_i \overset{i}{\xi}_{m_i} \right]}$$

$$\times \prod_{i=0}^{\ell} z_i^{\overset{i}{\eta}_0 + \overset{i}{\eta}_1 + ... + \overset{i}{\eta}_{m_i} + \overset{i}{\xi}_0 + \overset{i}{\xi}_1 + ... + \overset{i}{\xi}_{m_i}}.$$

Presenting this sum as

$$\prod_{i=0}^{\ell} \sum (u z_i)^{\overset{i}{\eta}_0} \sum (u^2 z_i)^{\overset{i}{\eta}_1} ... \sum (u^{m_i+1} z_i)^{\overset{i}{\eta}_{m_i}}$$

$$\times \sum z_i^{\overset{i}{\xi}_0} \sum (u z_i)^{\overset{i}{\xi}_1} ... \sum (u^{m_i} z_i)^{\overset{i}{\xi}_{m_i}},$$

we obtain that it is equivalent to the product

$$\prod_{i=0}^{\ell} \frac{1}{\prod_{p=1}^{m_i+1}(1-u^p z_i)} \cdot \frac{1}{\prod_{q=0}^{m_i}(1-u^q z_i)}. \tag{3.14}$$

For more convenient use, the expression (3.14) can be written as

$$\prod_{i=0}^{\ell} \Psi_{m_i}(u), \tag{3.15}$$

where

$$\Psi_{m_i}(u) = \begin{cases} \frac{1}{(1-z_i)(1-u z_i)} & \text{for } m_i = 0, \\ \frac{1}{(1-z_i)(1-u^{m_i+1} z_i) \prod_{k=1}^{m_i}(1-u^k z_i)^2} & \text{for } m_i \neq 0 \end{cases} \tag{3.16}$$

for each $i = 0, 1, 2, ..., \ell$. From the above said, we obtain that the number N_g is equal to coefficient at $u^{g+d} z_0^{d_0} z_1^{d_1} z_2^{d_2} ... z_\ell^{d_\ell}$ in decomposition of function (3.15) using (3.16) by degrees $u, z_0, z_1, z_2, ..., z_\ell$.

Similarly, taking into account Theorem 3.1, it can be shown that the number N_{g-1} is equal to coefficient at $u^{g+d-1} z_0^{d_0} z_1^{d_1} ... z_\ell^{d_\ell}$ in the decomposition by degrees $u, z_0, z_1, ..., z_\ell$ of expression (3.15), taking into account (3.16), or, which is the same, to coefficient at $u^{g+d} z_0^{d_0} z_1^{d_1} ... z_\ell^{d_\ell}$ in the expression $u \prod_{i=0}^{\ell} \Psi_{m_i}(u)$. From here, we have that difference $N_g - N_{g-1}$ at various g and d is a coefficient of the function

$$\Psi_\Gamma(u) = (1-u) \prod_{i=0}^{\ell} \Psi_{m_i}(u), \tag{3.17}$$

which we call *generating function* for centro-affine comitants of system (1.1)–(1.2) with $\Gamma = \{m_i\}_{i=0}^{\ell}$, where $\Psi_{m_i}(u)$ has the form (3.16).

In this way, we conclude that the following is true:

Theorem 3.2. *$\dim_{\mathbb{R}} V_\Gamma^{(d)}$ is equal to coefficient at the monomial $u^{g+d} z_0^{d_0} z_1^{d_1} ... z_\ell^{d_\ell}$ in the decomposition of generating function (3.17) for centro-affine comitants of system (1.1)–(1.2) by positive degrees u, z_0, $z_1, ..., z_\ell$.*

Following A. Cayley (see, e.g., [1]), write expression (3.17), by replacing u by u^{-2} and z_i by $u^{m_i+1} z_i$ in it, in the form

$$\varphi_\Gamma^{(0)}(u) = (1 - u^{-2}) \psi_{m_0}^{(0)}(u) \psi_{m_1}^{(0)}(u)...\psi_{m_\ell}^{(0)}(u), \tag{3.18}$$

where, according to (3.16), we have

$$\psi_{m_i}^{(0)}(u) = \begin{cases} \frac{1}{(1-u z_i)(1-u^{-1} z_i)} & \text{for } m_i = 0, \\ \frac{1}{(1-u^{m_i+1} z_i)(1-u^{-m_i-1} z_i) \prod_{k=1}^{m_i}(1-u^{m_i-2k+1} z_i)^2} & \text{for } m_i \neq 0 \end{cases} \tag{3.19}$$

for each $\Gamma = \{m_i\}_{i=0}^{\ell}$.

Following Sylvester (see, e.g., [1]), we will call expression (3.18) *as initial form of the generating function* for centro-affine comitants of system (1.1)–(1.2).

Note that with the indicated substitutions A. Cayley, the monomial $u^{g+d} z_0^{d_0} z_1^{d_1} ... z_\ell^{d_\ell}$, taking into account equality (3.3), goes to the monomial $u^\delta z_0^{d_0} z_1^{d_1} ... z_\ell^{d_\ell}$.

Taking into account the last affirmation, as well as Theorem 3.2 and equalities (3.18)–(3.19), it is proved

Theorem 3.3. $\dim_{\mathbb{R}} V_\Gamma^{(d)}$ *is equal to coefficient at the monomial* $u^\delta z_0^{d_0} z_1^{d_1} ... z_\ell^{d_\ell}$ *in the decomposition of initial form of generating function* (3.18)–(3.19) *for centro-affine comitants of system* (1.1)–(1.2) *by positive degrees* $u, z_0, z_1, ..., z_\ell$.

Example 3.3. If in system (1.1)–(1.2) we take $\Gamma = \{0, 1\}$, then we obtain $\ell = 1$, $m_0 = 0$, $m_1 = 1$. Then, from (3.18)–(3.19) for centro-affine comitants of system (1.18), we have an initial form of the generating function of the form

$$\varphi_{0,1}^{(0)}(u) = \frac{1 - u^{-2}}{(1 - uz_0)(1 - u^{-1}z_0)(1 - u^2 z_1)(1 - z_1)^2(1 - u^{-2}z_1)}. \tag{3.20}$$

3.3   Examples of Reduced Forms of Generating Functions for Centro-Affine Comitants of Differential Systems

In the papers of A. Cayley (see, for example, [1]) for binary forms, it is shown that if function (3.18)–(3.19) is represented as

$$\varphi_\Gamma(u) - u^{-2}\varphi_\Gamma(u^{-1}) = \varphi_\Gamma^{(0)}(u), \tag{3.21}$$

then we can limit ourselves to studying only rational function $\varphi_\Gamma(u)$, because the second term on the left side of (3.21) contains negative degrees of u, and according to Theorem 3.3 we are only interested in terms with positive degrees in decomposition of the function $\varphi_\Gamma^{(0)}(u)$.

Note that from Theorem 3.3 and equality (3.21), it follows

Proposition 3.1.

$$\varphi_\Gamma(u) = \sum_{(d)} \dim_R V_\Gamma^{(d)} u^\delta z_0^{d_0} z_1^{d_1} ... z_\ell^{d_\ell}. \tag{3.22}$$

Example 3.4. For function (3.20) we find

$$\varphi_{0,1}(u) - u^{-2}\varphi_{0,1}(u^{-1}) = \varphi_{0,1}^{(0)}(u), \tag{3.23}$$

where

$$\varphi_{0,1}(u) = \frac{1 + u z_0 z_1}{(1 - u z_0)(1 - z_1)(1 - z_1^2)(1 - z_0^2 z_1)(1 - u^2 z_1)}. \tag{3.24}$$

However, the question arises, how to obtain the function $\varphi_{\Gamma}(u)$ from (3.21) for more complicated Γ. This problem is solved by applying the improved Sylvester's method [1] in decomposition of the function $\varphi_{\Gamma}^{(0)}(u)$ in elementary fractions [14,33]. The resulting function $\varphi_{\Gamma}(u)$ at Sylvester's suggestion is named as *reduced form of generating function*.

We illustrate this method with one example.

Example 3.5. Suppose that in system (1.1)–(1.2), we have $\Gamma = \{2\}$, i.e. $\ell = 0$, $m_0 = 2$. If in this case to denote $z_0 = c$ in (3.19), then from (3.18) for comitants of this system, we obtain a reduced form of generating function in the form

$$\varphi_2^{(0)}(u) = \frac{1 + u^{-2}}{(1 - u^3 c)(1 - uc)^2(1 - u^{-1}c)^2(1 - u^{-3}c)}. \tag{3.25}$$

If you write the right part (3.25) in the form of elementary fractions with respect to u, then we obtain

$$\varphi_2^{(0)}(u) = \varphi_2(u) + \varphi_2^{(1)}(u), \tag{3.26}$$

where $\varphi_2(u)$ $(\varphi_2^{(1)}(u))$ is the sum of fractions corresponding to factors in the denominator of (3.25) with positive (negative) degrees of u. Then, according to this, we have

$$\varphi_2(u) = A + B, \tag{3.27}$$

$$A = \sum_{i=0}^{2} \frac{A_i}{1 - \alpha_i u}, \quad B = \sum_{j=0}^{1} \frac{B_j}{(1 - uc)^{2-j}}, \tag{3.28}$$

and $\alpha_i = \varrho_i c^{1/3}$ $(i = 0, 1, 2)$ and ϱ_i are roots of the equation

$$\varrho^3 - 1 = 0. \tag{3.29}$$

Multiplying both sides of equality (3.26) by $1 - u^3 c$ and taking into account (3.27) and (3.28), we obtain

$$\varphi_2(u)(1 - u^3 c) = \sum_{i=0}^{2} \frac{A_i}{1 - \alpha_i u}(1 - u^3 c) + [B + \varphi_2^{(1)}(u)](1 - u^3 c). \tag{3.30}$$

Since $1 - u^3c = (1 - \alpha_0 u)(1 - \alpha_1 u)(1 - \alpha_2 u)$, then (3.30) can be written as

$$\varphi_2^{(0)}(u)(1 - u^3c) = \sum_{i=0}^{2} A_i \prod_{\substack{j=0 \\ j \neq i}}^{2} (1 - \alpha_j u) + [B + \varphi_2^{(1)}(u)](1 - u^3c). \quad (3.31)$$

Substituting $u = \alpha_i^{-1}$ into this equality, we obtain

$$A_i \prod_{\substack{j=0 \\ j \neq i}}^{2} (1 - \alpha_j \alpha_i^{-1}) = \varphi_2^{(0)}(u)(1 - u^3c)|_{u = \alpha_i^{-1}} \ (i = 0, 1, 2). \quad (3.32)$$

Taking into account that ϱ_i are the roots of equation (3.29), it is easy to show that

$$\prod_{\substack{j=0 \\ j \neq i}}^{2} (1 - \alpha_j \alpha_i^{-1}) = 3 \ (i = 0, 1, 2).$$

Considering the last equality, from (3.32) and (3.25), we find

$$3A_i = \frac{1 - \alpha_i^2}{(1 - \alpha_i^{-1}c)^2(1 - \alpha_i c)(1 - \alpha_i^3 c)} \ (i = 0, 1, 2).$$

From here, taking into account that $\alpha_i = \varrho_i c^{1/3} \ (i = 0, 1, 2)$ and $\varrho_i^3 = 1$, we have

$$3A_i = \frac{1}{(1 - c^2)(1 - \alpha_i^2)(1 - \alpha_i^4)^2} \ (i = 0, 1, 2).$$

Since

$$1 - \alpha_i^2 = \frac{1 - \alpha_i^6}{1 + \alpha_i^2 + \alpha_i^4}, \ 1 - \alpha_i^4 = \frac{1 - \alpha_i^{12}}{1 + \alpha_i^4 + \alpha_i^8},$$

then from the last equality, we obtain

$$A_i = \frac{(1 + \alpha_i^2 + \alpha_i^4)(1 + \alpha_i^4 + \alpha_i^8)^2}{3(1 - c^2)^2(1 - c^4)^2}. \quad (3.33)$$

Substituting (3.33) into the first equality from (3.28), we find

$$A = \frac{1}{3(1 - c^2)^2(1 - c^4)^2} \left[\sum_{i=0}^{2} \frac{N_i}{1 - \alpha_i u} \right], \quad (3.34)$$

where

$$N_i = (1 + \alpha_i^2 + \alpha_i^4)(1 + \alpha_i^4 + \alpha_i^8)^2 \ (i = 0, 1, 2). \quad (3.35)$$

Bringing the expression in square brackets in (3.34) to a common denominator and taking into account that

$$\alpha_0 + \alpha_1 + \alpha_2 = c^{1/3}(\varrho_0 + \varrho_1 + \varrho_2) = 0,$$

we will obtain

$$A = \frac{\sum_{i=0}^{2} N_i + u \sum_{i=0}^{2} \alpha_i N_i + u^2(\alpha_1 \alpha_2 N_0 + \alpha_0 \alpha_2 N_1 + \alpha_0 \alpha_1 N_2)}{3(1 - c^2)^2(1 - c^4)^2(1 - u^3 c)}. \quad (3.36)$$

Taking into account that

$$\sum_{i=0}^{2} \alpha_i^m = \begin{cases} 3c^n & \text{for } m = 3n, \\ 0 & \text{for } m \neq 3n, \end{cases}$$

using (3.35), we find

$$\sum_{i=0}^{2} N_i = 3 + 6c^2 + 15c^4 + 3c^6,$$

$$\sum_{i=0}^{2} \alpha_i N_i = 3c + 15c^3 + 6c^5 + 3c^7,$$

$$\alpha_1 \alpha_2 N_0 + \alpha_0 \alpha_2 N_1 + \alpha_0 \alpha_1 N_2 = 9c^2 + 9c^4 + 9c^6.$$

From the last equalities and (3.36) after reduction by 3, we obtain

$$A = \frac{1}{(1 - c^2)^2(1 - c^4)^2(1 - u^3 c)}\Big[1 + 2c^2 + 5c^4 + c^6 + u(c + 5c^3$$

$$+ 2c^5 + c^7) + u^2(3c^2 + 3c^4 + 3c^6)\Big]. \quad (3.37)$$

If you multiply both parts (3.26) by $(1 - uc)^2$, then using (3.25) and the second equality from (3.28), we will obtain

$$B_0 + B_1(1 - uc) + [A + \varphi_2^{(1)}(u)](1 - uc)^2 = \frac{1 - u^{-2}}{(1 - u^3 c)(1 - u^{-1} c)^2(1 - u^{-3} c)}. \quad (3.38)$$

Substituting $u = c^{-1}$ into this equality, we find

$$B_0 = \frac{c^2}{(1 - c^2)^2(1 - c^4)}. \quad (3.39)$$

Taking the derivative with respect to u from both parts of (3.38) and taking $u = c^{-1}$, we have

$$-cB_1 = \frac{d}{du}\left[\frac{1 - u^{-2}}{(1 - u^3 c)(1 - u^{-1} c)^2(1 - u^{-3} c)}\right]_{u=c^{-1}},$$

from where we obtain

$$B_1 = -\frac{3c^2(1 + c^2 + c^4)}{(1 - c^2)^2(1 - c^4)^2}. \quad (3.40)$$

Substituting (3.39) in (3.40), in the second equality of (3.28), we have

$$B = \frac{1}{(1-c^2)^2(1-c^4)^2(1-uc)^2}\left[-4c^2 - 3c^4 - 2c^6 + u(3c^3 + 3c^5 + 3c^7)\right]. \quad (3.41)$$

From (3.27) with the help of (3.40) and (3.41) after reducing by $1 - c^2$, we obtain the following form of the generating function:

$$\varphi_2(u) = \frac{N_2(u,c)}{D_2(u,c)}, \quad (3.42)$$

where

$$D_2(u,c) = (1-c^2)(1-c^4)^2(1-uc)^2(1-u^3c),$$
$$N_2(u,c) = 1 - c^2 + c^4 + u(-c + 3c^3 - 2c^5) + u^2(2c^2 - 3c^4$$
$$+ c^6) + u^3(-c^3 + c^5 - c^7). \quad (3.43)$$

Example 3.6. Suppose that in system (1.1)–(1.2), we have $\Gamma = \{3\}$, i.e. $\ell = 0$, $m_0 = 3$. If in this case to denote $z_1 = d$ in (3.19), then from (3.18) for comitants of this system, we obtain the reduced form of generating function of the form

$$\varphi_3^{(0)}(u) = \frac{1 - u^{-2}}{(1 - u^4 d)(1 - u^2 d)^2(1 - d)^2(1 - u^{-2}d)^2(1 - u^{-4}d)}. \quad (3.44)$$

Analogously to Example 3.5, using this function, we obtain the following reduced form of generating function:

$$\varphi_3(u) = \frac{N_3(u,d)}{D_3(u,d)}, \quad (3.45)$$

where

$$D_3(u,d) = (1-d^2)^3(1-d^3)^2(1-u^2d)^2(1-u^4d),$$
$$N_3(u,d) = 1 - d^2 + d^4 + u^2(-d + d^2 + 3d^3 - 2d^5) + u^4(2d^2$$
$$- 3d^4 - d^5 + d^6) + u^6(-d^3 + d^5 - d^7). \quad (3.46)$$

Observation 3.3. *In the papers [17], [18], [33], [34], the reduced forms of generating functions for comitants of differential systems (1.1)–(1.2) for $\Gamma = \{0\}$, $\{1\}$, $\{0,1\}$, $\{2\}$, $\{0,2\}$, $\{1,2\}$, $\{0,1,2\}$, $\{3\}$, $\{0,3\}$, $\{1,3\}$, $\{2,3\}$, $\{4\}$, $\{1,4\}$, $\{5\}$, $\{1,5\}$ are found.*

Observation 3.4. *The reduced form of generating function for invariants of differential system (1.1)–(1.2) with fixed Γ is obtained from the reduced form of generating function for comitants $\varphi_\Gamma(u)$ of the same system for $u = 0$.*

3.4 Hilbert Series for Graded Algebras of Unimodular Comitants and Invariants of Differential Systems

Let $SL(2,\mathbb{R}) \subseteq GL(2,\mathbb{R})$ be a subgroup of unimodular transformations, i.e. $SL(2,\mathbb{R})$ consists of transformations of the form (1.27), for which $\Delta = 1$.

Definition 3.2. *We say that an integer rational function*

$$L(x,y,\overset{0}{a}_0^1,\overset{0}{a}_1^1,...,\overset{0}{a}_{m_0}^1,...,\overset{\ell}{a}_0^2,\overset{\ell}{a}_1^2,...,\overset{\ell}{a}_{m_\ell}^2)$$

with respect to the phase variables x,y and coefficients of system (1.1)–(1.2) is called a unimodular comitant of this system if the following equality takes place:

$$L(\overline{x},\overline{y},\overset{0}{b}_0^1,\overset{0}{b}_1^1,...,\overset{0}{b}_{m_0}^1,...,\overset{\ell}{b}_0^2,\overset{\ell}{b}_1^2,...,\overset{\ell}{b}_{m_\ell}^2)=L(x,y,\overset{0}{a}_0^1,\overset{0}{a}_1^1,...,\overset{0}{a}_{m_0}^1,...,\overset{\ell}{a}_0^2,\overset{\ell}{a}_1^2,...,\overset{\ell}{a}_{m_\ell}^2)$$

for any coefficients of system (1.1)–(1.2), phase variables x,y and transformations $q \in SL(2,\mathbb{R})$ from (1.27).

Similarly to (3.9) we denote the linear space of unimodular comitants of system (1.1)–(1.2) of type (3.1) by

$$S_\Gamma^{(d)}. \tag{3.47}$$

Lemma 3.3. $V_\Gamma^{(d)} \cong S_\Gamma^{(d)}.$
Proof. Let comitant

$$K(x,y,\overset{0}{a}_0^1,\overset{0}{a}_1^1,...,\overset{0}{a}_{m_0}^1,...,\overset{\ell}{a}_0^2,\overset{\ell}{a}_1^2,...,\overset{\ell}{a}_{m_\ell}^2)$$

belong to space (3.9). Then, taking into account the inclusion $SL(2,\mathbb{R}) \subseteq GL(2,\mathbb{R})$, the mentioned comitant will be an element of spaces (3.47), which indicates a one-to-one correspondence in one direction between the elements of spaces (3.9) and (3.47).

Let us show that such correspondence exists in the contrary direction. To do this, use Remark 1.2. If in system (1.1)–(1.2) one realizes transformation (1.28), then for coefficients of system (2.19) we will have $\overset{i}{b}_k^j = \Delta^{-\frac{1}{2}(m_i-1)}\overset{i}{a}_k^j$. Then, according to the fact that K is a homogeneous comitant of type (3.1) for unimodular comitant L that belongs to space (3.47), with this transformation we will have equality

$$L(\Delta^{\frac{1}{2}}x,\Delta^{\frac{1}{2}}y,\Delta^{-\frac{1}{2}(m_0-1)}\overset{0}{a}_0^1,\Delta^{-\frac{1}{2}(m_0-1)}\overset{0}{a}_1^1,...,\Delta^{-\frac{1}{2}(m_0-1)}\overset{0}{a}_{m_0}^0,$$
$$\Delta^{-\frac{1}{2}(m_0-1)}\overset{1}{a}_0^2,\Delta^{-\frac{1}{2}(m_0-1)}\overset{0}{a}_1^2,..,\Delta^{-\frac{1}{2}(m_0-1)}\overset{0}{a}_{m_0}^2,...,\Delta^{-\frac{1}{2}(m_\ell-1)}\overset{\ell}{a}_{m_\ell}^2) =$$
$$= \Delta^{-g}K(x,y,\overset{0}{a}_0^1,\overset{0}{a}_1^1,...,\overset{0}{a}_{m_0}^1,...,\overset{\ell}{a}_0^2,\overset{\ell}{a}_1^2,...,\overset{\ell}{a}_{m_\ell}^2), \tag{3.48}$$

where for g we have equality (3.13).

If in the resulting system after transformation (1.28) one realizes unimodular transformation (1.29), then for unimodular comitant from the left side of (3.48) we will have

$$L(\overline{x}, \overline{y}, \overset{0}{b_0^1}, \overset{0}{b_1^1}, ..., \overset{0}{b_{m_0}^1}, ..., \overset{\ell}{b_0^2}, \overset{\ell}{b_1^2}, ..., \overset{\ell}{b_{m_\ell}^2}) =$$

$$= L\Big(\Delta^{\frac{1}{2}}x, \Delta^{\frac{1}{2}}y, \Delta^{-\frac{1}{2}(m_0-1)}\overset{0}{a_0^1}, \Delta^{-\frac{1}{2}(m_0-1)}\overset{0}{a_1^1}, ..., \Delta^{-\frac{1}{2}(m_0-1)}\overset{0}{a_{m_0}^1},$$

$$\Delta^{-\frac{1}{2}(m_0-1)}\overset{0}{a_0^2}, \Delta^{-\frac{1}{2}(m_0-1)}\overset{0}{a_1^2}, ..., \Delta^{-\frac{1}{2}(m_0-1)}\overset{0}{a_{m_0}^2}, ..., \Delta^{-\frac{1}{2}(m_\ell-1)}\overset{\ell}{a_{m_\ell}^2}\Big).$$

Then, using the last equality and taking into account (3.48), we have the identity

$$L(\overline{x}, \overline{y}, \overset{0}{b_0^1}, \overset{0}{b_1^1}, ..., \overset{0}{b_{m_0}^1}, ..., \overset{\ell}{b_0^2}, \overset{\ell}{b_1^2}, ..., \overset{\ell}{b_{m_\ell}^2}) =$$

$$= \Delta^{-g}K(x, y, \overset{0}{a_0^1}, \overset{0}{a_1^1}, ..., \overset{0}{a_{m_0}^1}, ..., \overset{\ell}{a_0^2}, \overset{\ell}{a_1^2}, ..., \overset{\ell}{a_{m_\ell}^2})$$

for any transformation (1.28), for any coefficients of system (1.1)–(1.2) and any phase variables x, y. Note that g is defined using (3.3) and, according to Remark 2.1, it is an integer.

Thus, it was established that a unimodular comitant of type (3.1) is uniquely associated with a centro-affine comitant of system (1.1)–(1.2), which coincides with it. Lemma 3.3 is proved.

The proof of Lemma 3.3 contains a criterion for the centro-affine invariance of any homogeneous polynomial in the coefficients of system (1.1)–(1.2) and the phase variables x, y, which is formulated as

Consequence 3.2. *In order that an integer rational and homogeneous function of type (3.1) in coefficients of system (1.1)–(1.2) to be a centro-affine comitant of this system, it is necessary and sufficient that it be a unimodular comitant of the same type (3.1) of the mentioned system.*

Consequence 3.3. *The following equality takes place:*

$$dim_{\mathbb{R}}V_\Gamma^{(d)} = dim_{\mathbb{R}}S_\Gamma^{(d)}.$$

Consider a linear space

$$S_\Gamma = \sum_{(d)} S_\Gamma^{(d)}, \tag{3.49}$$

which is a graded algebra of comitants, in which its components satisfy the inclusion $S_\Gamma^{(d)}S_\Gamma^{(e)} \subseteq S_\Gamma^{(d+e)}$.

Following the paper [41], under a generalized Hilbert series of algebra (3.49), we will understand

$$H(S_\Gamma, u, z_0, z_1, ..., z_\ell) = \sum_{(d)} dim_{\mathbb{R}}S_\Gamma^{(d)} u^\delta z_0^{d_0} z_1^{d_1} ... z_\ell^{d_\ell}. \tag{3.50}$$

From equalities (3.22) and (3.50) with the help of Consequence 3.3 we obtain

$$H(S_\Gamma, u, z_0, z_1, ..., z_\ell) = \varphi_\Gamma(u). \tag{3.51}$$

Note that (according to the same paper [41]) an ordinary Hilbert series is obtained in an obvious way from the generalized

$$H_{S_\Gamma}(u) = H(S_\Gamma, u, u, u, ..., u). \tag{3.52}$$

Thus, using (3.51), we have

Conclusion 3.1. *The reduced form of generating function for comitants of system* (1.1)–(1.2) *with a fixed* Γ *is a generalized Hilbert series for algebra of unimodular comitants* (3.49) *with the same* Γ.

Observation 3.5. *If we denote the algebra of invariants for a fixed* Γ *for system* (1.1)–(1.2) *by* SI_Γ, *then according to Observation 3.4 and equality* (3.51) *for generalized Hilbert series of this algebra, we have*

$$H(SI_\Gamma, z_0, z_1, ..., z_\ell) = H(S_\Gamma, 0, z_0, z_1, ..., z_\ell) = \varphi_\Gamma(0), \tag{3.53}$$

and for the ordinary Hilbert series, we obtain

$$H_{SI_\Gamma}(z) = H(SI_\Gamma, z, z, ..., z). \tag{3.54}$$

Important Remark 3.1. *Note that centro-affine comitants and invariants of the systems of the form* (1.1)–(1.2) *were first studied in the works of Academician K. S. Sibirsky* [37–39] *and further developed in the works of his students. Since in the present section it is shown that the mentioned comitants and invariants form a basis of the graded algebras of comitants* S_Γ *and invariants* SI_Γ, *then these algebras will be called Sibirsky graded algebras of comitants* S_Γ *and Sibirsky graded algebras of invariants* SI_Γ *for the system* $s(\Gamma)$. *In the future we will use the abbreviation "Sibirsky algebras"* S_Γ *and* SI_Γ.

3.5 Comments to Chapter Three

It is worth noting that the method for generating functions is hundreds of years old. It was used in the papers of I. Newton (1642–1727), D. Bernoulli (1700–1782), L. Euler (1707–1783), K. Gauss (1777–1855), R. Riemann (1826–1866), A. Cayley (1821–1895), D. Sylvester (1814–1897), D. Hilbert (1862–1943) and other scientists to prove unexpected results.

Probably, the first manifestation of this method is the Newton binomial formula, which says that the number

$$\binom{n}{k} = \frac{n!}{k!(n-k)!}$$

is the coefficient of t^k in the polynomial $(1+t)^n$, i.e.

$$(1+t)^n = \sum_{k=0}^{n} \binom{n}{k} t^k.$$

In modern language, we can say that the function $(1+t)^n$ is a *generating function* for numbers

$$\binom{n}{0}, \binom{n}{1}, ..., \binom{n}{n}.$$

From these considerations, such numbers are also called *binomial coefficients*.

The method of generating functions is based on a very simple idea. A sequence of real numbers $a_0, a_1, a_2, ...$ is associated with an expression of the form

$$a(t) = a_0 + a_1 t + a_2 t^2 + ...,$$

which we will call a series, or a *generating function* of this sequence. This function can be represented as a polynomial of infinite degree. Such an expression is called a formal power series, since we are not interested in its convergence.

These series often have simple forms that allow us to draw certain conclusions about the sequence $\{a_n\}_{n\geq 0}$, which are very difficult to obtain in another way.

Let V be a vector space represented as a direct sum of finite-dimensional subspaces

$$V = \bigoplus_{n=0}^{\infty} V_n, \quad V_n \bigcap_{(n\neq m)} V_m = \{0\}.$$

Such expansion will be called a *graduation*. A generating function for V, or sequence $dim_{\mathbb{R}} V_n$ $(n = 0, 1, 2, 3, ...)$, will be called the formal series

$$\Phi_V(t) = \sum_{n=0}^{\infty} (dim_{\mathbb{R}} V_n) t^n. \tag{3.55}$$

A remarkable effect for generating functions is that the corresponding formal series can converge in some neighborhood of zero to a concrete function. Thus, by studying its properties (e.g., poles, zeros), we can obtain additional information on the structure of the space V, in particular, on the asymptotic behavior of the sequence $\{dim_{\mathbb{R}} V_n\}_{n=0}^{\infty}$.

If $V = A$ is a graded algebra, then (3.55) is called the Hilbert series for this algebra and is denoted by $H_A(t)$, which carries profound information on the nature of asymptotic behavior of the algebra A.

By studying the space V or the algebra A, the generating functions or the Hilbert series, which depend on several variables, can be introduced in some cases. This fact reflects a more detailed graduation of these objects. As a result, these functions are called the *generalized* generating functions and the Hilbert series, respectively, and those having the form (3.55), are called *ordinary*.

The application of generating functions and Hilbert series in the theory of two-dimensional autonomous first-order polynomial differential systems has its origin in the papers [33,34].

We note that (see [41]) the term Hilbert series arises from the classical Hilbert results relating to the commutative case. Sometimes it is also called the Poincaré series, but today this term should be considered to be settled, linking the name of Poincaré only to the homologous series. Despite the fact that the algebras S_Γ and SI_Γ for system of the form (1.1)–(1.2) have their origin and are thoroughly studied in the papers [33,34], they received the name of "Sibirsky algebra" only in the article [29]. This was because one of the most important problems in the qualitative theory of differential systems, the problem of the number of algebraically independent focus quantities, could be solved using these algebras, which are involved in solving the center and focus problem for any differential system of the form (1.1)–(1.2) with polynomial nonlinearities.

4

Hilbert Series for Sibirsky Algebras S_Γ (SI_Γ) and Krull Dimension for Them

---

4.1 Krull Dimension for Sibirsky Graded Algebras

In the following, we will limit ourselves to studying systems of the form $s(1, m_1, m_2, ..., m_\ell)$ from (1.1)–(1.2) and therefore the Sibirsky algebras $S_{1,m_1,m_2,...,m_\ell}$ and $SI_{1,m_1,m_2,...,m_\ell}$.

From theory of invariants of differential systems [37] and tensors [16], it follows that Sibirsky graded algebras $S_{1,m_1,m_2,...,m_\ell}$ and $SI_{1,m_1,m_2,...,m_\ell}$ are commutative and finitely defined algebras. If for these algebras, we introduce a single notation A, then the last statement for them is written in the form

$$A = < a_1, a_2, ..., a_m | f_1 = 0, f_2 = 0, ..., f_n = 0 > \quad (m, n < \infty), \tag{4.1}$$

where a_i are the generators of this algebra, f_j – defining relations (syzygies) between these generators.

It is known, for example, from [33], that for the simplest differential system $s(0, 1)$ from (1.3) of the form

$$\dot{x} = a + cx + dy, \quad \dot{y} = b + ex + fy$$

the finitely defined graded algebras of comitants $S_{0,1}$ and invariants $SI_{0,1}$ will be written, respectively, as

$$S_{0,1} = < i_1, i_2, i_3, k_1, k_2, k_3 | (i_1 k_1 - k_3)^2 + k_3^2 - i_2 k_1^2 - 2i_3 k_2 = 0 >,$$
$$SI_{0,1} = < i_1, i_2, i_3 >, \tag{4.2}$$

where according to (2.12), we have

$$i_1 = c + f, \quad i_2 = c^2 + 2de + f^2, \quad i_3 = -ea^2 + (c - f)ab + db^2,$$
$$k_1 = -bx + ay, \quad k_2 = -ex^2 + (c - f)xy + dy^2,$$
$$k_3 = -(ea + fb)x + (ca + db)y.$$

Note that in [28] on example of the system $s(0, 1)$ from (1.3), some experience was gained in the application of classical groups, Lie algebras and theory

DOI: 10.1201/9781003193074-5   61 of invariants and comitants in the qualitative study of autonomous polynomial differential systems, accumulated in the Chişinău school of differential equations.

The work [33] also contains examples of Sibirsky algebras of invariants for systems (1.7) and (1.10), which are written, respectively, as

$$SI_2 =< I_7, I_8, I_9, I_{15}|f_1 = 0 >,$$

where $I_7 - I_9$, I_{15} are from (2.65) and f_1 is from (2.67), and

$$SI_3 =< J_1, J_2, J_3, J_4, J_5, J_6|\varphi_1 = 0 >,$$

where $J_1 - J_6$ are from (2.71) and φ_1 is from (2.73).

Definition 4.1. [3] *The elements $a_1, a_2, ..., a_r$ of the algebra A are called algebraically independent, if for any nontrivial polynomial F in r variables, the following inequality holds:*

$$F(a_1, a_2, ..., a_r) \neq 0.$$

Definition 4.2. *Maximal number of algebraically independent elements of a graded algebra A is called the Krull dimension of this algebra, which is denoted by $\varrho(A)$.*

It is known that for the algebra A, given in the form (4.1), the equality $n = m - \varrho(A)$ is true. However, this equality is not very effective, since it is impossible to determine the numbers m and n for the majority of algebras of invariants and comitants of systems of the form (1.1)–(1.2).

In the classical theory of invariants [1] the set of elements $a_1, a_2, ..., a_{\varrho(A)}$ from A, which reflect the Krull dimension of the algebra A, are called *algebraic basis* of this set. This means (similar to Remark 2.2) that for $\forall a \in A$ ($a \neq a_j$) there is such a natural number p, corresponding to a, that the following equality holds:

$$P_0 a^p + P_1 a^{p-1} + ... + P_p = 0, \tag{4.3}$$

where P_k ($k = \overline{0, p}$ are polynomials on a_j ($j = \overline{1, \varrho(A)}$). Note that generally speaking $P_0 \not\equiv 1$.

If for all $a \in A$ in (4.3), we have $P_0 \equiv 1$, then such basis is called *integral algebraic*. Its existence is shown by Hilbert (see, e.g. [1]). The number of its elements is denoted by $\varrho'(A)$.

Note that the number of elements in the mentioned bases does not always coincide between themselves. So in [33], we have that for the system $s(4)$ the Krull dimension $\varrho(SI_4) = 7$, and in [42] for the same system, we find that the number of elements in integral algebraic basis of the same algebra is $\varrho'(SI_4) = 9$, i.e. $\varrho(SI_4) < \varrho'(SI_4)$. From [33], we have that for the system $s(0, 1)$ the equalities $\varrho(S_{0,1}) = \varrho'(S_{0,1}) = 5$ and $\varrho(SI_{0,1}) = = \varrho'(SI_{0,1}) = 3$ are true. From the papers [14,33,44] it follows that for the systems $s(2)$ and $s(3)$

we have $\varrho(SI_2) = \varrho'(SI_2) = 3$, $\varrho(SI_3) = = \varrho'(SI_3) = 5$. In the papers [21,33] we find that for the system $s(1,2)$ the equality $\varrho(SI_{1,2}) = \varrho'(SI_{1,2}) = 7$ is true. However, for the system $s(1,2,3)$, according to [33] and [4], we have that $\varrho(SI_{1,2,3}) = 15$ and $\varrho'(SI_{1,2,3}) = 21$.

These examples lead us to

$$\varrho(A) \leq \varrho'(A).$$

This inequality underlines that an integral algebraic basis contains an algebraic basis of the algebra A. The proof of this fact is easily obtained by contradiction.

Remark 4.1. *The main property of integral algebraic basis of the algebra of invariants A is that this is the smallest number of elements of the algebra A, the equity to zero of which turns to zero all elements of the algebra A.*

Further, we will need obvious statements

Proposition 4.1. *If B is a graded subalgebra of the algebra A, then between the Krull dimensions of these algebras, the inequality holds*

$$\varrho(B) \leq \varrho(A).$$

Proposition 4.2. *If the Krull dimension of the algebra A is $\varrho(A)$, then on any variety $V = \{a = 0, b < 0\}$ with fixed $a, b \in A$ (b does not affect the mentioned variety) in the algebra A the number of algebraically independent elements of this algebra is not greater then $\varrho(A)$ (maybe the number of elements that form an integral algebraic basis is not greater than $\varrho(A)$).*

From Theorem 2.4, Remark 2.2, and Definition 4.2, it follows

Conclusion 4.1. *Krull dimension for Sibirsky graded algebras $S_{1,m_1,m_2,...,m_\ell}$ and $SI_{1,m_1,m_2,...,m_\ell}$ are expressed by the formulas*

$$\varrho(S_{1,m_1,m_2,...,m_\ell}) = 2\left(\sum_{i=1}^{\ell} m_i + \ell\right) + 3, \qquad (4.4)$$

$$\varrho(SI_{1,m_1,m_2,...,m_\ell}) = 2\left(\sum_{i=1}^{\ell} m_i + \ell\right) + 1. \qquad (4.5)$$

4.2 Hilbert Series for Sibirsky Graded Algebras $S_{1,m_1,m_2,...,m_\ell}$, $SI_{1,m_1,m_2,...,m_\ell}$

According to Theorem 3.3 and Consequence 3.3, we obtain that in Sibirsky algebra $S_{1,m_1,m_2,...,m_\ell}$ for its spaces, we have $dim_{\mathbb{R}} S_{1,m_1,m_2,...,m_\ell}^{(d)} < \infty$.

Then, the following (3.50), by the *generalized Hilbert series* of the algebra $S_{1,m_1,m_2,...,m_\ell}$, we will understand the formal series

$$H(S_{1,m_1,m_2,...,m_\ell}, u, z_0, z_1, ..., z_\ell)$$
$$= \sum_{(d)} dim_{\mathbb{R}} S^{(d)}_{1,m_1,m_2,...,m_\ell} u^\delta z_0^{d_0} z_1^{d_1} ... z_\ell^{d_\ell}, \qquad (4.6)$$

about which it is said that it reflects u, z – graduation of the mentioned algebra.

According to Observation 3.5 for the algebra of invariants $SI_{1,m_1,m_2,...,m_\ell}$, we obtain the equality

$$H(SI_{1,m_1,m_2,...,m_\ell}, z_0, z_1, ..., z_\ell) = H(S_{1,m_1,m_2,...,m_\ell}, 0, z_0, z_1, ..., z_\ell), \qquad (4.7)$$

and *usual Hilbert series* are written, respectively, as

$$\begin{aligned} H_{S_{1,m_1,m_2,...,m_\ell}}(u) &= H(S_{1,m_1,m_2,...,m_\ell}, u, u, u, ..., u), \\ H_{SI_{1,m_1,m_2,...,m_\ell}}(z) &= H(SI_{1,m_1,m_2,...,m_\ell}, z, z, ..., z). \end{aligned} \qquad (4.8)$$

The last series carry important information about the nature of asymptotic behavior of these algebras.

The method of constructing generalized Hilbert series (4.6)–(4.8) for Sibirsky algebras $S_{1,m_1,m_2,...,m_\ell}$ and $SI_{1,m_1,m_2,...,m_\ell}$ is developed in the papers [33,34] and shown in simple examples from Section 3.3. As it was shown in this section, the generalized Hilbert series for algebras $S_{0,1}$ and $SI_{0,1}$ of unimodular comitants and invariants of the system $s(0,1)$ have the form, respectively,

$$H(S_{0,1}, u, z_0, z_1) = \frac{1 + uz_0z_1}{(1 - uz_0)(1 - z_1)(1 - z_1^2)(1 - z_0^2 z_1)(1 - u^2 z_1)},$$
$$H(SI_{0,1}, z_0, z_1) = \frac{1}{(1 - z_1)(1 - z_1^2)(1 - z_0^2 z_1)}.$$

Then, according to (4.8), with their help, an ordinary Hilbert series is written as

$$H_{S_{0,1}}(u) = \frac{1 - u + u^2}{(1 - u)^2(1 - u^2)(1 - u^3)^2}, \quad H_{SI_{0,1}}(z) = \frac{1}{(1 - z)(1 - z^2)(1 - z^3)}.$$

Remark 4.2. *Following* [40], *note that the Krull dimension* $\varrho(S_{1,m_1,m_2,...,m_\ell})$ *(*$\varrho(SI_{1,m_1,m_2,...,m_\ell})$*) for Sibirsky graded algebras* $S_{1,m_1,m_2,...,m_\ell}$ *(*$SI_{1,m_1,m_2,...,m_\ell}$*) equals to order of pole of the ordinary Hilbert series* $H_{S_{1,m_1,m_2,...,m_\ell}}(u)$ *(*$H_{SI_{1,m_1,m_2,...,m_\ell}}(z)$*) at the unit.*

For example, taking into account the above-mentioned Hilbert series $H_{S_{0,1}}(u)$ and $H_{SI_{0,1}}(z)$, we obtain for Krull dimension of the algebras $S_{0,1}$ and $SI_{0,1}$, respectively, $\varrho(S_{0,1}) = 5$ and $\varrho(SI_{0,1}) = 3$.

In the other cases, when there is no explicit form of ordinary Hilbert series, but a power series decomposition is known, the following can be used:

Remark 4.3. *We agree that the comparison of series with non-negative coefficients occurs by coefficients* $(\sum a_n t^n \leq \sum b_n t^n \Leftrightarrow a_n \leq b_n, \forall n)$. *Taking this into account, if for commutative graded algebras A and B, we have*

$$H_A(t) \leq H_B(t), \tag{4.9}$$

then for their Krull's dimensions we also obtain $\varrho(A) \leq \varrho(B)$.

It is also clear that if for ordinary Hilbert series of the commutative graded algebra A, we have

$$H_A(t) \leq \frac{C}{(1-t)^m}, \tag{4.10}$$

where C – some fixed constant, then we obtain $\varrho(A) \leq m$.

The proof of Remark 4.3 is obtained using the Macaulay theorem from the paper [13].

Consequence 4.1. *Note that formulas* (4.4)–(4.5) *contain an explicit form of Krull dimensions for Sibirsky graded algebras* $S_{1,m_1,m_2,\ldots,m_\ell}$, $SI_{1,m_1,m_2,\ldots,m_\ell}$ *of the differential system* $s(1, m_1, m_2, \ldots, m_\ell)$.

However, knowledge of Hilbert series for these algebras gives additional information about the mentioned algebras, which will be used further. At the same time, using these series once again, an information about Krull dimensions of Sibirsky algebras for concrete systems $s(1, m_1, m_2, \ldots, m_\ell)$ is confirmed.

4.3 Hilbert Series for Sibirsky Algebras $S_{1,2}$, $SI_{1,2}$ and Their Krull Dimensions

From (3.18)–(3.19) for $\Gamma = \{1, 2\}$, putting $\ell = 1$ and $m_0 = 1$, $m_1 = 2$ and introducing for convenience the notation $z_0 = b$, $z_1 = c$, we find for comitants of the differential system with quadratic nonlinearities $s(1, 2)$ an initial form of generating function

$$\varphi_{1,2}^{(0)}(u) = (1 - u^{-2})\psi_1^{(0)}(u)\psi_2^{(0)}(u), \tag{4.11}$$

where

$$\psi_1^{(0)}(u) = \frac{1}{(1 - u^2 b)(1 - b)^2(1 - u^{-2}b)},$$

$$\psi_2^{(0)}(u) = \frac{1}{(1 - u^3 c)(1 - uc)^2(1 - u^{-1}c)^2(1 - u^{-3}c)}. \tag{4.12}$$

Using the advanced Sylvester method for decomposition of the function $\varphi_{1,2}^{(0)}(u)$ on elementary fractions by analogy with the Example 3.5 and taking into account Cayley functional equation (3.21) and Conclusion 3.1, we obtain

Theorem 4.1. *A generalized Hilbert series $H(S_{1,2}, u, b, c)$ for Sibirsky algebra $S_{1,2}$ of the system $s(1,2)$ from (1.5) is a rational function of u, b, c and has the form*

$$H(S_{1,2}, u, b, c) = \frac{N_{1,2}(u, b, c)}{D_{1,2}(u, b, c)}, \qquad (4.13)$$

where

$$D_{1,2}(u, b, c) = (1-b)(1-b^2)(1-c^2)(1-c^4)^2(1-bc^2)^2(1-b^3c^2) \cdot \\ \cdot (1-u^2 b)(1-uc)^2(1-u^3 c), \qquad (4.14)$$

$$\begin{aligned}
N_{1,2}(u, b, c) &= 1 - c^2 + c^4 + b(c^2 + 2c^4 - 2c^6) + b^2(c^2 + c^4 - c^6 - c^8) \\
&+ b^3(2c^4 - 2c^6 - c^8) + b^4(-c^6 + c^8 - c^{10}) + u[-c + 3c^3 - 2c^5 \\
&+ b(2c - 5c^5 + 3c^7) + b^2(c - 2c^5 + c^9) + b^3(c^3 - 3c^5 + 2c^7) + b^4(c^7 \\
&- 3c^9 + 2c^{11})] + u^2[2c^2 - 3c^4 + c^6 + b(-3c^4 + 4c^6 - 2c^8) + b^2(-2c^4 \\
&+ c^{10}) + b^3(c^2 - 3c^4 + 2c^{10}) + b^4(-2c^4 + 2c^6 - c^8 + 3c^{10} - c^{12}) \\
&+ b^5(c^6 - c^8 + c^{10})] + u^3[-c^3 + c^5 - c^7 + b(c - 3c^3 + c^5 - 2c^7 + 2c^9) \\
&+ b^2(-2c^3 + 3c^9 - c^{11}) + b^3(-c^3 + 2c^9) + b^4(2c^5 - 4c^7 + 3c^9) \\
&+ b^5(-c^7 + 3c^9 - 2c^{11})] + u^4[b(-2c^2 + 3c^4 - c^6) + b^2(-2c^6 + 3c^8 - c^{10}) + \\
&+ b^3(-c^4 + 2c^8 - c^{12}) + b^4(-3c^6 + 5c^8 - 2c^{12}) + b^5(2c^8 - 3c^{10} + c^{12})] \\
&+ u^5[b(c^3 - c^5 + c^7) + b^2(c^5 + 2c^7 - 2c^9) + b^3(c^5 + c^7 - c^9 - c^{11}) \\
&+ b^4(2c^7 - 2c^9 - c^{11}) + b^5(-c^9 + c^{11} - c^{13})].
\end{aligned}$$

$$(4.15)$$

Proof of Theorem 4.1 follows from the validity of the Cayley functional equation

$$H(S_{1,2}, u, b, c) - u^{-2} H(S_{1,2}, u^{-1}, b, c) = \varphi_{1,2}^{(0)}(u),$$

where $H(S_{1,2}, u, b, c)$ is from (4.13)–(4.15), and $\varphi_{1,2}^{(0)}(u)$ is from (4.11)–(4.12).

According to Observation 3.5 of Theorem 4.1, we have

Consequence 4.2. *A generalized Hilbert series $H(SI_{1,2}, b, c)$ for Sibirsky algebra $SI_{1,2}$ has the form*

$$H(SI_{1,2}, b, c) = \frac{NI_{1,2}(b, c)}{DI_{1,2}(b, c)}, \qquad (4.16)$$

where

$$DI_{1,2}(b,c) = (1-b)(1-b^2)(1-c^2)(1-c^4)^2(1-bc^2)^2(1-b^3c^2),$$

$$NI_{1,2}(b,c) = 1 - c^2 + c^4 + b(c^2 + 2c^4 - 2c^6) + b^2(c^2 + c^4 - c^6) \qquad (4.17)$$

$$-c^8) + b^3(2c^4 - 2c^6 - c^8) + b^4(-c^6 + c^8 - c^{10}).$$

Using equalities (3.52) and (3.54) from expressions (4.13)–(4.17), we obtain that there takes place

Theorem 4.2. *Ordinary Hilbert series* $H_{S_{1,2}}(u)$ *and* $H_{SI_{1,2}}(z)$ *for Sibirsky algebras* $S_{1,2}$ *and* $SI_{1,2}$ *of the system* $s(1,2)$ *from* (1.5) *have the form*

$$H_{S_{1,2}}(u) = \frac{1}{(1-u^2)^2(1-u^3)^3(1-u^4)^3(1-u^5)}(1 + u + u^2$$
$$+ 4u^3 + 11u^4 + 20u^5 + 29u^6 + 33u^7 + 39u^8 + 41u^9 + 39u^{10}$$
$$+ 33u^{11} + 29u^{12} + 20u^{13} + 11u^{14} + 4u^{15} + u^{16} + u^{17} + u^{18}), \quad (4.18)$$

$$H_{SI_{1,2}}(z) = \frac{1 + z^3 + 2z^4 + 3z^5 + 3z^6 + 3z^7 + 2z^8 + z^9 + z^{12}}{(1-z)(1-z^2)(1-z^3)^2(1-z^4)^2(1-z^5)}. \qquad (4.19)$$

Theorem 4.3. *Krull dimensions of Sibirsky algebras* $S_{1,2}$ *and* $SI_{1,2}$ *are equal to* $\varrho(S_{1,2}) = 9$ *and* $\varrho(SI_{1,2}) = 7$, *respectively.*
   Proof. From the paper [22], it is known that in order that a point $u = 1$ ($z = 1$), to be a pole of the function $H_{S_\Gamma}(u) H_{SI_\Gamma}(z)$ with order multiplicities k ($k \geq 1$), it is necessary and sufficient that it be a zero of multiplicity k for the function $\frac{1}{H_{S_\Gamma}(u)} \left(\frac{1}{H_{SI_\Gamma}(z)} \right)$.
   We illustrate this with the ordinary Hilbert series $H_{S_{1,2}}(u)$ and $H_{SI_{1,2}}(z)$ from (4.18) and (4.19). It is easy to see that

$$\frac{1}{H_{S_{1,2}}(u)} = \frac{(1-u)^9}{(1+u)^{-2}(1+u+u^2)^{-3}(1+u+u^2+u^3)^{-3}}$$
$$\cdot \frac{1}{(1+u+u^2+u^3+u^4)^{-1}}(1 + u + u^2 + 4u^3 + 11u^4 + 20u^5 + 29u^6$$
$$+ 33u^7 + 39u^8 + 41u^9 + 39u^{10} + 33u^{11} + 29u^{12} + 20u^{13} + 11u^{14} + 4u^{15}$$
$$+ u^{16} + u^{17} + u^{18})^{-1}$$

and

$$\frac{1}{H_{SI_{1,2}}(z)} = \frac{(1-z)^7}{(1+z)^{-1}(1+z+z^2)^{-2}(1+z+z^2+z^3)^{-2}}$$
$$\cdot \frac{1}{(1+z+z^2+z^3+z^4)^{-1}}(1 + z^3 + 2z^4 + 3z^5 + 3z^6 + 3z^7 + 2z^8$$
$$+ z^9 + z^{12})^{-1},$$

from where we find that $lim_{u\to1}(1-u)^9 H_{S_{1,2}}(u) \neq 0$ and $lim_{z\to1}(1-z)^7 \cdot$ $\cdot H_{SI_{1,2}}(z) \neq 0$. Hence we have that at the point $u = 1$ ($z = 1$) the function $H_{S_{1,2}}(u)$ ($H_{SI_{1,2}}(z)$) has a pole of multiplicity 9 (7). According to Remark 4.2, Theorem 4.3 is proved.

4.4 Hilbert Series for Sibirsky Algebras $S_{1,3}$, $SI_{1,3}$ and Their Krull Dimensions

From (3.18)–(3.19) for $\Gamma = \{1, 3\}$, by accepting $\ell = 1$ and $m_0 = 1$, $m_1 = 3$ and introducing for convenience the notation $z_0 = b$, $z_1 = d$, we find for comitants of the differential system $s(1,3)$ the initial form of the generating function

$$\varphi_{1,3}^{(0)}(u) = (1 - u^{-2})\psi_1^{(0)}(u)\psi_3^{(0)}(u), \tag{4.20}$$

where

$$\psi_1^{(0)}(u) = \frac{1}{(1 - u^2 b)(1 - b)^2(1 - u^{-2} b)}, \tag{4.21}$$

$$\psi_3^{(0)}(u) = \frac{1}{(1 - u^4 d)(1 - u^2 d)^2(1 - d)^2(1 - u^{-2} d)^2(1 - u^{-4} d)}.$$

Using Sylvester's advanced method of decomposition of the function $\varphi_{1,3}^{(0)}(u)$ to elementary fractions, by analogy with Example 3.5 and taking into account Cayley functional equation (3.21) and Conclusion 3.1, we obtain that the following statement is true

Theorem 4.4. *A generalized Hilbert series $H(S_{1,3}, u, b, d)$ for Sibirsky algebra $S_{1,3}$ of the system $s(1,3)$ from (1.8) is a rational function of u, b, d and has the form*

$$H(S_{1,3}, u, b, d) = \frac{N_{1,3}(u, b, d)}{D_{1,3}(u, b, d)}, \tag{4.22}$$

where

$$D_{1,3}(u, b, d) = (1 - b)(1 - b^2)(1 - u^2 b)(1 - bd)^2(1 - b^2 d)(1 - d^2)^3$$
$$\cdot(1 - d^3)^2(1 - u^2 d)^2(1 - u^4 d), \tag{4.23}$$

$$N_{1,3}(u, b, d) = \sum_{k=0}^{4} R_{2k}(b, d)u^{2k},$$

and

$$R_0(b,d) = 1 + b(-d + d^2 + 3d^3 - 2d^5) + b^2(2d^2 - 3d^4 - d^5 + d^6)$$
$$+b^3(-d^3 + d^5 - d^7) - d^2 + d^4,$$
$$R_2(b,d) = b(2d + 4d^2 - 2d^3 - 8d^4 + 4d^6) + b^2(d - d^2 - 6d^3 + 7d^5 - 3d^7)$$
$$+b^3(-2d^2 + 4d^4 - 4d^6 + 2d^8) + b^4(d^3 - d^5 + d^7) - d + d^2 + 3d^3 - 2d^5,$$
$$R_4(b,d) = b(d - d^2 - 6d^3 + 7d^5 - 3d^7) + b^2(-3d^2 - 2d^3 + 6d^4 - 6d^6$$
$$+2d^7 + 3d^8) + b^3(3d^3 - 7d^5 + 6d^7 + d^8 - d^9) + b^4(-d^4 + d^5 + 3d^6 - 2d^8)$$
$$+2d^2 - 3d^4 - d^5 + d^6,$$
$$R_{8-2k}(b,d) = -b^4 d^{10} R_{2k}(b^{-1}, d^{-1}) \ (k = 0, 1).$$
$$(4.24)$$

According to Observation 3.5 from Theorem 4.4, we have

Consequence 4.3. *A generalized Hilbert series* $H(SI_{1,3}, b, d)$ *for Sibirsky algebra* $SI_{1,3}$ *has the form*

$$H(SI_{1,3}, b, d) = \frac{NI_{1,3}(b,d)}{DI_{1,3}(b,d)}, \qquad (4.25)$$

where

$$DI_{1,3}(b,d) = (1-b)(1-b^2)(1-bd)^2(1-b^2d)(1-d^2)^3(1-d^3)^2,$$
$$NI_{1,3}(b,d) = 1 - d^2 + d^4 + b(-d + d^2 + 3d^3 - 2d^5)$$
$$+ b^2(2d^2 - 3d^4 - d^5 + d^6) + b^3(-d^3 + d^5 - d^7). \qquad (4.26)$$

Using equalities (3.52) and (3.54) from expressions (4.22)–(4.26), we obtain that there takes place the following

Theorem 4.5. *Ordinary Hilbert series* $H_{S_{1,3}}(u)$ *and* $H_{SI_{1,3}}(z)$ *for Sibirsky algebra* $S_{1,3}$ *and* $SI_{1,3}$ *of the system* $s(1,3)$ *from (1.8) have the form*

$$H_{S_{1,3}}(u) = \frac{1}{(1-u^2)^5(1-u^3)^5(1-u^5)}(1 + u + u^3 + 9u^4 + 16u^5$$
$$+ 19u^6 + 15u^7 + 14u^8 + 15u^9 + 19u^{10} + 16u^{11} + 9u^{12} + u^{13}$$
$$+ u^{15} + u^{16}), \qquad (4.27)$$

$$H_{SI_{1,3}}(z) = \frac{1}{(1-z)(1-z^2)^5(1-z^3)^3}(1 - z^2 + z^3 + 5z^4 + z^5$$
$$- z^6 + z^8). \qquad (4.28)$$

Theorem 4.6. *Krull dimensions of Sibirsky algebras* $S_{1,3}$ *and* $SI_{1,3}$ *are equal to* $\varrho(S_{1,3}) = 11$ *and* $\varrho(SI_{1,3}) = 9$, *respectively*.

The proof of this theorem is similar to the proof of Theorem 4.3.

4.5 Hilbert Series for Sibirsky Algebras $S_{1,4}$, $SI_{1,4}$ and Their Krull Dimensions

From (3.18)–(3.19) for $\Gamma = \{1,4\}$, by accepting $\ell = 1$ and $m_0 = 1$, $m_1 = 4$ and introducing for convenience the notation $z_0 = b$, $z_1 = e$, we find for comitants of the differential system $s(1,4)$ the initial form of the generating function

$$\varphi_{1,4}^{(0)}(u) = (1 - u^{-2})\psi_1^{(0)}(u)\psi_4^{(0)}(u), \qquad (4.29)$$

where

$$\psi_1^{(0)}(u) = \frac{1}{(1 - u^2 b)(1 - b)^2(1 - u^{-2}b)},$$

$$\psi_4^{(0)}(u) = \frac{1}{(1 - u^5 e)(1 - u^3 e)^2(1 - ue)^2(1 - u^{-1}e)^2(1 - u^{-3}e)^2(1 - u^{-5}e)}.$$
$$(4.30)$$

Using Sylvester's advanced method of decomposition of the function $\varphi_{1,4}^{(0)}(u)$ to elementary fractions, by analogy with Example 3.5 and taking into account Cayley functional equation (3.21) and Conclusion 3.1, we obtain that the following statement is true

Theorem 4.7. *Ordinary Hilbert series $H(S_{1,4}, u, b, e)$ for Sibirsky algebra $S_{1,4}$ of the system $s(1,4)$ is a rational function of u, b, e and has the form*

$$H(S_{1,4}, u, b, e) = \frac{N_{1,4}(u, b, e)}{D_{1,4}(u, b, e)}, \qquad (4.31)$$

where

$$D_{1,4}(u, b, e) = (1 - b)(1 - b^2)(1 - bu^2)(1 - be^2)^2(1 - b^3 e^2)^2(1 - b^5 e^2) \\
\cdot (1 - e^2)(1 - e^4)^2(1 - e^6)^2(1 - e^8)^2(1 - eu)^2(1 - eu^3)^2(1 - eu^5), \qquad (4.32)$$

$$N_{1,4}(u, b, e) = \sum_{k=0}^{13} R_k(b, e)u^k,$$

and R_k is from Appendix 1.

According to Observation 3.5 from Theorem 4.7, we have

Consequence 4.4. *Ordinary Hilbert series $H(SI_{1,4}, b, e)$ for Sibirsky algebra $SI_{1,4}$ has the form*

$$H(SI_{1,4}, b, e) = \frac{NI_{1,4}(b, e)}{DI_{1,4}(b, e)}, \qquad (4.33)$$

where

$$DI_{1,4}(b, e) = (1 - b)(1 - b^2)(1 - be^2)^2(1 - b^3 e^2)^2(1 - b^5 e^2)(1 - e^2)(1 - e^4)^2 \cdot \\
(1 - e^6)^2(1 - e^8)^2, NI_{1,4}(b, e) = R_0(b, e), \qquad (4.34)$$

and $R_0(b, e)$ is from Appendix 1.

Using equalities (3.52) and (3.54) from expressions (4.31)–(4.34), we obtain that there takes place

Theorem 4.8. *Ordinary Hilbert series $H_{S_{1,4}}(u)$ and $H_{SI_{1,4}}(z)$ for Sibirsky algebras $S_{1,4}$ and $SI_{1,4}$ of system $s(1,4)$ have the form*

$$H_{S_{1,4}}(u) = \frac{N_{1,4}(u)}{D_{1,4}(u)}, \tag{4.35}$$

where

$$D_{1,4}(u) = (1 - u^2)(1 - u^3)(1 - u^4)^3(1 - u^5)^2(1 - u^6)^3(1 - u^7)(1 - u^8)^2, \tag{4.36}$$

$$\begin{aligned}
N_{1,4}(u) = {}& 1 + u + u^2 + 5u^3 + 17u^4 + 39u^5 + 100u^6 + 218u^7 + 467u^8 \\
&+ 865u^9 + 1586u^{10} + 2685u^{11} + 4467u^{12} + 6889u^{13} + 10423u^{14} \\
&+ 14934u^{15} + 20921u^{16} + 27849u^{17} + 36293u^{18} + 45278u^{19} + 55254u^{20} \\
&+ 64697u^{21} + 74134u^{22} + 81782u^{23} + 88328u^{24} + 91866u^{25} + 93539u^{26} \\
&+ 91866u^{27} + 88328u^{28} + 81782u^{29} + 74134u^{30} + 64697u^{31} + 55254u^{32} \\
&+ 45278u^{33} + 36293u^{34} + 27849u^{35} + 20921u^{36} + 14934u^{37} + 10423u^{38} \\
&+ 6889u^{39} + 4467u^{40} + 2685u^{41} + 1586u^{42} + 865u^{43} + 467u^{44} + 218u^{45} \\
&+ 100u^{46} + 39u^{47} + 17u^{48} + 5u^{49} + u^{50} + u^{51} + u^{52}.
\end{aligned} \tag{4.37}$$

$$H_{SI_{1,4}}(z) = \frac{NI_{1,4}(z)}{DI_{1,4}(z)}, \tag{4.38}$$

where

$$DI_{1,4}(z) = (1 - z^3)(1 - z^4)^3(1 - z^5)^2(1 - z^6)^2(1 - z^7)(1 - z^8)^2,$$

$$\begin{aligned}
NI_{1,4}(z) = {}& 1 + z + z^2 + 3z^3 + 8z^4 + 15z^5 + 32z^6 + 67z^7 + 129z^8 + 217z^9 \\
&+ 355z^{10} + 546z^{11} + 812z^{12} + 1122z^{13} + 1511z^{14} + 1948z^{15} + 2447z^{16} \\
&+ 2923z^{17} + 3410z^{18} + 3827z^{19} + 4183z^{20} + 4375z^{21} + 4461z^{22} + 4375z^{23} \\
&+ 4183z^{24} + 3827z^{25} + 3410z^{26} + 2923z^{27} + 2447z^{28} + 1948z^{29} + 1511z^{30} \\
&+ 1122z^{31} + 812z^{32} + 546z^{33} + 355z^{34} + 217z^{35} + 129z^{36} + 67z^{37} + 32z^{38} \\
&+ 15z^{39} + 8z^{40} + 3z^{41} + z^{42} + z^{43} + z^{44}.
\end{aligned} \tag{4.39}$$

Theorem 4.9. *Krull dimensions of Sibirsky algebras $S_{1,4}$ and $SI_{1,4}$ are equal to $\varrho(S_{1,4}) = 13$ and $\varrho(SI_{1,4}) = 11$, respectively.*

4.6 Hilbert Series for Sibirsky Algebras $S_{1,5}$, $SI_{1,5}$ and Their Krull Dimensions

From (3.18)–(3.19) for $\Gamma = \{1,5\}$, by accepting $\ell = 1$ and $m_0 = 1$, $m_1 = 5$ and introducing for convenience the notation $z_0 = b$, $z_1 = f$, we find for comitants of the differential system $s(1,5)$ the initial form of the generating function

$$\varphi_{1,5}^{(0)}(u) = (1 - u^{-2})\psi_1^{(0)}(u)\psi_5^{(0)}(u), \tag{4.40}$$

where

$$\psi_1^{(0)}(u) = \frac{1}{(1 - u^2 b)(1 - b)^2(1 - u^{-2}b)},$$

$$\psi_5^{(0)}(u) = \frac{1}{(1 - u^6 f)(1 - u^4 f)^2(1 - u^2 f)(1 - f^2)^2(1 - u^{-2}f)^2} \tag{4.41}$$

$$\cdot \frac{1}{(1 - u^{-4}f)^2(1 - u^{-6}f)}.$$

Using Sylvester's advanced method of decomposition of the function $\varphi_{1,5}^{(0)}(u)$ to elementary fractions, by analogy with Example 3.5 and taking into account Cayley functional equation (3.21) and Conclusion 3.1, we obtain that the following statement is true

Theorem 4.10. *Ordinary Hilbert series $H(S_{1,5}, u, b, f)$ for Sibirsky algebra $S_{1,5}$ of system $s(1,5)$ is a rational function on u, b, f and has the form*

$$H(S_{1,5}, u, b, f) = \frac{N_{1,5}(u, b, f)}{D_{1,5}(u, b, f)}, \tag{4.42}$$

where

$$D_{1,5}(u, b, f) = (1 - b)(1 - b^2)(1 - bu^2)(1 - bf)^2(1 - b^2 f)^2(1 - b^3 f)(1 + f)$$
$$\cdot (1 - f^2)^2(1 - f^3)^3(1 - f^4)^2(1 - f^5)^2(1 - fu^2)^2(1 - fu^4)^2(1 - fu^6), \tag{4.43}$$

$$N_{1,5}(u, b, f) = \sum_{k=0}^{8} R_{2k}(b, f)u^{2k}, $$

and $R_{2k}(b, f)$ is from Appendix 2.

According to Observation 3.5 from Theorem 4.10, we have

Consequence 4.5. *A generalized Hilbert series $H(SI_{1,5}, b, f)$ of Sibirsky algebra $SI_{1,5}$ has the form*

$$H(SI_{1,5}, b, f) = \frac{NI_{1,5}(b, f)}{DI_{1,5}(b, f)}, \tag{4.44}$$

where

$$DI_{1,5}(b,f) = (1-b)(1-b^2)(1-bf)^2(1-b^2f)^2(1-b^3f)(1+f)$$
$$\cdot(1-f^2)^2(1-f^3)^3(1-f^4)^2(1-f^5)^2, \tag{4.45}$$
$$NI_{1,5}(b,f) = R_0(b,f),$$

and $R_0(b,e)$ is from Appendix 2.

Using equalities (3.52) and (3.54) from expressions (4.42)–(4.45), we obtain that there takes place

Theorem 4.11. *Ordinary Hilbert series* $H_{S_{1,5}}(u)$ *and* $H_{SI_{1,5}}(z)$ *for Sibirsky algebras* $S_{1,5}$ *and* $SI_{1,5}$ *of system* $s(1,5)$ *from* (1.5) *have the form*

$$H(S_{1,5}, u) = \frac{N_{1,5}(u)}{D_{1,5}(u)}, \tag{4.46}$$

where

$$D_{1,5}(u) = (1-u^4)^4(1-u^6)^5(1-u^8)^4(1-u^{10})^2, \tag{4.47}$$

$$N_{1,5}(u) = 1 + u^2 + u^4 + 3u^6 + 27u^8 + 70u^{10} + 177u^{12} + 338u^{14}$$
$$+644u^{16} + 1090u^{18} + 1800u^{20} + 2640u^{22} + 3689u^{24} + 4658u^{26}$$
$$+5555u^{28} + 6063u^{30} + 6317u^{32} + 6063u^{34} + 5555u^{36} + 4658u^{38} \tag{4.48}$$
$$+3689u^{40} + 2640u^{42} + 1800u^{44} + 1090u^{46} + 644u^{48} + 338u^{50}$$
$$+177u^{52} + 70u^{54} + 27u^{56} + 3u^{58} + u^{60} + u^{62} + u^{64}.$$

$$H_{SI_{1,5}}(z) = \frac{NI_{1,5}(z)}{DI_{1,5}(z)}, \tag{4.49}$$

where

$$DI_{1,5}(z) = (1-z^2)^4(1-z^3)^4(1-z^4)^3(1-z^5)^2,$$
$$NI_{1,5}(z) = 1 + z + 9z^4 + 22z^5 + 50z^6 + 79z^7 + 120z^8 + 160z^9$$
$$+221z^{10} + 269z^{11} + 325z^{12} + 339z^{13} + 325z^{14} + 269z^{15} \tag{4.50}$$
$$+221z^{16} + 160z^{17} + 120z^{18} + 79z^{19} + 50z^{20} + 22z^{21} + 9z^{22}$$
$$+z^{25} + z^{26}.$$

Theorem 4.12. *Krull dimensions of Sibirsky algebras* $S_{1,5}$ *and* $SI_{1,5}$ *are equal to* $\varrho(S_{1,5}) = 15$ *and* $\varrho(SI_{1,5}) = 13$, *respectively.*

4.7   Obtaining Ordinary Hilbert Series for Sibirsky Algebras $S_{1,2,3}$, $SI_{1,2,3}$ Using, Residues, and Calculating Krull Dimensions for Them

Let G be a linear reductive group on algebraically closed field K and V– an n–dimensional rational representation. Hilbert series of the ring of invariants $K[V]^G$ are denoted by $H(K[V]^G, t)$. [12].

Theorem 4.13. (Molien's formula [12]**).** Let G be a finite group acting on a finite-dimensional vector space V over a field K of characteristic, not divisible by $|G|$. Then

$$H(K[V]^G, t) = \frac{1}{|G|} \sum_{\sigma \in G} \frac{1}{det_V^0 (1 - t\sigma)}.$$

If K has the characteristic 0, then $det_V^0 (1 - t\sigma)$ can be taken as $det_V (1 - t\sigma)$.

Assume that $char(K) = 0$. From Theorem 4.13, it follows that for a finite group, Hilbert series of the ring of invariants can be easily calculated. If G is a finite group, and V is a finite-dimensional representation, then according to [12], we have

$$H(K[V]^G, t) = \frac{1}{|G|} \sum_{\sigma \in G} \frac{1}{det_V (1 - t\sigma)}. \tag{4.51}$$

This idea can be generalized to arbitrary reductive groups. Assume that K is a field of complex numbers $\mathbb{C}$. Then, we can choose the Haar measure $d\mu$ on C with the norm $\int_C d\mu = 1$. Let V be a finite-dimensional rational representation for G. Then, according to [12], the following expression is the generalization of (4.51):

$$H(\mathbb{C}[V]^G, t) = \int_C \frac{d\mu}{det_V (1 - t\sigma)}. \tag{4.52}$$

Note that the Hilbert series $H(\mathbb{C}[V]^G, t)$ converges for $|t| < 1$, because this series is a rational function with poles only in $t = 1$. Since C is compact, there exists a constant $A > 0$ such that for every $\sigma \in C$ and every eigenvalue λ, for σ we have $|\lambda| \leq A$. Since λ^ℓ is an eigenvalue of σ^ℓ, it follows that $|\lambda^\ell| \leq A$ for all ℓ. This means that $|\lambda| \leq 1$. Obviously, the integral of the right-hand side of (4.52) is also defined for $|t| < 1$ [12].

Assume that G is a connected group. Let T be a maximal torus for G, and D the maximal compact subgroup for T. We may assume that C contains D. The torus can be identified with $(\mathbb{C}^*)^r$, where r is the rank of G and D can be identified with the subgroup $(S^1)^r$ of $(\mathbb{C}^*)^r$, where $\mathbb{C}^* \supset S^1$ is a unit circle. We can choose a Haar measure $d\nu$ on D such that the equality $\int_D d\nu = 1$ holds. Suppose that f is a continuous class of functions on C. An integral like $\int_C f(\sigma) d\mu$ can be considered as an integral over D, since f is constant on conjugacy classes. More precisely, there exists a weight function $\varphi : D \to \mathbb{R}$, such that for every continuous class of functions, f we have [12]

$$\int_C f(\sigma)d\mu = \int_D \varphi(\sigma)f(\sigma)d\nu.$$

Then we obtain [12]

$$H(\mathbb{C}[V]^G, t) = \int_C \frac{d\mu}{det_V(1 - t\sigma)} = \int_D \frac{\varphi(\sigma)d\nu}{det_V(1 - t\sigma)}. \qquad (4.53)$$

Choosing the appropriate bases in V and its adjoint space V^*, we can achieve that the compact torus D acts diagonally on V and V^*. Then the action $(z_1, ..., z_r) \in D$ on V^* is defined by the matrix

$$\begin{pmatrix} m_1(z) & 0 & \cdots & 0 \\ 0 & m_2(z) & \cdots & 0 \\ \vdots & \vdots & \ddots & \vdots \\ 0 & 0 & \cdots & m_n(z) \end{pmatrix},$$

where $m_1, m_2, ..., m_n$ are Laurent monomials of $z_1, ..., z_r$.

In these notations, we have $det_V(1 - t \cdot (z_1, ..., z_n)) = (1 - m_1(z)t) \cdot (1 - m_2(z)t) \cdots (1 - m_n(z)t)$. Hence, according to [12], we obtain

$$H(\mathbb{C}[V]^G, t) = \int_D \frac{\varphi(z)d\nu}{(1 - m_1(z)t)(1 - m_2(z)t) \cdots (1 - m_n(z)t)}. \qquad (4.54)$$

It follows that $H(K[V]^G, t)$ is the coefficient z^ρ (in the form of series in $z_1, ..., z_r$ with coefficients at $K(t)$) in the expression

$$\frac{\sum_{w \in W} sgn(w)z^{w(\rho)}}{(1 - m_1(z)t)(1 - m_2(z)t) \cdots (1 - m_n(z)t)},$$

or coefficient $z^0 = 1$ in

$$\frac{z^{-\rho} \sum_{w \in W} sgn(w)z^{w(\rho)}}{(1 - m_1(z)t)(1 - m_2(z)t) \cdots (1 - m_n(z)t)}.$$

Recall Residue Theorem from the theory of functions of a complex variable, which can be applied to calculate Hilbert series for rings of invariants.

Suppose that $f(z)$ is a meromorphic function on $\mathbb{C}$. If $a \in \mathbb{C}$, then f can be written as a Laurent series:

$$f(z) = \sum_{k=-d}^{\infty} c_k(z - a)^k$$

at the point $z = a$.

If $d > 0$ and $c_{-d} \neq 0$, then f has a pole in $z = a$ of order d.

Residue of the function f in $z = a$ is denoted by $Res(f, a)$ and is determined by the equality

$$Res(f, a) = c_{-1}.$$

If the order k of the pole $z = a$ of the function f satisfies the inequality $k \geq 1$, then residue can be calculated from the formula

$$Res(f, a) = \frac{1}{(k-1)!} \lim_{z \to a} \frac{d^{k-1}}{dz^{k-1}} ((z-a)^k f(z)).$$

Choose D such that $\gamma : [0, 1] \to \mathbb{C}$ is a smooth curve. Then the integral over the curve γ is determined by the equality

$$\int_\gamma f(z) dz = \int_0^1 f(\gamma(t)) \gamma'(t) dt.$$

Theorem 4.14. (Residue Theorem [12]**).** Suppose that D is a closed, simply connected compact domain in $\mathbb{C}$, whose border is ∂D, and $\gamma : [0, 1] \to \mathbb{C}$ is such a smooth curve, that $\gamma([0, 1]) = \partial D$, $\gamma(0) = \gamma(1)$, surrounding D exactly once counterclockwise. Suppose that f is a meromorphic function on $\mathbb{C}$ without poles in ∂D. Then we have the formula

$$\frac{1}{2\pi i} \int_\gamma f(z) dz = \sum_{a \in D} Res(f, a).$$

Note that in D there is a finite number of points at which f has a nonzero residue.

Example 4.1 [12]. Let $T = \mathbb{G}_m$ is a one-dimensional torus acting on a three-dimensional space V with the matrix

$$\varrho = \begin{pmatrix} z & 0 & 0 \\ 0 & z & 0 \\ 0 & 0 & z^{-2} \end{pmatrix}.$$

The action $\mathbb{G}_m$ on V^* is given by the matrix

$$\begin{pmatrix} z^{-1} & 0 & 0 \\ 0 & z^{-1} & 0 \\ 0 & 0 & z^2 \end{pmatrix}.$$

Thus, we obtain the equality

$$H_T(K[V], z, t) = \frac{1}{(1 - z^{-1}t)^2 (1 - z^2 t)}. \tag{4.55}$$

In order that the Hilbert series converge, it is necessary that $|z^{-1}t| < 1$ and $|z^2t| < 1$. Suppose that $|z| = 1$ and $|t| < 1$. To find the coefficient of z^0, divide (4.55) by $2\pi i z$ and integrate on the unit circle S^1 in $\mathbb{C}$. Then we have

$$H(K[V]^{G_m}, t) = \frac{1}{2\pi i} \int_{S^1} \frac{dz}{z(1 - z^{-1}t)^2(1 - z^2t)}. \qquad (4.56)$$

According to Residue Theorem 4.14, the expression from (4.56) can be written as

$$\frac{1}{2\pi i} \int_{S^1} f(z)dz = \sum_{a \in D^1} Res(f(z), a),$$

where D^1 is a unit circle and $f(z) = z^{-1}(1 - z^{-1}t)^{-2}(1 - z^2t)^{-1}$. The poles of the function $f(z)$ are only $z = t$ and $z = \pm t^{-1/2}$. Since $|t| < 1$, then the only pole in the unit circle is $z = t$. Calculating the residue, we have

$$\frac{1}{z(1 - z^{-1}t)^2(1 - z^2t)} = \frac{1}{(z - t)^2} \frac{z}{1 - z^2t}.$$

Power series for the $g(z) = \frac{z}{1 - z^2 t}$ in a neighborhood of $z = t$ is given by

$$g(z) = g(t) + g'(t)(z - t) + \frac{g''(t)(z - t)^2}{2} + \ldots$$

$$= \frac{t}{1 - t^3} + \frac{1 + t^3}{(1 - t^3)^2}(z - t) + \ldots . \qquad (4.57)$$

The Hilbert series $H(K[V]^{G_m}, t)$ is a residue of the function

$$\frac{1}{z(1 - z^{-1}t)^2(1 - z^2t)} = \frac{g(z)}{(z - t)^2}$$

in $z = t$. Then, from (4.57) it follows that the indicated series has the form

$$\frac{1 + t^3}{(1 - t^3)^2}.$$

From [11] it is known

Theorem 4.15.

$$H(K[V]^G, t) = \frac{1}{2\pi i} \int_{S^1} \frac{1}{det(I - t_{\rho v}(z))} \frac{dz}{z}$$

where $S^1 \subset \mathbb{C}$ is a unit circle $\{z : |z| = 1\}$.

Using Residue Theorem and the corresponding generating function (3.18)–(3.19), we conclude that the last theorem can be adapted to calculate an ordinary Hilbert series for algebra of comitants and invariants of differential systems.

Theorem 4.16.

$$H_{SI_\Gamma}(t) = \frac{1}{2\pi i} \int\limits_{S^1} \frac{\varphi_\Gamma^{(0)}(z)}{z} dz,$$

where $S^1 \subset \mathbb{C}$ is a unit circle $\{z : |z| = 1\}$, $\varphi_\Gamma^{(0)}(z)$ – corresponding generating function (3.18)–(3.19).

Using Theorem 4.16, the following statement is obtained

Theorem 4.17. *For a cubic differential system $s(1,2,3)$, the ordinary Hilbert series for Sibirsky algebras $S(1,2,3)$ and $SI(1,2,3)$ of comitants and invariants are as follows:*

$$H_{S_{1,2,3}}(t) = \frac{1}{(1-t)^2(1-t^2)^2(1-t^3)^6(1-t^4)^3(1-t^5)^3(1-t^7)}(1-t$$

$$+3t^2 + 9t^3 + 36t^4 + 90t^5 + 220t^6 + 459t^7 + 946t^8 + 1748t^9$$

$$+3032t^{10} + 4845t^{11} + 7302t^{12} + 10268t^{13} + 13749t^{14}$$

$$+17327t^{15} + 20781t^{16} + 23565t^{17} + 25460t^{18} + 26051t^{19}$$

$$+25460t^{20} + 23565t^{21} + 20781t^{22} + 17327t^{23} + 13749t^{24}$$

$$+10268t^{25} + 7302t^{26} + 4845t^{27} + 3032t^{28} + 1748t^{29}$$

$$+946t^{30} + 459t^{31} + 220t^{32} + 90t^{33} + 36t^{34} + 9t^{35}$$

$$+3t^{36} - t^{37} + t^{38}),$$

$$H_{SI_{1,2,3}}(t) = \frac{1}{(1-t)(1-t^2)^3(1-t^3)^5(1-t^4)^2(1-t^5)^3(1-t^7)}(1$$

$$+t^2 + 6t^3 + 24t^4 + 57t^5 + 128t^6 + 244t^7 + 447t^8 + 756t^9 + 1203t^{10}$$

$$+1760t^{11} + 2433t^{12} + 3124t^{13} + 3800t^{14} + 4351t^{15} + 4736t^{16}$$

$$+4854t^{17} + 4736t^{18} + 4351t^{19} + 3800t^{20} + 3124t^{21} + 2433t^{22}$$

$$+1760t^{23} + 1203t^{24} + 756t^{25} + 447t^{26} + 244t^{27}$$

$$+128t^{28} + 57t^{29} + 24t^{30} + 6t^{31} + t^{32} + t^{34}).$$

From this theorem it follows that a Krull dimension of the Sibirsky graded algebra $S_{1,2,3}$ ($SI_{1,2,3}$) is equal to 17 (15).

We note that the method of calculating the ordinary Hilbert series using Theorem 4.16 for Sibirsky algebras of different differential systems was confirmed on the following Hilbert series: H_{SI_1}, H_{S_2}, H_{SI_2}, $H_{SI_{0,2}}$, $H_{SI_{1,2}}$, $H_{SI_{1,3}}$, $H_{SI_{2,3}}$, H_{S_5}, H_{SI_5}, known from [33,34].

Remark 4.4. *Note that for Hilbert series of Sibirsky graded algebra of comitants of the system $s(\Gamma)$, where $\Gamma \not\ni 0$, the following equality holds: $H_{S_\Gamma}(t) = H_{SI_{\Gamma \cup \{0\}}}(t)$.*

4.8 Comments to Chapter Four

As it follows from the papers [33,34], the generalized and the ordinary Hilbert series for Sibirsky algebras of differential systems of the form (1.1)–(1.2) play an important role in studying the structures of these algebras. With the help of these series, one can get an idea of the upper bound of degrees of generators of these algebras. It is known from [40] that if the ordinary Hilbert series for commutative algebras has the form $H(t) = \frac{N(t)}{D(t)}$, where $D(t)$ and $N(t)$ are polynomials in t, then $deg D(t) > deg N(t)$. In our case, for Sibirsky algebras, this inequality also holds. If these algebras are written in the form (4.1) and their ordinary Hilbert series in the form $H_A(t) = \frac{N_A(t)}{D_A(t)}$, then from all examples concerning the construction of the generators of algebras A at the Chişinău school of differential equations, it was found that $deg\, a_i \leq deg D_A(t)$, where $1 \leq i \leq m$

$deg\, a_i$ denotes the degree of generators of these algebras with respect to the coefficients and phase variables of corresponding differential systems. In addition, the ordinary Hilbert series for Sibirsky algebra makes it possible to calculate the Krull dimension of the given algebra, and at the same time, to estimate the upper bound of these dimensions for their subalgebras for which these series are unknown.

5

About the Center and Focus Problem

5.1 On a New Formulation of the Center and Focus Problem for Differential Systems $s(1, m_1, m_2, ..., m_\ell)$

Consider the system $s(1, m_1, m_2, ..., m_\ell)$ of the form (1.1)–(1.2). Suppose that the roots of characteristic equation of the considered system are purely imaginary, i.e., a singular point $O(0,0)$ of this system is a center for it (surrounded by closed trajectories) or a focus (surrounded by spirals) [2], [20].

The center and focus problem can be formulated as follows: *Let for the system $s(1, m_1, m_2, ..., m_\ell)$ from (1.1)–(1.2) the origin of coordinates be a singular point of the second type (center or focus), then what will be the conditions which distinguish a center from a focus?* This problem was posed by H. Poincaré [26]. Fundamental results were obtained by A. Lyapunov [20] which showed that the conditions for center are the vanishing of an infinite sequence of polynomials (focus quantities)

$$L_1, L_2, ..., L_k, ... \tag{5.1}$$

in the coefficients of right side of the system $s(1, m_1, ..., m_\ell)$ of the form (1.1)–(1.2). If at least one of the quantities from (5.1) is nonzero, then the origin of coordinates for the system $s(1, m_1, ..., m_\ell)$ of the form (1.1)–(1.2) is a focus. These conditions are necessary and sufficient.

In the case of the system $s(1, m_1, ..., m_\ell)$ from Hilbert's theorem on the finiteness of basis of polynomial ideals, it follows that *essential conditions for center* in the indicated sequence are only a finite number, and the rest are consequences of them. Then the center and focus problem for the system $s(1, m_1, ..., m_\ell)$ of the form (1.1)–(1.2) takes the following formulation: *how many polynomials ω (essential conditions for center)*

$$L_{n_1}, L_{n_2}, ..., L_{n_\omega}, ... \ (n_i \in \{1, 2, ..., k, ...\}; \ i = \overline{1, \omega}; \ \omega < \infty) \tag{5.2}$$

from (5.1) is it necessary to attract so that their equality to zero annuls all other polynomials in (5.1)?

In the works of academician K. S. Sibirsky [37–39] and his disciples [6]–[8], [10], it was shown that for some systems of the form (1.1)–(1.2), essential conditions of center (5.2) are expressed through centro-affine comitants of these

DOI: 10.1201/9781003193074-6

systems. This prompted the authors not to look for an explicit form of center conditions but to determine the relationship between the number of focus
quantities (5.2) and some characteristics of the set of centro-affine comitants
and invariants. From here arose the *generalized center and focus problem* in
the following formulation: *to determine the upper bound of number of algebraically independent essential focus quantities involved in solving the center
and focus problem for any system $s(1, m_1, ..., m_\ell)$ of the form (1.1)–(1.2).*

5.2   Sibirsky Invariant Variety for Center and Focus

The center and focus problem for differential systems $s(1, m_1, ..., m_\ell)$ of the
form (1.1)–(1.2) has the following classical formulation: *for an infinite system
of polynomials*

$$\{(x^2 + y^2)^k\}_{k=1}^\infty \tag{5.3}$$

there exists such function

$$U(x, y) = x^2 + y^2 + \sum_{k=3}^\infty f_k(x, y), \tag{5.4}$$

*where $f_k(x, y)$ are homogeneous polynomials of degree k with respect to
variables x, y, and such constants*

$$L_1, L_2, ..., L_k, ..., \tag{27.1}$$

that the following identity holds:

$$\frac{dU}{dt} = \sum_{k=1}^\infty L_k(x^2 + y^2)^{k+1} \tag{5.5}$$

(with respect to x and y) along the trajectories of the Lyapunov system [20]

$$\dot{x} = y + \sum_{i=1}^\ell P_{m_i}(x, y), \quad \dot{y} = -x + \sum_{i=1}^\ell Q_{m_i}(x, y), \tag{5.6}$$

which we denote by $s\mathcal{L}(1, m_1, ..., m_\ell)$.

As it follows from the theory of invariants and comitants of differential systems [33,37,43], the algebra $S_{1,m_1,...,m_\ell}$ for any differential system
$s(1, m_1, ..., m_\ell)$, written as

$$\dot{x} = cx + dy + \sum_{i=1}^\ell P_{m_i}(x, y), \quad \dot{y} = ex + fy + \sum_{i=1}^\ell Q_{m_i}(x, y) \tag{5.7}$$

contains, among its generators, the polynomials

$$i_1 = c + f, \quad i_2 = c^2 + 2de + f^2, \quad k_2 = -ex^2 + (c-f)xy + dy^2, \qquad (5.8)$$

given earlier in (2.12).

Remark 5.1. *Note that the set*

$$\mathcal{V} = \{i_1 = c + f = 0, \ Discr(k_2) = 2i_2 - i_1^2 < 0\} \qquad (5.9)$$

will be called the Sibirsky invariant variety of center and focus for differential system (5.7), since the comitant k_2 from (5.8), using a real centro-affine transformation of the plane xOy on the variety $\mathcal{V}$, can be reduced to the form

$$x^2 + y^2, \qquad (5.10)$$

and system (5.7) (scalar change of time t is allowed here) can be reduced to the form (5.6) [37], for which the roots of the characteristic equation are purely imaginary, i.e., the origin of coordinates for this system is a singular point of the second group (center or focus).

According to Remark 5.1, we obtain

Observation 5.1. *If we consider the expression of the comitant k_2 from (5.8) and the fact that with the help of a real centro-affine transformation on Sibirsky invariant variety $\mathcal{V}$ its expression can be reduced to the form (5.10), then formally this variety for differential system (5.7) can be written as*

$$\mathcal{V} = \{f = -c\} \cup \{c = 0, \ d = -e = 1\}. \qquad (5.11)$$

5.3 Focus Quantities L_k and Constants G_k on Sibirsky Invariant Variety of the System $s(1, m_1, ..., m_\ell)$ and Null Focus Pseudo-Quantity

Consider the center and focus problem for differential system (5.7). Then, for this system, we write the identity

$$\left[cx + dy + \sum_{i=1}^{\ell} P_{m_i}(x,y) \right] \frac{\partial U}{\partial x} + \left[ex + fy + \sum_{i=1}^{\ell} Q_{m_i}(x,y) \right] \frac{\partial U}{\partial y} = \sum_{k=1}^{\infty} G_k k_2^{k+1},$$

$$(5.12)$$

where

$$U(x,y) = k_2 + \sum_{r=3}^{\infty} F_r(x,y), \qquad (5.13)$$

and $k_2 \not\equiv 0$ from (5.8) and $F_r(x,y)$ are homogeneous polynomials of degree r with respect to x, y. Equality (5.12) splits by powers of x and y into an infinite number of algebraic equations, where the quantities $G_1, G_2, ..., G_k, ...,$ are variables as well as coefficients of the polynomials $F_r(x, y)$.

For any differential system (5.7) from identity (5.12) with k_2 from (5.8), we find that the first three equations have the form

$$x^2 : \; e(c + f) = 0, \; xy : \; (c - f)(c + f) = 0, \; y^2 : \; d(c + f) = 0.$$

These equations are equivalent to one of two series of the conditions: 1) $c + f = 0$; 2) $e = c - f = d = 0$. Since $k_2 \not\equiv 0$, then, according to (5.8), these conditions are equivalent with the first equality $c + f = 0$, which is contained in the variety $\mathcal{V}$ from (5.9). Following the abovementioned, according to Remark 5.1 and the formulated center and focus problem for differential system (5.6), we conclude that *for focus quantities L_k from (5.1) and constants G_k from (5.12) the following equalities hold*:

$$L_k = G_k|_{\mathcal{V}} \; (k = 1, 2, ...), \tag{5.14}$$

where $\mathcal{V}$ is from (5.9).

From the abovementioned, it follows

Observation 5.2. *Identity (5.12) with function (5.13) on the variety $\mathcal{V}$ from (5.9) allows us to state that differential system (5.7) under these conditions has a singular point of the second group at the origin of coordinates (center or focus).*

We denote the expression $c + f = 0$, which is contained in the variety $\mathcal{V}$ from (5.9), by

$$G_0 \equiv c + f = 0, \tag{5.15}$$

and will call it *null focus pseudo-quantity*. We note that G_0 from (5.15) is a centro-affine (unimodular) invariant of the system $s(1, m_1, ..., m_\ell)$ of a type

$$(0, 1, \underbrace{0, ..., 0}_{\ell}).$$

To obtain a more clear idea about the quantities $G_1, G_2, ..., G_k, ...$ from identity (5.12) with function (5.13), let us study, in the next chapter, the remaining equations from the expansion of this identity in powers of $x^3, x^2 y, xy^2, y^3, ...$ for various differential systems $s(1, m_1, ..., m_\ell)$ without considering equality (5.15).

5.4 Polynomials in Coefficients of Differential Systems that Have Weight Isobarity (h, g)

Consider a special case of differential system (5.7) ($\ell = 1, m_1 = 2; s(1, 2)$) written in a tensor form (this form of writing differential systems of the form

(1.1)–(1.2) was introduced by academician K. S. Sibirsky [39] in the 60s of the last century):

$$\dot{x}^j = a^j_\alpha x^\alpha + a^j_{\alpha\beta} x^\alpha x^\beta \ (j, \alpha, \beta = 1, 2), \tag{5.16}$$

where the tensor coefficient $a^j_{\alpha\beta}$ is symmetric in the lower indices, by which a total convolution is carried out here.

Differential system (5.16) in an expanded form is written in the following form:

$$\begin{aligned}
\dot{x}^1 &= a^1_1 x^1 + a^1_2 x^2 + a^1_{11}(x^1)^2 + 2a^1_{12}x^1 x^2 + a^1_{22}(x^2)^2, \\
\dot{x}^2 &= a^2_1 x^1 + a^2_2 x^2 + a^2_{11}(x^1)^2 + 2a^2_{12}x^1 x^2 + a^2_{22}(x^2)^2.
\end{aligned} \tag{5.17}$$

Note that if we introduce the notation

$$\begin{aligned}
x = x^1, \ c = a^1_1, \ d = a^1_2, \ g = a^1_{11}, \ h = a^1_{12}, \ k = a^1_{22}, \\
y = x^2, \ e = a^2_1, \ f = a^2_2, \ l = a^2_{11}, \ m = a^2_{12}, \ n = a^2_{22},
\end{aligned} \tag{5.18}$$

then we obtain previously encountered system (1.5) and vice versa.

From the theory of invariants of differential systems (1.1)–(1.2) [43], the difference between the number of subscripts and superscripts that are equal to 1 (2), will be called as *a weight of any coordinate of the tensor coefficient* a^j_α or $a^j_{\alpha\beta}$ relative to the coordinate x^1 (x^2). For example:

(1) Weight of a^1_2 relative to x^1 is equal to -1, and relative to x^2 is equal to 1;

(2) Weight of a^1_1 relative to x^1 is equal to 0, and relative to x^2 is equal to 0;

(3) Weight of a^2_1 relative to x^1 is equal to 1, and relative to x^2 is equal to -1;

(4) Weight of a^1_{11} relative to x^1 is equal to 1, and relative to x^2 is equal to 0;

(5) Weight of a^1_{12} relative to x^1 is equal to 0, and relative to x^2 is equal to 1 etc.

If the polynomial S is considered, which depends on coefficients of differential system (5.17) of the form

$$(0, d_0, d_1), \tag{5.19}$$

i.e. of homogeneous degree d_0 relative to coordinates of the tensor a^j_α and of degree d_1 relative to coordinates of the tensor $a^j_{\alpha\beta}$, then the weight of each member of this polynomial relatively to x^1 or x^2 is equal to the sum of the weights corresponding to each factor of this member with respect to x^1 or x^2. Zero in (5.19) indicates that the expression S does not contain the phase variables x^1, x^2. For example, the monomial $a^1_1 a^2_1 a^1_{22}$ has type $(0, 2, 1)$, and weight is equal to 0 relative to x^1, and weight is equal to 1 relative to x^2.

If all members of the polynomial S of type (5.19) have the weight h relative to x^1 and the weight g relative to x^2, then we say that the polynomial S has *weight isobarity* (h, g) [43].

The isobaric property of polynomials is of great importance in the theory of invariants. For example, if we want to check if any polynomial S in coefficients of system (5.17) can be a coefficient in some comitant of type (δ, d_0, d_1) of this system, it is necessary that the polynomial S has isobarity of some weight (h, g).

From the work [37], it is known that a comitant K_{11} of the system (5.17) has type $(3, 1, 1)$ and the form

$$K_{11} = A_0(x^1)^3 + A_1(x^1)^2 x^2 + A_2 x^1(x^2)^2 + A_3(x^3)^3, \qquad (5.20)$$

where

$$\begin{aligned} A_0 &= -a_1^2 a_{11}^1 - a_2^2 a_{11}^2, \; A_1 = a_1^1 a_{11}^1 + a_2^1 a_{11}^2 - 2a_1^2 a_{12}^1 - 2a_2^2 a_{12}^2, \\ A_2 &= 2a_1^1 a_{12}^1 + 2a_2^1 a_{12}^2 - a_1^2 a_{22}^1 - a_2^2 a_{22}^2, \; A_3 = a_1^1 a_{22}^1 + a_2^1 a_{22}^2. \end{aligned} \qquad (5.21)$$

Note that in A_0 all terms have weight isobarity $(2, -1)$, in A_1–of weight $(1, 0)$, in A_2–of weight $(0, 1)$ and in A_3–of weight $(-1, 2)$.

In the above example, the expression A_0 from (5.21) is a *semi-invariant* of the comitant K_{11} from (5.20).

Further, we will see that *semi-invariant* of any comitant has a special meaning.

Note that for the semi-invariant *of weight isobarity* (h, g) in any comitant, the number g coincides with the *weight* of this comitant.

5.5   Comments to Chapter Five

In this chapter, we formulate the generalized center and focus problem and the arguments that prompted the authors to such formulation. We consider the Sibirsky invariant variety of center and focus, which is closely related to centro-affine classification of the quadratic form (comitant) k_2 (see [37], p. 31) of the system $s(1, m_1, ..., m_\ell)$.

The concepts of isobarity and semi-invariants, which play an important role in the construction of centro-affine comitants and invariants, are explained.

6

On the Upper Bound of the Number of Algebraically Independent Focus Quantities that Take Part in Solving the Center and Focus Problem for the System $s(1, m_1, ..., m_\ell)$

6.1 Applications of Generating Functions and Hilbert Series to the Center and Focus Problem for the Differential System $s(1, 2)$

Consider the differential system $s(1, 2)$, which we write as follows:

$$\dot{x} = cx + dy + gx^2 + 2hxy + ky^2,$$
$$\dot{y} = ex + fy + lx^2 + 2mxy + ny^2 \qquad (6.1)$$

with a finitely defined graded algebra of unimodular comitants $S_{1,2}$ [33]. For this system, we write function (5.13) in the form

$$U = k_2 + a_0 x^3 + 3a_1 x^2 y + 3a_2 xy^2 + a_3 y^3 + b_0 x^4 + 4b_1 x^3 y + 6b_2 x^2 y^2$$
$$+4b_3 xy^3 + b_4 y^4 + c_0 x^5 + 5c_1 x^4 y + 10c_2 x^3 y^2 + 10c_3 x^2 y^3 + 5c_4 xy^4$$
$$+c_5 y^5 + d_0 x^6 + 6d_1 x^5 y + 15d_2 x^4 y^2 + 20d_3 x^3 y^3 + 15d_4 x^2 y^4 + 6d_5 xy^5$$
$$+d_6 y^6 + e_0 x^7 + 7e_1 x^6 y + 21e_2 x^5 y^2 + 35e_3 x^4 y^3 + 21e_5 x^2 y^5 + 7e_6 xy^6 \qquad (6.2)$$
$$+e_7 y^7 + f_0 x^8 + 8f_1 x^7 y + 28f_2 x^6 y^2 + 56f_3 x^5 y^3 + 70f_4 y^4$$
$$+56f_5 x^3 y^5 + 28f_6 x^2 y^6 + 8f_7 xy^7 + f_8 y^8 + ...,$$

where $k_2 \not\equiv 0$ is from (5.8), and $a_0, a_1, ..., f_7, f_8, ...$ are unknown coefficients. Identity (5.12) along the trajectories of differential system (6.1) with function (6.2) splits into the following systems of equations (equality (5.15) is omitted):

$$x^3 : 3ca_0 + 3ea_1 = 2eg - (c - f)l,$$
$$x^2 y : 3da_0 + 3(2c + f)a_1 + 6ea_2 = (f - c)(g + 2m) - 2dl + 4eh,$$
$$xy^2 : 6da_1 + 3(2f + c)a_2 + 3ea_3 = (f - c)(2h + n) + 2ek - 4dm, \qquad (6.3)$$
$$y^3 : 3da_2 + 3fa_3 = (f - c)k - 2dn;$$

DOI: 10.1201/9781003193074-7

$$x^4 : 4cb_0 + 4eb_1 - e^2G_1 = -3ga_0 - 3la_1,$$

$$x^3y : 4db_0 + 4(f + 3c)b_1 + 12eb_2 + 2e(c - f)G_1 = -6ha_0$$
$$- 6(g + m)a_1 - 6la_2,$$

$$x^2y^2 : 12db_1 + 12(c + f)b_2 + 12eb_3 + [2de - (c - f)^2]G_1 =$$
$$= -3ka_0 - 3(4h + n)a_1 - 3(g + 4m)a_2 - 3la_3,$$

$$xy^3 : 12bd_2 + 4(3f + c)b_3 + 4eb_4 + 2d(f - c)G_1 = -6ka_1$$
$$- 6(h + n)a_2 - 6ma_3,$$

$$y^4 : 4db_3 + 4fb_4 - d^2G_1 = -3ka_2 - 3na_3; \qquad (6.4)$$

$$x^5 : 5cc_0 + 5ec_1 = -4gb_0 - 4lb_1,$$

$$x^4y : 5dc_0 + 5(4c + f)c_1 + 20ec_2 = -8hb_0 - 4(3g + 2m)b_1$$
$$- 12lb_2,$$

$$x^3y^2 : 20dc_1 + 10(3c + 2f)c_2 + 30ec_3 = -4kb_0 - 4(6h + n)b_1$$
$$- 12(g + 2m)b_2 - 12lb_3,$$

$$x^2y^3 : 30dc_2 + 10(2c + 3f)c_3 + 20ec_4 = -12kb_1 - 12(2h$$
$$+ n)b_2 - 4(g + 6m)b_3 - 4lb_4,$$

$$xy^4 : 20dc_3 + 5(c + 4f)c_4 + 5ec_5 = -12kb_2 - 4(2h + 3n)b_3$$
$$- 8mb_4,$$

$$y^5 : 5dc_4 + 5fc_5 = -4kb_3 - 4nb_4; \qquad (6.5)$$

$$x^6 : 6cd_0 + 6ed_1 + e^3G_2 = -5gc_0 - 5lc_1,$$

$$x^5y : 6dd_0 + 6(5c + f)d_1 + 30ed_2 + 3e^2(f - c)G_2 = -10hc_0$$
$$- 10(2g + m)c_1 - 20lc_2,$$

$$x^4y^2 : 30dd_1 + 30(2c + f)d_2 + 60ed_3 + 3e[(c - f)^2 - de]G_2 =$$
$$= -5kc_0 - 5(8h + n)c_1 - 10(3g + 4m)c_2 - 30lc_3,$$

$$x^3y^3 : 60dd_2 + 60(c + f)d_3 + 60ed_4 + (f - c)[(c - f)^2$$
$$- 6de]G_2 = -20kc_1 - 20(3h + n)c_2 - 20(g + 3m)c_3 - 20lc_4,$$

$$x^2y^4 : 60dd_3 + 30(c + 2f)d_4 + 30ed_5 + 3d[de - (c - f)^2]G_2 =$$
$$= -30kc_2 - 10(4h + 3n)c_3 - 5(g + 8m)c_4 - 5lc_5,$$

$$xy^5 : 30dd_4 + 6(c + 5f)d_5 + 6ed_6 + 3d^2(f - c)G_2 = -20kc_3$$
$$- 10(h + 2n)c_4 - 10mc_5,$$

$$y^6 : 6dd_5 + 6fd_6 - d^3G_2 = -5kc_4 - 5nc_5; \qquad (6.6)$$

$$x^7 : 7ce_0 + 7ee_1 = -6gd_0 - 6ld_1,$$

$$x^6y : 7de_0 + 7(6c + f)e_1 + 42ee_2 = -12hd_0 - 6(5g + 2m)d_1$$
$$- 30ld_2,$$

$$x^5y^2 : 42de_1 + 7(15c + 6f)e_2 + 105ee_3 = -6kd_0 - 6(10h$$
$$+ n)d_1 - 60(g + m)d_2 - 60ld_3,$$

$$x^4y^3 : 105de_2 + 5(28c + 21f)e_3 + 140ee_4 = -30kd_1 - 30(4h$$
$$+ n)d_2 - 60(g + 2m)d_3 - 60ld_4,$$

$$x^3y^4 : 140de_3 + 35(3c + 4f)e_4 + 105ee_5 = -60kd_2 - 60(2h$$
$$+ n)d_3 - 30(g + 4m)d_4 - 30ld_5,$$

$$x^2y^5 : 105de_4 + 7(6c + 15f)e_5 + 42ee_6 = -60kd_3 - 60(h + n)d_4$$
$$- 6(g + 10m)d_5 - 6ld_6,$$

$$xy^6 : 42de_5 + 7(c + 6f)e_6 + 7ee_7 = -30kd_4 - 6(2h + 5n)d_5$$
$$- 12md_6,$$

$$y^7 : 7de_6 + 7fe_7 = -6kd_5 - 6nd_6; \tag{6.7}$$

$$x^8 : 8cf_0 + 8ef_1 - e^4G_3 = -7ge_0 - 7le_1,$$

$$x^7y : 8df_0 + 8(7c + f)f_1 + 56ef_2 + 4e^3(c - f)G_3 =$$
$$= -14he_0 - 14(3g + m)e_1 - 42le_2,$$

$$x^6y^2 : 56df_1 + 56(3c + f)f_2 + 168ef_3 + 2e^2[2de - 3(c - f)^2]G_3 =$$
$$= -7ke_0 - 7(12h + n)e_1 - 21(5g + 4m)e_2 - 105le_3,$$

$$x^5y^3 : 168df_2 + 56(5c + 3f)f_3 + 280ef_4 + 4e(f - c)[3de - (c$$
$$- f)^2]G_3 = -42ke_1 - 42(5h + n)e_2 - 70(2g + 3m)e_3$$
$$- 140le_4,$$

$$x^4y^4 : 280df_3 + 280(c + f)f_4 + 280ef_5 + [12de(c - f)^2 - 6d^2e^2$$
$$- (c - f)^4]G_3 = -105ke_2 - 35(8h + 3n)e_3 - 35(3g$$
$$+ 8m)e_4 - 105le_5,$$

$$x^3y^5 : 280df_4 + 56(3c + 5f)f_5 + 168ef_6 + 4d(f - c)[(c - f)^2$$
$$- 3de]G_3 = -140ke_3 - 70(3h + 2n)e_4 - 42(g + 5m)e_5$$
$$- 42le_6,$$

$$x^2y^6 : 168df_5 + 56(c + 3f)f_6 + 56ef_7 + 2d^2[2de - 3(c - f)^2]G_3 =$$
$$= -105ke_4 - 21(4h + 5n)e_5 - 7(g + 12m)e_6 - 7le_7,$$

$$xy^7 : 56df_6 + 8(c + 7f)f_7 + 8ef_8 + 4d^3(f - c)G_3 = -42ke_5$$
$$- 14(h + 3n)e_6 - 14me_7,$$

$$y^8 : 78df_7 + 8ff_8 - d^4G_3 = -7ke_6 - 7ne_7. \tag{6.8}$$

It is evident that linear systems of equations (6.3)–(6.8) in the variables a_0, $a_1, a_2, a_3, b_0, b_1,...,b_4, c_0, c_1,...,c_5, d_0, d_1,...,d_6, e_0, e_1,...,e_7, f_0, f_1,...,f_8,..., G_1$, $G_2, G_3, ...$ can be extended by adding, after the last equation from (6.8), an infinite number of equations obtained from the equality of coefficients of $x^\alpha y^\beta$ for $\alpha + \beta > 8$ in identity (5.12).

For a clearer reflection of the process of obtaining G_1, we write systems (6.3), (6.4) in the matrix form

$$A_1 B_1 = C_1, \tag{6.9}$$

where

$$A_1 = \begin{pmatrix} 3c & 3e & 0 & 0 & 0 & 0 & 0 \\ 3d & 3(2c+f) & 6e & 0 & 0 & 0 & 0 \\ 0 & 6d & 3(2c+f) & 3e & 0 & 0 & 0 \\ 0 & 0 & 3d & 3f & 0 & 0 & 0 \\ 3g & 3l & 0 & 0 & 4c & 4e & 0 \\ 6h & 6(g+m) & 6l & 0 & 4d & 4(f+3c) & 12e \\ 3k & 3(4h+n) & 3(g+4m) & 3l & 0 & 12d & 12(c+f) \\ 0 & 6k & 6(h+n) & 6m & 0 & 0 & 12d \\ 0 & 0 & 3k & 3n & 0 & 0 & 0 \end{pmatrix}$$

$$\begin{pmatrix} 0 & 0 & 0 \\ 0 & 0 & 0 \\ 0 & 0 & 0 \\ 0 & 0 & 0 \\ 0 & 0 & -e^2 \\ 0 & 0 & 2e(c-f) \\ 12e & 0 & 2de-(c-f)^2 \\ 4(3f+c) & 4l & 2d(f-c) \\ 4d & 4f & -d^2 \end{pmatrix},$$

$$B_1 = \begin{pmatrix} a_0 \\ a_1 \\ a_2 \\ a_3 \\ b_0 \\ b_1 \\ b_2 \\ b_3 \\ b_4 \\ G_1 \end{pmatrix}, C_1 = \begin{pmatrix} 2eg+(f-c)l \\ (f-c)(g+2m)-2dl+4eh \\ (f-c)(2h+n)+3ek-4dm \\ (f-c)k-2dn \\ 0 \\ 0 \\ 0 \\ 0 \\ 0 \\ 0 \end{pmatrix}. \tag{6.10}$$

Since the dimension of the matrix A_1 is 9×10, then it is clear that we have at least one free variable. Therefore, choosing as a free variable, one of b_i ($i \in \{0, ..., 4\}$) and using the Cramer's rule for system (6.9), for each fixed i, we obtain

$$G_1 = \frac{G_{1,i} + B_{1,i} b_i}{\sigma_{1,i}}, \tag{6.11}$$

where $G_{1,i}, B_{1,i}, \sigma_{1,i}$ (see, Appendix 3) are polynomials in coefficients of system (6.1), and b_i are undetermined coefficients of the function $U(x,y)$ from (6.2).

In the future, we will need an explicit form of the operators $X_1, ..., X_4$ of the Lie algebra L_4 for system (6.1), expressions of which are obtained from Section 1.5:

$$X_1 = x\frac{\partial}{\partial x} + D_1, \ X_2 = y\frac{\partial}{\partial x} + D_2, \ X_3 = x\frac{\partial}{\partial y} + D_3, \ X_4 = y\frac{\partial}{\partial y} + D_4,$$

$$(6.12)$$

where

$$D_1 = d\frac{\partial}{\partial d} - e\frac{\partial}{\partial e} - g\frac{\partial}{\partial g} + k\frac{\partial}{\partial k} - 2l\frac{\partial}{\partial l} - m\frac{\partial}{\partial m},$$

$$D_2 = e\frac{\partial}{\partial c} + (f-c)\frac{\partial}{\partial d} - e\frac{\partial}{\partial f} + l\frac{\partial}{\partial g} + (m-g)\frac{\partial}{\partial h}$$

$$+(n-2h)\frac{\partial}{\partial k} - l\frac{\partial}{\partial m} - 2m\frac{\partial}{\partial n},$$

$$(6.13)$$

$$D_3 = -d\frac{\partial}{\partial c} + (c-f)\frac{\partial}{\partial e} + d\frac{\partial}{\partial f} - 2h\frac{\partial}{\partial g} - k\frac{\partial}{\partial h}$$

$$+(g-2m)\frac{\partial}{\partial l} + (h-n)\frac{\partial}{\partial m} + k\frac{\partial}{\partial n},$$

$$D_4 = -d\frac{\partial}{\partial d} + e\frac{\partial}{\partial e} - h\frac{\partial}{\partial h} - 2k\frac{\partial}{\partial k} + l\frac{\partial}{\partial l} - n\frac{\partial}{\partial n}.$$

By studying matrices (6.10) of system (6.9), we conclude that $G_{1,i}$ from (6.11) are homogeneous polynomials of degree 8 with respect to the linear part, and of degree 2 with respect to the quadratic part of system (6.1).

Note that $G_{1,i}$ from (6.11) for $i = 0, 1, 2, 3, 4$ are homogeneous polynomials of isobarities with weights, respectively (see, Appendix 3):

$$(3,-1), \ (2,0), \ (1,1), \ (0,2), \ (-1,3). \tag{6.14}$$

According to formula (3.3) for differential system (6.1) and theory of invariants of differential systems [37,43], it follows that the numerators of fractions (6.11) can be coefficients in comitants of the weight -1 of type $(4,8,2)$. This means that according to (2.56) with the help of Lie differential operator D_3 from (6.13) for system (6.1) and numerator of fraction (6.11), we obtain a system of four linear nonhomogeneous partial differential equations

$$D_3(G_{1,0} + B_{1,0}b_0) = G_{1,1} + B_{1,1}b_1, \ D_3(G_{1,1} + B_{1,1}b_1) = -G_{1,2} - B_{1,2}b_2,$$

$$-D_3(G_{1,2} + B_{1,2}b_2) = G_{1,3} + B_{1,3}b_3, \ D_3(G_{1,3} + B_{1,3}b_3) = -G_{1,4} - B_{1,4}b_4,$$

$$(6.15)$$

with five unknowns b_0, b_1, b_2, b_3, b_4. According to Lemma 2.4, system (6.15) has an infinite number of solutions. Note that a particular solution of this system is $b_0 = b_1 = b_2 = b_3 = b_4 = 0$, for which the polynomial

$$f_4'(x,y) = G_{1,0}x^4 + 4G_{1,1}x^3y + 2G_{1,2}x^2y^2 + 4G_{1,3}xy^3 + G_{1,4}y^4 \tag{6.16}$$

is a centro-affine comitant of differential system (6.1). This fact is also confirmed by Theorem 2.2 with the operators $X_1 - X_4$ from (6.12)–(6.13) for differential system (6.1), for which we have the equalities

$$X_1(f_4') = X_4(f_4') = f_4', \; X_2(f_4') = X_3(f_4') = 0.$$

Another particular solution for system (6.15) is given by the following expressions

$$b_0 = \frac{-e(g^2 + 2hl + m^2)}{3c^2 - 4de + 10cf + 3f^2},$$

$$b_1 = \frac{(c-f)(g^2 + 2hl + m^2) - 2e(gh + kl + hm + mn)}{4(3c^2 - 4de + 10cf + 3f^2)},$$

$$b_2 = \frac{2(c-f)(gh + kl + hm + mn) - e(h^2 + 2km + n^2) + d(g^2 + 2hl + m^2)}{6(3c^2 - 4de + 10cf + 3f^2)},$$

$$b_3 = \frac{(c-f)(h^2 + 2km + n^2) + 2d(gh + kl + hm + mn)}{4(3c^2 - 4de + 10cf + 3f^2)},$$

$$b_4 = \frac{d(h^2 + 2km + n^2)}{3c^2 - 4de + 10cf + 3f^2},$$

whose denominators are different from zero on the invariant variety $\mathcal{V}$ from (5.11). This solution determines the centro-affine comitant

$$f_4''(x,y) = (G_{1,0} + B_{1,0}b_0)x^4 + 4(G_{1,1} + B_{1,1}b_1)x^3y + 2(G_{1,2} \\ + B_{1,2}b_2)x^2y^2 + 4(G_{1,3} + B_{1,3}b_3)xy^3 + (G_{1,4} + B_{1,4}b_4)y^4. \tag{6.17}$$

It is evident that differential system (6.15) has infinite number of solutions b_0, b_1, b_2, b_3, b_4, which define a centro-affine comitants of type $(4, 8, 2)$.

In view of the abovementioned, comitants (6.16), (6.17) belong to the space

$$S_{1,2}^{(4,8,2)}, \tag{6.18}$$

of Sibirsky algebra $S_{1,2}$.

Note that comitants (6.16), (6.17) on the invariant variety $\mathcal{V}$ from (5.11) for differential system (6.1) have the following form

$$f_4'(x,y)|_\mathcal{V} = f_4''(x,y)|_\mathcal{V} = -8L_1(x^2 + y^2)^2 \; (G_1|_\mathcal{V} = -8L_1), \tag{6.19}$$

where

$$L_1 = \frac{1}{2}\left[g(l - h) - k(h + n) + m(l + n)\right]$$

is the first focus quantity of differential system (6.1) on the invariant variety $\mathcal{V}$ and coincides with the focus quantity from [35, p. 110] for system (6.1), received after substitution $f = -c = 0$, $d = -e = 1$.

Similarly to the previous case, for determining the quantity G_2, we write system of equations (6.3)–(6.6) in the matrix form (see, Appendix 4)

$$A_2B_2 = C_2, \tag{6.20}$$

from where we find

$$G_2 = \frac{G_{2,i,j} + B_{2,i,j} b_i + D_{2,i,j} d_j}{\sigma_{2,i,j}}, \quad (i = \overline{0,4}, \ j = \overline{0,6}). \tag{6.21}$$

By studying matrix equality (6.20), we obtain that $deg G_{2,i,j} = 24$, and using system (6.3)–(6.6), we obtain that $G_{2,i,j}$ from (6.21) has type $(0, 20, 4)$, i.e. $G_{2,i,j}$ are homogeneous polynomials of degree 20 in coefficients of the linear part and of degree 4 in coefficients of the quadratic part of the system $s(1, 2)$ from (6.1).

Computing the expressions $G_{2,i,j}$ for each $i = \overline{0,4}$ and $j = \overline{0,6}$, we obtain for their isobarities the following Table 6.1 (see, Appendix 4):

Table 6.1. *Isobarities with weight of polynomials $G_{2,i,j}$ for the system $s(1, 2)$*

$G_{2,i,j}$	d_0	d_1	d_2	d_3	d_4	d_5	d_6
b_0	$(7,-3)$	$(6,-2)$	$(5,-1)$	$(4,0)$	$(3,1)$	$(2,2)$	$(1,3)$
b_1	$(6,-2)$	$(5,-1)$	$(4,0)$	$(3,1)$	$(2,2)$	$(1,3)$	$(0,4)$
b_2	$(5,-1)$	$(4,0)$	$(3,1)$	$(2,2)$	$(1,3)$	$(0,4)$	$(-1,5)$
b_3	$(4,0)$	$(3,1)$	$(2,2)$	$(1,3)$	$(0,4)$	$(-1,5)$	$(-2,6)$
b_4	$(3,1)$	$(2,2)$	$(1,3)$	$(0,4)$	$(-1,5)$	$(-2,6)$	$(-3,7)$

Note that for $j = \overline{0,6}$, we have

$$\frac{G_{2,0,j}}{\sigma_{2,0,j}}\big|_V = \frac{1}{24}(38g^3 h + 46gh^3 + 71g^2 hk + 46h^3 k + 38ghk^2 + 5hk^3$$

$$-38g^3 l + 3gh^2 l - 39g^2 kl + 53h^2 kl - 15gk^2 l - 32ghl^2 + 15hkl^2 - 5gl^3$$

$$+29g^2 hm + 42ghkm + 13hk^2 m - 79g^2 lm - 54h^2 lm - 68gklm - 15k^2 lm$$

$$-37hl^2 m - 5l^3 m + 6ghm^2 + 6hkm^2 - 39glm^2 - 29klm^2 + 2lm^3 + 6g^3 n$$

$$+109gh^2 n + 48g^2 kn + 159h^2 kn + 33gk^2 n + 5k^3 n - 34ghln + 116hkln$$

$$-57gl^2 n + 15kl^2 n - 48g^2 mn - 54h^2 mn - 14gkmn + 8k^2 mn - 138hlmn$$

$$-62l^2 mn - 37gm^2 n - 27km^2 n + 2m^3 n + 72ghn^2 + 175hkn^2 - 72gln^2$$

$$+63kln^2 - 101hmn^2 - 119lmn^2 - 6gn^3 + 62kn^3 - 62mn^3), \tag{6.22}$$

$$\frac{G_{2,2,j}}{\sigma_{2,2,j}}\big|_V = \frac{1}{24}(62g^3 h - 2gh^3 + 95g^2 hk - 2h^3 k + 38ghk^2 + 5hk^3 - 62g^3 l$$

$$+27gh^2 l - 39g^2 kl + 29h^2 kl - 15gk^2 l - 8ghl^2 + 15hkl^2 - 5gl^3 + 53g^2 hm$$

$$+66ghkm + 13hk^2 m - 127g^2 lm - 6h^2 lm - 68gklm - 15k^2 lm - 13hl^2 m$$

$$-5l^3 m + 6ghm^2 + 6hkm^2 - 63glm^2 - 29klm^2 + 2lm^3 + 6g^3 n$$

$$+61gh^2 n + 72g^2 kn + 63h^2 kn + 33gk^2 n + 5k^3 n - 10ghln + 68hkln$$

$$-33gl^2 n + 15kl^2 n - 72g^2 mn - 6h^2 mn + 10gkmn + 8k^2 mn - 66hlmn$$

$$-38l^2 mn - 61gm^2 n - 27km^2 n + 2m^3 n + 72ghn^2 + 127hkn^2 - 72gln^2$$

$$+39kln^2 - 53hmn^2 - 95lmn^2 - 6gn^3 + 62kn^3 - 62mn^3), \tag{6.23}$$

$$\frac{G_{2,4,j}}{\sigma_{2,4,j}}|_{\mathcal{V}} = \frac{1}{24}(62g^3h - 2gh^3 + 119g^2hk - 2h^3k + 62ghk^2 + 5hk^3 - 62g^3l$$

$$+27gh^2l - 63g^2kl + 29h^2kl - 15gk^2l - 8ghl^2 + 15hkl^2 - 5gl^3 + 101g^2hm$$

$$+138ghkm + 37hk^2m - 175g^2lm - 6h^2lm - 116gklm - 15k^2lm - 13hl^2m$$

$$-5l^3m + 54ghm^2 + 54hkm^2 - 159glm^2 - 53klm^2 - 46lm^3 + 6g^3n$$

$$+37gh^2n + 72g^2kn + 39h^2kn + 57gk^2n + 5k^3n + 14ghln + 68hkln$$

$$-33gl^2n + 15kl^2n - 72g^2mn - 6h^2mn + 34gkmn + 32k^2mn - 42hlmn$$

$$-38l^2mn - 109gm^2n - 3km^2n - 46m^3n + 48ghn^2 + 79hkn^2 - 48gln^2$$

$$+39kln^2 - 29hmn^2 - 71lmn^2 - 6gn^3 + 38kn^3 - 38mn^3)$$

$$(6.24)$$

and $\dfrac{G_{2,1,j}}{\sigma_{2,1,j}}$ and $\dfrac{G_{2,3,j}}{\sigma_{2,3,j}}$ on the invariant variety $\mathcal{V}$ give uncertainties.

By studying the isobarities of $G_{2,i,j}$ top-down for each line from Table 6.1, according to the theory of invariants of differential systems [37,43], we find that the numerators of fraction (6.21) can be coefficients in a centro-affine comitants with the corresponding weights $-3, -2, -1, 0, 1$. Using these weights and formula (3.3) for differential system (6.1), as well as the fact that $G_{2,i,j}$ have type $(0, 20, 4)$, we obtain that the mentioned comitants correspond to the types

$$(10, 20, 4), \ (8, 20, 4), \ (6, 20, 4), \ (4, 20, 4), \ (2, 20, 4). \quad (6.25)$$

As the quantity G_2 in (5.12) is a coefficient of homogeneity of degree 6 in the phase variables x and y, then it is logical to choose from (6.25) the type

$$(6, 20, 4), \quad (6.26)$$

which corresponds to the expression $G_{2,2,j}$ $(j = \overline{0, 6})$ from Table 6.1.

This means that according to (2.56), using Lie differential operator D_3 from (6.13) for differential system (6.1) and the numerator of fraction (6.21) for the fixed $i = 2$, we obtain a system of six linear nonhomogeneous partial differential equations:

$$
\begin{aligned}
D_3(G_{2,2,0} + B_{2,2,0}b_0 + D_{2,2,0}d_0) &= -(G_{2,2,1} + B_{2,2,1}b_1 + D_{2,2,1}d_1), \\
-D_3(G_{2,2,1} + B_{2,2,1}b_1 + D_{2,2,1}d_1) &= G_{2,2,2} + B_{2,2,2}b_2 + D_{2,2,2}d_2, \\
D_3(G_{2,2,2} + B_{2,2,2}b_2 + D_{2,2,2}d_2) &= -(G_{2,2,3} + B_{2,2,3}b_3 + D_{2,2,3}d_3), \\
-D_3(G_{2,2,3} + B_{2,2,3}b_3 + D_{2,2,3}d_3) &= G_{2,2,4} + B_{2,2,4}b_4 + D_{2,2,4}d_4, \\
D_3(G_{2,2,4} + B_{2,2,4}b_4 + D_{2,2,4}d_4) &= -(G_{2,2,5} + B_{2,2,5}b_5 + D_{2,2,5}d_5), \\
-D_3(G_{2,2,5} + B_{2,2,5}b_5 + D_{2,2,5}d_5) &= G_{2,2,6} + B_{2,2,6}b_6 + D_{2,2,6}d_6,
\end{aligned}
\quad (6.27)
$$

with seven unknowns $d_0, d_1, ..., d_6$, where b_i $(i = \overline{0, 4})$ are defined from system (6.15). According to Lemma 2.4, system (6.27) has infinite number of solutions that define comitants of type $(6, 20, 4)$. Note that obtaining explicit solutions of system (6.27) is a complicated enough procedure. We will show the importance of homogeneities of the expressions $G_{2,2,j}$ from (6.21) in obtaining a focus quantities of differential system (6.1) on the invariant variety of center and focus $\mathcal{V}$ from (5.11). According to (6.26), system (6.27) defines centro-affine comitants belonging to the space

$$S_{1,2}^{(6,20,4)} \tag{6.28}$$

of Sibirsky algebra $S_{1,2}$.

According to (2.56) and (6.27), each comitant, belonging to this space, can be written as

$$f_6'(x,y) = (G_{2,2,0} + B_{2,2,0}b_2 + D_{2,2,0}d_0)x^6 - (G_{2,2,1} + B_{2,2,1}b_2$$
$$+D_{2,2,1}d_1)x^5y + \frac{1}{2!}(G_{2,2,2} + B_{2,2,2}b_2 + D_{2,2,2}d_2)x^4y^2 - \frac{1}{3!}(G_{2,2,3}$$
$$+B_{2,2,3}b_2 + D_{2,2,3}d_3)x^3y^3 + \frac{1}{4!}(G_{2,2,4} + B_{2,2,4}b_2 + D_{2,2,4}d_4)x^2y^4$$
$$-\frac{1}{5!}(G_{2,2,5} + B_{2,2,5}b_2 + D_{2,2,5}d_5)xy^5 + \frac{1}{6!}(G_{2,2,6} + B_{2,2,6}b_2 + D_{2,2,6}d_6)y^6.$$

Note that on the invariant variety $\mathcal{V}$ from (5.11) for differential system (6.1) the expressions $G_{2,2,j}$ $(j = \overline{0,6})$ get the form

$$G_{2,2,0}|_{\mathcal{V}} = G_{2,2,2}|_{\mathcal{V}} = G_{2,2,4}|_{\mathcal{V}} = G_{2,2,6}|_{\mathcal{V}} = -2304L_2,$$
$$G_{2,2,1}|_{\mathcal{V}} = G_{2,2,3}|_{\mathcal{V}} = G_{2,2,5}|_{\mathcal{V}} = 0, \tag{6.29}$$

where

$$24L_2 = 62g^3h - 2gh^3 + 95g^2hk - 2h^3k + 38ghk^2 + 5hk^3 - 62g^3l$$
$$+27gh^2l - 39g^2kl + 29h^2kl - 15gk^2l - 8ghl^2 + 15hkl^2 - 5gl^3$$
$$+53g^2hm + 66ghkm + 13hk^2m - 127g^2lm - 6h^2lm - 68gklm$$
$$-15k^2lm - 13hl^2m - 5l^3m + 6ghm^2 + 6hkm^2 - 63glm^2 - 29klm^2$$
$$+2lm^3 + 6g^3n + 61gh^2n + 72g^2kn + 63h^2kn + 33gk^2n + 5k^3n$$
$$-10ghln + 68hkln - 33gl^2n + 15kl^2n - 72g^2mn - 6h^2mn$$
$$+10gkmn + 8k^2mn - 66hlmn - 38l^2mn - 61gm^2n - 27km^2n$$
$$+2m^3n + 72ghn^2 + 127hkn^2 - 72gln^2 + 39kln^2 - 53hmn^2$$
$$-95lmn^2 - 6gn^3 + 62kn^3 - 62mn^3$$

is a second focus quantity of system (6.1) on the invariant variety $\mathcal{V}$.

For the second focus quantity corresponding to $G_{2,4,j}$, we obtain

$$G_{2,4,0}|_{\mathcal{V}} = G_{2,4,2}|_{\mathcal{V}} = G_{2,4,4}|_{\mathcal{V}} = G_{2,4,6}|_{\mathcal{V}} = -2304LS_2,$$
$$G_{2,4,1}|_{\mathcal{V}} = G_{2,4,3}|_{\mathcal{V}} = G_{2,4,5}|_{\mathcal{V}} = 0,$$

$$24LS_2 = 62g^3h - 2gh^3 + 119g^2hk - 2h^3k + 62ghk^2 + 5hk^3 - 62g^3l$$

$$+27gh^2l - 63g^2kl + 29h^2kl - 15gk^2l - 8ghl^2 + 15hkl^2 - 5gl^3 + 101g^2hm$$

$$+138ghkm + 37hk^2m - 175g^2lm - 6h^2lm - 116gklm - 15k^2lm$$

$$-13hl^2m - 5l^3m + 54ghm^2 + 54hkm^2 - 159glm^2 - 53klm^2 - 46lm^3$$

$$+6g^3n + 37gh^2n + 72g^2kn + 39h^2kn + 57gk^2n + 5k^3n + 14ghln$$

$$+68hkln - 33gl^2n + 15kl^2n - 72g^2mn - 6h^2mn + 34gkmn + 32k^2mn$$

$$-42hlmn - 38l^2mn - 109gm^2n - 3km^2n - 46m^3n + 48ghn^2 + 79hkn^2$$

$$-48gln^2 + 39kln^2 - 29hmn^2 - 71lmn^2 - 6gn^3 + 38kn^3 - 38mn^3.$$

This expression coincides with the focus quantity from [35 p. 110] for system (6.1), received after substitution $f = -c = 0$, $d = -e = 1$.

Consider determination of G_3 of homogeneity of degree 8 with respect to the phase variables x and y in (6.21). Writing system (6.3)–(6.8) in the matrix form

$$A_3B_3 = C_3,$$

we obtain

$$G_3 = \frac{G_{3,i,j,k} + B_{3,i,j,k}b_i + D_{3,i,j,k}d_j + F_{3,i,j,k}f_k}{\sigma_{3,i,j,k}} \tag{6.30}$$

$$(i = \overline{0,4}; \; j = \overline{0,6}; \; k = \overline{0,8}).$$

Similarly to the previous case, we choose a comitant of the weight -1 of the differential system $s(1,2)$ from (6.1), which contains the expressions $G_{3,2,j,k} + B_{3,2,j,k}b_2 + D_{3,2,j,k}d_j + F_{3,2,j,k}f_k$ $(k = \overline{0,8})$ as a semi-invariant, and we find that it belongs to the space

$$S_{1,2}^{(8,37,6)} \tag{6.31}$$

of Sibirsky algebra $S_{1,2}$.

Lets consider the extension of system (6.3)–(6.8) obtained from identity (5.12) for differential system (6.1) and function (6.2), which contains the quantity G_k, which we write in a matrix form as follows $A_kB_k = C_k$. We denote by m_{G_k} the number of equations and by n_{G_k} the number of unknowns of this system. Note that these numbers are written as

$$m_{G_k} = \underbrace{4+5}_{G_1} + \underbrace{6+7}_{G_2} + \underbrace{8+9}_{G_3} + \cdots + \underbrace{(2k+2)+(2k+3)}_{G_k}, \; (k = 1,2,3,...),$$

$$n_{G_k} = \underbrace{4+6}_{G_1} + \underbrace{6+8}_{G_2} + \underbrace{8+10}_{G_3} + \cdots + \underbrace{(2k+2)+(2k+4)}_{G_k}.$$

Hence we obtain

$$m_{G_k} = k(2k+7), n_{G_k} = m_{G_k} + k > m_{G_k}. \tag{6.32}$$

Similarly to the previous cases, from this system, we have

$$G_k = \frac{G_{k,i_1,i_2,...,i_k} + B_{k,i_1,i_2,...,i_k}b_{i_1} + \cdots + Z_{k,i_1,i_2,...,i_k}z_{i_k}}{\sigma_{k,i_1,i_2,...,i_k}}. \tag{6.33}$$

Now it is important to determine the degree of the polynomial $G_{k,i_1,i_2,...,i_k} +$ $+B_{k,i_1,i_2,...,i_k} b_{i_1} + \cdots + Z_{k,i_1,i_2,...,i_k} z_{i_k}$ in coefficients of differential system (6.1).

Note that the degree of nonzero polynomial coefficient of G_i $(i = \overline{1,k})$ in coefficients of system (6.1) in Cramer's determinant of the order m_{G_k}, when the last column corresponding to the quantity G_k is replaced with the column corresponding to free members, forms the following diagram (coefficients of the last quantity G_k have the degree 2 according to the substitution):

$$G_1, G_2, G_3, ..., G_{k-1}, G_k.$$

$$\downarrow \quad \downarrow \quad \downarrow \quad \quad \downarrow \quad \quad \downarrow$$

$$2 \quad 3 \quad 4 \quad \quad k \quad \quad 2$$

Then the degree of the polynomial $G_{k,i_1,i_2,...,i_k} + B_{k,i_1,i_2,...,i_k} b_{i_1} + \cdots +$ $Z_{k,i_1,i_2,...,i_k} z_{i_k}$ in coefficients of differential system (6.1), denoted by N_{G_k}, will be written as

$$N_{G_k} = m_{G_k} - k + \frac{k(k+1)}{2} + 1,$$

hence according to (6.32), we obtain

$$N_{G_k} = \frac{1}{2}(5k^2 + 13k + 2). \tag{6.34}$$

It is the degree of homogeneity of $G_{k,i_1,i_2,...,i_k} + B_{k,i_1,i_2,...,i_k} b_{i_1} + \cdots$ $+Z_{k,i_1,i_2,...,i_k} z_{i_k}$ in coefficients of linear and quadratic parts of differential system (6.1), which is contained in a polynomial of type $(d) = (\delta, d_0, d_1)$. Since $\delta = 2(k+1)$, and $d_1 = 2k$, then $d_0 = N_{G_k} - 2k$. So we obtain that a comitant of the weight -1 of the differential system $s(1,2)$ from (6.1), containing the semi-invariant $G_{k,i_1,i_2,...,i_k} + B_{k,i_1,i_2,...,i_k} b_{i_1} + \cdots$ $+Z_{k,i_1,i_2,...,i_k} z_{i_k}$ which corresponds to the quantity G_k for $k = 1,2,3,...,$ belongs to the type

$$\left(2(k+1), \frac{1}{2}(5k^2 + 9k + 2), 2k\right), \tag{6.35}$$

where $2(k+1)$ is the degree of homogeneity of the comitant in phase variables x, y; $\frac{1}{2}(5k^2 + 9k + 2)$ is the degree of homogeneity of the comitant in coefficients of linear part, and $2k$ is the degree of homogeneity of the comitant in coefficients of the quadratic part of the differential system $s(1,2)$ from (6.1).

Hereafter the expressions $G_{k,i_1,i_2, ..., i_k} + B_{k,i_1,i_2, ..., i_k} b_{i_1} + \cdots$ $+Z_{k,i_1,i_2,...,i_k} z_{i_k}$, which determine comitants of types (6.35), corresponding to the quantity G_k $(k = 1,2,3,...)$, will be called *generalized focus pseudo-quantities*, and comitants of the type (6.35) for $k = 1,2,3,...$ will be called *comitants which contain the generalized focus pseudo-quantities* $G_{k,i_1,i_2,...,i_k} +$ $B_{k,i_1,i_2,...,i_k} b_{i_1} + \cdots + Z_{k,i_1,i_2,...,i_k} z_{i_k}$ as coefficients.

Note that the spaces $S^{(2(k+1),\frac{1}{2}(5k^2+9k+2),2k)}$ are generalized records of spaces (6.18), (6.28), (6.31) for $k = 1,2,3,...$ of Sibirsky algebra $S_{1,2}$.

According to the paper [33], using Theorem 4.1, there takes place

Theorem 6.1. *Dimension of linear space of centro-affine comitants of type* $(d) = (\delta, d_0, d_1)$ *for the differential system* $s(1,2)$ *from* (6.1), *denoted by* $dim_{\mathbb{R}} V_{1,2}^{(d)}$, *is equal to the coefficient of the monomial* $u^{\delta} b^{d_0} c^{d_1}$ *in decomposition of generalized Hilbert series from* (4.13)–(4.15) *for Sibirsky algebra* $S_{1,2}$ *of comitants of the considered system.*

Consider the subalgebra $S'_{1,2} \subset S_{1,2}$, which we write in the form

$$S'_{1,2} = \bigoplus_{(d)} S_{1,2}^{(d')}, \tag{6.36}$$

where by $S_{1,2}^{(d')}$ the following linear spaces are denoted:

$$S_{1,2}^{(0,0,0)} = \mathbb{R}, \ S_{1,2}^{(0,1,0)}, ..., S_{1,2}^{(2(k+1), \frac{1}{2}(5k^2+9k+2), 2k)}, \ k = 1, 2, ..., \tag{6.37}$$

as well as spaces from $S_{1,2}$, which contain all kinds of their products.

Since the algebra $S'_{1,2}$ is a graded subalgebra in a finitely defined algebra $S_{1,2}$, then according to Proposition 4.1, we obtain $\varrho(S'_{1,2}) \leq \varrho(S_{1,2})$. From this inequality and from the fact that $\varrho(S_{1,2}) = 9$ [33], according to Remark 2.3 on semi-invariants and the fact that generalized focus pseudo-quantities are coefficients of some comitants, there takes place

Theorem 6.2. *Maximal number of algebraically independent generalized focus pseudo-quantities in the center and focus problem for differential system* (6.1) *does not exceed 9.*

According to Proposition 4.2, Remark 5.1, and equality (5.14), there follows that the maximal number of algebraically independent focus quantities L_k ($k = \overline{1, \infty}$) cannot exceed the maximal number of algebraically independent generalized focus pseudo-quantities $G_{k,i_1,i_2,...,i_k} + B_{k,i_1,i_2,...,i_k} b_{i_1} + \cdots + Z_{k,i_1,i_2,...,i_k} z_{i_k}$.

Hence, according to equalities (5.5), we have

Consequence 6.1. *Upper bound of the number of algebraically independent focus quantities that take part in solving the center and focus problem for differential system* (6.1) *does not exceed 9.*

Consider the types of spaces $S_{1,2}^{(d')}$ from (6.37) for $d' = \delta' + d'_0 + d'_1 \leq 60$, which are obtained from expansion in a power series of the fraction

$$\frac{1}{(1-b)(1-u^4 b^8 c^2)(1-u^6 b^{20} c^4)(1-u^8 b^{37} c^6)}, \tag{6.38}$$

where $u^{\delta'} b^{d'_0} c^{d'_1}$ shows a type of space $S_{1,2}^{(d')}$ for $(d') = (\delta', d'_0, d'_1)$. In consideration of these types and the generalized Hilbert series (4.13)–(4.15) of the algebra $S_{1,2}$, we can write expansion of Hilbert series of the algebra $S'_{1,2}$ for $d' = \delta' + d'_0 + d'_1 \leq 60$, which has the form

$$H(S'_{1,2}, u, b, c) = 1 + b + 2b^2 + 2b^3 + 3b^4 + 3b^5 + 4b^6 + 4b^7 + 5b^8 + 5b^9$$
$$+6b^{10} + 6b^{11} + 7b^{12} + 7b^{13} + 8b^{14} + 8b^{15} + 9b^{16} + 9b^{17} + 10b^{18} + 10b^{19}$$
$$+11b^{20} + 11b^{21} + 12b^{22} + 12b^{23} + 13b^{24} + 13b^{25} + 14b^{26} + 14b^{27} + 15b^{28}$$
$$+15b^{29} + 16b^{30} + 16b^{31} + 17b^{32} + 17b^{33} + 18b^{34} + 18b^{35} + 19b^{36} + 19b^{37}$$
$$+20b^{38} + 20b^{39} + 21b^{40} + 21b^{41} + 22b^{42} + 22b^{43} + 23b^{44} + 23b^{45} + 24b^{46}$$
$$+24b^{47} + 25b^{48} + 25b^{49} + 26b^{50} + 26b^{51} + 27b^{52} + 27b^{53} + 28b^{54} + 28b^{55}$$
$$+29b^{56} + 29b^{57} + 30b^{58} + 30b^{59} + 31b^{60} + u^4(68b^8c^2 + 79b^9c^2 + 87b^{10}c^2$$
$$+98b^{11}c^2 + 106b^{12}c^2 + 117b^{13}c^2 + 125b^{14}c^2 + 136b^{15}c^2 + 144b^{16}c^2$$
$$+155b^{17}c^2 + 163b^{18}c^2 + 174b^{19}c^2 + 182b^{20}c^2 + 193b^{21}c^2 + 201b^{22}c^2$$
$$+212b^{23}c^2 + 220b^{24}c^2 231b^{25}c^2 + 239b^{26}c^2 + 250b^{27}c^2 + 258b^{28}c^2$$
$$+269b^{29}c^2 + 277b^{30}c^2 + 288b^{31}c^2 + 296b^{32}c^2 + 307b^{33}c^2 + 315b^{34}c^2$$
$$+326b^{35}c^2 + 334b^{36}c^2 + 345b^{37}c^2 + 353b^{38}c^2 + 364b^{39}c^2 + 372b^{40}c^2$$
$$+383b^{41}c^2 + 391b^{42}c^2 + 402b^{43}c^2 + 410b^{44}c^2 + 421b^{45}c^2 + 429b^{46}c^2$$
$$+440b^{47}c^2 + 448b^{48}c^2 + 459b^{49}c^2 + 467b^{50}c^2 + 478b^{51}c^2 + 486b^{52}c^2$$
$$+497b^{53}c^2 + 505b^{54}c^2) + u^6(988b^{20}c^4 + 1046b^{21}c^4u^6 + 1098b^{22}c^4$$
$$+1156b^{23}c^4 + 1208b^{24}c^4 + 1266b^{25}c^4 + 1318b^{26}c^4 + 1376b^{27}c^4 + 1428b^{28}c^4$$
$$+1486b^{29}c^4 + 1538b^{30}c^4 + 1596b^{31}c^4 + 1648b^{32}c^4 + 1706b^{33}c^4 + 1758b^{34}c^4$$
$$+1816b^{35}c^4 + 1868b^{36}c^4 + 1926b^{37}c^4 + 1978b^{38}c^4 + 2036b^{39}c^4 + 2088b^{40}c^4$$
$$+2146b^{41}c^4 + 2198b^{42}c^4 + 2256b^{43}c^4 + 2308b^{44}c^4 + 2366b^{45}c^4 + 2418b^{46}c^4$$
$$+2476b^{47}c^4 + 2528b^{48}c^4 + 2586b^{49}c^4 + 2638b^{50}c^4) + u^8(798b^{16}c^4$$
$$+855b^{17}c^4u^8 + 918b^{18}c^4 + 975b^{19}c^4 + 1038b^{20}c^4 + 1095b^{21}c^4 + 1158b^{22}c^4$$
$$+1215b^{23}c^4 + 1278b^{24}c^4 + 1335b^{25}c^4 + 1398b^{26}c^4 + 1455b^{27}c^4 + 1518b^{28}c^4$$
$$+1575b^{29}c^4 + 1638b^{30}c^4 + 1695b^{31}c^4 + 1758b^{32}c^4 + 1815b^{33}c^4 + 1878b^{34}c^4$$
$$+1935b^{35}c^4 + 1998b^{36}c^4 + 2055b^{37}c^4 + 2118b^{38}c^4 + 2175b^{39}c^4 + 2238b^{40}c^4$$
$$+2295b^{41}c^4 + 2358b^{42}c^4 + 2415b^{43}c^4 + 2478b^{44}c^4 + 2535b^{45}c^4 + 2598b^{46}c^4$$
$$+2655b^{47}c^4 + 2718b^{48}c^4 + 6685b^{37}c^6 + 6878b^{38}c^6 + 7081b^{39}c^6 + 7274b^{40}c^6$$
$$+7477b^{41}c^6 + 7670b^{42}c^6 + 7873b^{43}c^6 + 8066b^{44}c^6 + 8269b^{45}c^6 + 8462b^{46}c^6)$$
$$+u^{10}(5152b^{28}c^6 + 5361b^{29}c^6u^{10} + 5580b^{30}c^6 + 5789b^{31}c^6 + 6008b^{32}c^6$$
$$+6217b^{33}c^6 + 6436b^{34}c^6 + 6645b^{35}c^6 + 6864b^{36}c^6 + 7073b^{37}c^6 + 7292b^{38}c^6$$
$$+7501b^{39}c^6 + 7720b^{40}c^6 + 7929b^{41}c^6 + 8148b^{42}c^6 + 8357b^{43}c^6$$
$$+8576b^{44}c^6) + u^{12}(4294b^{24}c^6 + 4522b^{25}c^6u^{12} + 4740b^{26}c^6$$
$$+4968b^{27}c^6 + 5186b^{28}c^6 + 5414b^{29}c^6 + 5632b^{30}c^6 + 5860b^{31}c^6$$
$$+6078b^{32}c^6 + b^{33}c^6 + 6524b^{34}c^6 + 6752b^{35}c^6 + 6970b^{36}c^6$$
$$+7198b^{37}c^6 + 7416b^{38}c^6 + 7644b^{39}c^6 + 7862b^{40}c^6 + 8090b^{41}c^6$$

The Center and Focus Problem

$$+8308b^{42}c^6 + 20412b^{40}c^8) + u^{14}(18369b^{36}c^8 + 18987b^{37}c^8$$
$$+19590b^{38}c^8) + u^{16}(15835b^{32}c^8 + 16454b^{33}c^8 + 17088b^{34}c^8 \qquad (6.39)$$
$$+17707b^{35}c^8 + 18341b^{36}c^8) + \cdots$$

There from, an ordinary Hilbert series of the algebra $S'_{1,2}$ has the form (the first 61 terms):

$$H_{S'_{1,2}}(t) = H(S'_{1,2}, t, t, t) = 1 + t + 2t^2 + 2t^3 + 3t^4 + 3t^5$$
$$+4t^6 + 4t^7 + 5t^8 + 5t^9 + 6t^{10} + 6t^{11} + 7t^{12} + 7t^{13} + 76t^{14}$$
$$+87t^{15} + 96t^{16} + 107t^{17} + 116t^{18} + 127t^{19} + 136t^{20} + 147t^{21}$$
$$+156t^{22} + 167t^{23} + 176t^{24} + 187t^{25} + 196t^{26} + 207t^{27}$$
$$+1014t^{28} + 1082t^{29} + 2142t^{30} + 2268t^{31} + 2392t^{32} + 2518t^{33}$$
$$+2642t^{34} + 2768t^{35} + 2892t^{36} + 3018t^{37} + 3142t^{38} + 3268t^{39} \qquad (6.40)$$
$$+3392t^{40} + 3518t^{41} + 7936t^{42} + 8290t^{43} + 13784t^{44}$$
$$+14347t^{45} + 14908t^{46} + 15471t^{47} + 16032t^{48} + 16595t^{49}$$
$$+17156t^{50} + 24404t^{51} + 25158t^{52} + 25924t^{53} + 26678t^{54}$$
$$+27444t^{55} + 44033t^{56} + 45418t^{57} + 65175t^{58}$$
$$+67178t^{59} + 89581t^{60} + \cdots$$

We consider the first 61 terms in expansion of Hilbert series of the algebra $SI_{1,2}$, which according to [33] are obtained from (4.13)–(4.15) in the following way:

$$H_{SI_{1,2}}(t) = H(S_{1,2}, 0, t, t) = 1 + t + 2t^2 + 5t^3 + 10t^4 + 17t^5 + 30t^6 + 50t^7$$
$$+81t^8 + 125t^9 + 188t^{10} + 276t^{11} + 399t^{12} + 559t^{13} + 772t^{14} + 1051t^{15}$$
$$+1409t^{16} + 1859t^{17} + 2428t^{18} + 3133t^{19} + 4004t^{20} + 5064t^{21} + 6350t^{22}$$
$$+7897t^{23} + 9752t^{24} + 11947t^{25} + 14544t^{26} + 17597t^{27} + 21168t^{28}$$
$$+25315t^{29} + 30127t^{30} + 35673t^{31} + 42051t^{32} + 49345t^{33} + 57668t^{34}$$
$$+67127t^{35} + 77855t^{36} + 89960t^{37} + 103603t^{38} + 118928t^{39} + 136102t^{40}$$
$$+155281t^{41} + 176675t^{42} + 200462t^{43} + 226870t^{44} + 256104t^{45}$$

$$+288419t^{46} + 324057t^{47} + 363307t^{48} + 406419t^{49} + 453726t^{50}$$
$$+505532t^{51} + 562185t^{52} + 624013t^{53} + 691426t^{54} + 764788t^{55}$$
$$+844540t^{56} + 931088t^{57} + 1024916t^{58} + 1126484t^{59} \qquad (6.41)$$
$$+1236327t^{60} + \cdots$$

Since for series (6.40) and (6.41) there is an inequality

$$H_{S'_{1,2}}(t) \le H_{SI_{1,2}}(t),$$

then in assumption that this inequality holds for the remaining terms of the considered series, we obtain the inequality

$$\varrho(S'_{1,2}) \leq \varrho(SI_{1,2}).$$

Note that $S'_{1,2}$ is not a subalgebra in $SI_{1,2}$.

Since from [33] we have $\varrho(SI_{1,2}) = 7$, then according to the last inequality we obtain

Remark 6.1. *One of the ways to improve the upper bound of a number of algebraically independent generalized focus pseudo-quantities (as well as focus quantities) for differential system (6.1), that take part in solving the center and focus problem for a given differential system, is in the supposed inequality* $\varrho(S'_{1,2}) \leq 7$.

But, on the other hand, you can easily check with (6.40), that for the first 61 terms from (6.40), we have

$$H_{S'_{1,2}}(t) < \frac{1}{(1-t)^5}.$$

If we assume that this inequality is true for all terms of series $H_{S'_{1,2}}(t)$ and $(1-t)^{-5}$, then perhaps there is an improvement in the majorant assessment of the maximal number of algebraically independent focus quantities that take part in solving the center and focus problem for the differential system $s(1,2)$ from (6.1), which is expressed by the inequality $\varrho(S'_{1,2}) < 5$.

6.2 Type of Generalized Focus Pseudo-Quantities for the Differential System $s(1,3)$

Consider the differential system $s(1,3)$, which we write in the form

$$\dot{x} = cx + dy + px^3 + 3qx^2y + 3rxy^2 + sy^3,$$
$$\dot{y} = ex + fy + tx^3 + 3ux^2y + 3vxy^2 + wy^3 \tag{6.42}$$

with finitely defined graded algebra of unimodular comitants $S_{1,3}$ [33]. For this system, we write function (5.13) in the form

$$\begin{aligned}
U = {} & k_2 + a_0x^3 + 3a_1x^2y + 3a_2xy^2 + a_3y^3 + b_0x^4 + 4b_1x^3y \\
& + 6b_2x^2y^2 + 4b_3xy^3 + b_4y^4 + c_0x^5 + 5c_1x^4y + 10c_2x^3y^2 \\
& + 10c_3x^2y^3 + 5c_4xy^4 + c_5y^5 + d_0x^6 + 6d_1x^5y + 15d_2x^4y^2 \\
& + 20d_3x^3y^3 + 15d_4x^2y^4 + 6d_5xy^5 + d_6y^6 + e_0x^7 + 7e_1x^6y \\
& + 21e_2x^5y^2 + 35e_3x^4y^3 + 21e_5x^2y^5 + 7e_6xy^6 + e_7y^7 + f_0x^8 \\
& + 8f_1x^7y + 28f_2x^6y^2 + 56f_3x^5y^3 + 70f_4y^4 + 56f_5x^3y^5 \\
& + 28f_6x^2y^6 + 8f_7xy^7 + f_8y^8 + ...,
\end{aligned} \tag{6.43}$$

where $k_2 \not\equiv 0$ is from (5.8), and $a_0, a_1, ..., f_7, f_8, ...$ are unknown coefficients.

Identity (5.12) along the trajectories of differential system (6.42) with function (6.43) splits into the following systems of equations (equality (5.15) is omitted):

$$
\begin{aligned}
x^3 &: 3a_0c + 3a_1e = 0, \\
x^2y &: 6a_1c + 3a_0d + 6a_2e + 3a_1f = 0, \\
xy^2 &: 3a_2c + 6a_1d + 3a_3e + 6a_2f = 0, \\
y^3 &: 3a_2d + 3a_3f = 0;
\end{aligned}
\tag{6.44}
$$

$$
\begin{aligned}
x^4 &: 4b_0c + 4b_1e - e^2G_1 = 2ep - ct + ft, \\
x^3y &: 12b_1c + 4b_0d + 12b_2e + 4b_1f + 2ceG_1 - 2efG_1 = -cp \\
      &\quad + fp + 6eq - 2dt - 3cu + 3fu, \\
x^2y^2 &: 12b_2c + 12b_1d + 12b_3e + 12b_2f - c^2G_1 + 2deG_1 \\
       &\quad + 2cfG_1 - f^2G_1 = -3cq + 3fq + 6er - 6du - 3cv + 3fv, \\
xy^3 &: 4b_3c + 12b_2d + 4b_4e + 12b_3f - 2cdG_1 + 2dfG_1 = -3cr \\
     &\quad + 3fr + 2es - 6dv - cw + fw, \\
y^4 &: 4b_3d + 4b_4f - d^2G_1 = -cs + fs - 2dw;
\end{aligned}
\tag{6.45}
$$

$$
\begin{aligned}
x^5 &: 5cc_0 + 5c_1e + 3a_0p + 3a_1t = 0, \\
x^4y &: 20cc_1 + 5c_0d + 20c_2e + 5c_1f + 6a_1p + 9a_0q + 6a_2t \\
      &\quad + 9a_1u = 0, \\
x^3y^2 &: 30cc_2 + 20c_1d + 30c_3e + 20c_2f + 3a_2p + 18a_1q + 9a_0r \\
       &\quad + 3a_3t + 18a_2u + 9a_1v = 0,
\end{aligned}
$$

$$
\begin{aligned}
x^2y^3 &: 20cc_3 + 30c_2d + 20c_4e + 30c_3f + 9a_2q + 18a_1r + 3a_0s \\
       &\quad + 9a_3u + 18a_2v + 3a_1w = 0, \\
xy^4 &: 5cc_4 + 20c_3d + 5c_5e + 20c_4f + 9a_2r + 6a_1s + 9a_3v \\
     &\quad + 6a_2w = 0, \\
y^5 &: 5c_4d + 5c_5f + 3a_2s + 3a_3w = 0; \\
x^6 &: 6cd_0 + 6d_1e + 4b_0p + 4b_1t + e^3G_2 = 0, \\
x^5y &: 6dd_0 + 30cd_1 + 30d_2e + 6d_1f + 12b_1p + 12b_0q \\
     &\quad + 12b_2t + 12b_1u - 3ce^2G_2 + 3e^2fG_2 = 0, \\
x^4y^2 &: 30dd_1 + 60cd_2 + 60d_3e + 30d_2f + 12b_2p + 36b_1q + 12b_0r \\
       &\quad + 12b_3t + 36b_2u + 12b_1v + 3c^2eG_2 - 3de^2G_2 - 6cefG_2 \\
       &\quad + 3ef^2G_2 = 0,
\end{aligned}
\tag{6.46}
$$

$x^3y^3 : 60dd_2 + 60cd_3 + 60d_4e + 60d_3f + 4b_3p + 36b_2q + 36b_1r$
$\qquad + 4b_0s + 4b_4t + 36b_3u + 36b_2v + 4b_1w - c^3G_2 + 6cdeG_2$
$\qquad + 3c^2fG_2 - 6defG_2 - 3cf^2G_2 + f^3G_2 = 0,$

$x^2y^4 : 60dd_3 + 30cd_4 + 30d_5e + 60d_4f + 12b_3q + 36b_2r + 12b_1s$
$\qquad + 12b_4u + 36b_3v + 12b_2w - 3c^2dG_2 + 3d^2eG_2 + 6cdfG_2 \qquad (6.47)$
$\qquad - 3df^2G_2 = 0,$

$xy^5 : 30dd_4 + 6cd_5 + 6d_6e + 30d_5f + 12b_3r + 12b_2s$
$\qquad + 12b_4v + 12b_3w - 3cd^2G_2 + 3d^2fG_2 = 0,$

$y^6 : 6dd_5 + 6d_6f + 4b_3s + 4b_4w - d^3G_2 = 0;$

$\qquad x^7 : 7ce_0 + 7ee_1 + 5c_0p + 5c_1t = 0,$
$\qquad x^6y : 7de_0 + 42ce_1 + 42ee_2 + 7e_1f + 20c_1p + 15c_0q + 20c_2t$
$\qquad\qquad + 15c_1u = 0,$
$\qquad x^5y^2 : 42de_1 + 105ce_2 + 105ee_3 + 42e_2f + 30c_2p + 60c_1q$
$\qquad\qquad + 15c_0r + 30c_3t + 60c_2u + 15c_1v = 0,$
$\qquad x^4y^3 : 105de_2 + 140ce_3 + 140ee_4 + 105e_3f + 20c_3p + 90c_2q$
$\qquad\qquad + 60c_1r + 5c_0s + 20c_4t + 90c_3u + 60c_2v + 5c_1w = 0,$
$\qquad x^3y^4 : 140de_3 + 105ce_4 + 105ee_5 + 140e_4f + 5c_4p + 60c_3q$
$\qquad\qquad + 90c_2r + 20c_1s + 5c_5t + 60c_4u + 90c_3v + 20c_2w = 0,$

$x^2y^5 : 105de_4 + 42ce_5 + 42ee_6 + 105e_5f + 15c_4q + 60c_3r$
$\qquad + 30c_2s + 15c_5u + 60c_4v + 30c_3w = 0,$
$xy^6 : 42de_5 + 7ce_6 + 7ee_7 + 42e_6f + 15c_4r + 20c_3s + 15c_5v$
$\qquad + 20c_4w = 0,$
$y^7 : 7de_6 + 7e_7f + 5c_4s + 5c_5w = 0;$
$x^8 : 8cf_0 + 8ef_1 + 6d_0p + 6d_1t - e^4G_3 = 0,$
$x^7y : 8df_0 + 56cf_1 + 8ff_1 + 56ef_2 + 30d_1p + 18d_0q + 30d_2t$
$\qquad + 18d_1u + 4ce^3G_3 - 4e^3fG_3 = 0,$
$x^6y^2 : 56df_1 + 168cf_2 + 56ff_2 + 168ef_3 + 60d_2p + 90d_1q + 18d_0r$
$\qquad + 60d_3t + 90d_2u + 18d_1v - 6c^2e^2G_3 + 4de^3G_3$
$\qquad + 12ce^2fG_3 - 6e^2f^2G_3 = 0,$
$x^5y^3 : 168df_2 + 280cf_3 + 168ff_3 + 280ef_4 + 60d_3p + 180d_2q + 90d_1r$
$\qquad + 6d_0s + 60d_4t + 180d_3u + 90d_2v + 6d_1w + 4c^3eG_3$
$\qquad - 12cde^2G_3 - 12c^2efG_3 + 12de^2fG_3 + 12cef^2G_3 - 4ef^3G_3 = 0,$
$\qquad\qquad\qquad\qquad\qquad\qquad\qquad\qquad\qquad\qquad\qquad (6.48)$

$x^4 y^4 : 280df_3 + 280cf_4 + 280ff_4 + 280ef_5 + 30d_4p + 180d_3q + 180d_2r$

$\qquad + 30d_1s + 30d_5t + 180d_4u + 180d_3v + 30d_2w - c^4 G_3$

$\qquad + 12c^2 deG_3 - 6d^2 e^2 G_3 + 4c^3 fG_3 - 24cdefG_3 - 6c^2 f^2 G_3$

$\qquad + 12def^2 G_3 + 4cf^3 G_3 - f^4 G_3 = 0,$

$x^3 y^5 : 280df_4 + 168cf_5 + 280ff_5 + 168ef_6 + 6d_5p + 90d_4q + 180d_3r$

$\qquad + 60d_2s + 6d_6t + 90d_5u + 180d_4v + 60d_3w - 4c^3 dG_3$

$\qquad + 12cd^2 eG_3 + 12c^2 dfG_3 - 12d^2 efG_3 - 12cdf^2 G_3 + 4df^3 G_3 = 0,$

$x^2 y^6 : 168df_5 + 56cf_6 + 168ff_6 + 56ef_7 + 18d_5q + 90d_4r + 60d_3s$

$\qquad + 18d_6u + 90d_5v + 60d_4w - 6c^2 d^2 G_3 + 4d^3 eG_3 + 12cd^2 fG_3$

$\qquad - 6d^2 f^2 G_3 = 0,$

$xy^7 : 56df_6 + 8cf_7 + 56ff_7 + 8ef_8 + 18d_5r + 30d_4s + 18d_6v$

$\qquad + 30d_5w - 4cd^3 G_3 + 4d^3 fG_3 = 0,$

$y^8 : 8df_7 + 8ff_8 + 6d_5s + 6d_6w - d^4 G_3 = 0.$

$$\text{(6.49)}$$

It is evident that linear systems of equations (6.44)–(6.49) in the variables a_0, a_1, a_2, a_3, b_0, b_1,...,b_4, c_0, c_1,...,c_5, d_0, d_1,...,d_6, e_0, e_1,...,e_7, f_0, f_1,...,f_8,..., G_1, G_2, G_3, ... can be considered as a single system that can be extended by adding, after the last equation from (6.49), an infinite number of equations obtained from the equality of coefficients of $x^\alpha y^\beta$ for $\alpha + \beta > 8$ in identity (5.12).

For obtaining the quantity G_1, we write system (6.45) in the matrix form

$$A_1 B_1 = C_1, \qquad\qquad (6.50)$$

where

$$A_1 = \begin{pmatrix} 4c & 4e & 0 & 0 & 0 & -e^2 \\ 4d & 12c+4f & 12e & 0 & 0 & 2ce-2ef \\ 0 & 12d & 12c+12f & 12e & 0 & -c^2+2de+2cf-f^2 \\ 0 & 0 & 12d & 4c+12f & 4e & -2cd+2df \\ 0 & 0 & 0 & 4d & 4f & -d^2 \end{pmatrix},$$

$$B_1 = \begin{pmatrix} b_0 \\ b_1 \\ b_2 \\ b_3 \\ b_4 \\ G_1 \end{pmatrix}, C_1 = \begin{pmatrix} 2ep-ct+ft \\ cp+fp+6eq-2dt-3cu+3fu \\ -3cq+3fq+6er-6du-3cv+3fv \\ -3cr+3fr+2es-6dv-cw+fw \\ -cs+fs-2dw \end{pmatrix}. \quad (6.51)$$

Since the dimension of the matrix A_1 is 5×6, then it is clear that we have at least one free variable. Therefore, choosing one of b_i ($i \in \{0, , 1, ..., 4\}$) as a free variable and using the Cramer's rule for system (6.50), for each fixed i, we obtain

$$G_1 = \frac{G_{1,i} + B_{1,i} b_i}{\sigma_{1,i}}, \qquad\qquad (6.52)$$

where $G_{1,i}, B_{1,i}, \sigma_{1,i}$ (see, Appendix 5) are polynomials in coefficients of system (6.42), and b_i are undetermined coefficients of the function $U(x,y)$ from (6.49).

In the future, we will need an explicit form of the operators $X_1, ..., X_4$ of the Lie algebra L_4 for system (6.42), the expressions of which are obtained from Section 1.5:

$$X_1 = x\frac{\partial}{\partial x} + D_1, \ X_2 = y\frac{\partial}{\partial x} + D_2, \ X_3 = x\frac{\partial}{\partial y} + D_3, \ X_4 = y\frac{\partial}{\partial y} + D_4,$$

$$(6.53)$$

where

$$D_1 = d\frac{\partial}{\partial d} - e\frac{\partial}{\partial e} - 2p\frac{\partial}{\partial p} - q\frac{\partial}{\partial q} + s\frac{\partial}{\partial s} - 3t\frac{\partial}{\partial t} - 2u\frac{\partial}{\partial u} - v\frac{\partial}{\partial v},$$

$$D_2 = e\frac{\partial}{\partial c} + (f-c)\frac{\partial}{\partial d} - e\frac{\partial}{\partial f} + t\frac{\partial}{\partial p} + (u-p)\frac{\partial}{\partial q} + (v-2q)\frac{\partial}{\partial r}$$

$$+(w-3r)\frac{\partial}{\partial s} - t\frac{\partial}{\partial u} - 2u\frac{\partial}{\partial v} - 3v\frac{\partial}{\partial w},$$

$$D_3 = -d\frac{\partial}{\partial c} + (c-f)\frac{\partial}{\partial e} + d\frac{\partial}{\partial f} - 3q\frac{\partial}{\partial p} - 2r\frac{\partial}{\partial q} - s\frac{\partial}{\partial r} + (p-3u)\frac{\partial}{\partial t}$$

$$+(q-2v)\frac{\partial}{\partial u} + (r-w)\frac{\partial}{\partial v} + s\frac{\partial}{\partial w},$$

$$D_4 = -d\frac{\partial}{\partial d} + e\frac{\partial}{\partial e} - q\frac{\partial}{\partial q} - 2r\frac{\partial}{\partial r} - 3s\frac{\partial}{\partial s} + t\frac{\partial}{\partial t} - v\frac{\partial}{\partial v} - 2w\frac{\partial}{\partial w}.$$

$$(6.54)$$

By studying matrices (6.51) of system (6.50), we conclude that $G_{1,i}$ from (6.52) are homogeneous polynomials of degree 5 with respect to the linear part, and of degree 1 with respect to the cubic part of system (6.42).

Note that $G_{1,i}$ from (6.52) are homogeneous polynomials in coefficients of system (6.42), where for $i = 0, 1, 2, 3, 4$ they are polynomials of isobarities with weights, respectively (see, Appendix 5):

$$(3,-1), \ (2,0), \ (1,1), \ (0,2), \ (-1,3). \tag{6.55}$$

According to formula (3.3) (for differential system (6.42) and the theory of invariants of differential systems [37,43]), it follows that the numerators of fractions (6.52) can be coefficients in comitants of the weight -1 of the type $(4,5,1)$. This means that according to (2.56) with the help of Lie differential operator D_3 from (6.54) for differential system (6.42) and numerator of fraction (6.52), we obtain a system of four linear nonhomogeneous partial differential equations:

$$D_3(G_{1,0} + B_{1,0}b_0) = G_{1,1} + B_{1,1}b_1, \ D_3(G_{1,1} + B_{1,1}b_1) = -G_{1,2} - B_{1,2}b_2,$$

$$-D_3(G_{1,2} + B_{1,2}b_2) = G_{1,3} + B_{1,3}b_3, \ D_3(G_{1,3} + B_{1,3}b_3) = -G_{1,4} - B_{1,4}b_4,$$

$$(6.56)$$

with five unknowns b_0, b_1, b_2, b_3, b_4. According to Lemma 2.4, system (6.56) has an infinite number of solutions. Note that a particular solution of this system is $b_0 = b_1 = b_2 = b_3 = b_4 = 0$, for which the polynomial

$$f_4'(x,y) = G_{1,0}x^4 + 4G_{1,1}x^3y + 2G_{1,2}x^2y^2 + 4G_{1,3}xy^3 + G_{1,4}y^4 \qquad (6.57)$$

is a centro-affine comitant of differential system (6.42). This fact is also confirmed by Theorem 2.2 with the operators $X_1 - X_4$ from (6.54) for differential system (6.42), for which we have the equalities

$$X_1(f_4') = X_4(f_4') = f_4', \; X_2(f_4') = X_3(f_4') = 0.$$

It is evident that differential system (6.56) has an infinite number of solutions b_0, b_1, b_2, b_3, b_4, which define centro-affine comitants of type

$$f_4''(x,y) = (G_{1,0} + B_{1,0}b_0)x^4 + 4(G_{1,1} + B_{1,1}b_1)x^3y + 2(G_{1,2}$$
$$+ B_{1,2}b_2)x^2y^2 + 4(G_{1,3} + B_{1,3}b_3)xy^3 + (G_{1,4} + B_{1,4}b_4)y^4. \qquad (6.58)$$

In view of the abovementioned, comitant (6.57) or (6.58) belongs to the linear space

$$S_{1,3}^{(4,5,1)}, \qquad (6.59)$$

of Sibirsky algebra $S_{1,3}$.

Note that comitants (6.57) on the invariant variety $\mathcal{V}$ from (5.11) for differential system (6.42) have the form

$$f_4'(x,y)|_\mathcal{V} = 6(p+r+u+w)(x^2+y^2)^2 \; (G_1|_\mathcal{V} = 6(p+r+u+w)). \qquad (6.60)$$

Similarly to the previous case, for determining the quantity G_2, we write systems of equations (6.45), (6.47) in the matrix form (see, Appendix 6):

$$A_2 B_2 = C_2, \qquad (6.61)$$

from where we obtain

$$G_2 = \frac{G_{2,i,j} + B_{2,i,j}b_i + D_{2,i,j}d_j}{\sigma_{2,i,j}}, \; (i = \overline{0,4}, \; j = \overline{0,6}). \qquad (6.62)$$

By studying matrix equality (6.61), we obtain that $deg G_{2,i,j} = 14$, and using systems (6.45), (6.47), we obtain that $G_{2,i,j}$ from (6.62) has the type $(0, 12, 2)$, i.e. $G_{2,i,j}$ are homogeneous polynomials of degree 12 in coefficients of the linear part and of degree 2 in coefficients of the cubic part of the differential system $s(1,3)$ from (6.42).

Computing the expressions $G_{2,i,j}$ for each $i = \overline{0,4}$ and $j = \overline{0,6}$, we obtain for their isobarities the following Table 6.2 (see, Appendix 6).
Table 6.2. *Isobarities with weight of polynomials* $G_{2,i,j}$ *for the system* $s(1,3)$
Note that for $j = \overline{0,6}$ we have

$G_{2,i,j}$	d_0	d_1	d_2	d_3	d_4	d_5	d_6
b_0	$(7,-3)$	$(6,-2)$	$(5,-1)$	$(4,0)$	$(3,1)$	$(2,2)$	$(1,3)$
b_1	$(6,-2)$	$(5,-1)$	$(4,0)$	$(3,1)$	$(2,2)$	$(1,3)$	$(0,4)$
b_2	$(5,-1)$	$(4,0)$	$(3,1)$	$(2,2)$	$(1,3)$	$(0,4)$	$(-1,5)$
b_3	$(4,0)$	$(3,1)$	$(2,2)$	$(1,3)$	$(0,4)$	$(-1,5)$	$(-2,6)$
b_4	$(3,1)$	$(2,2)$	$(1,3)$	$(0,4)$	$(-1,5)$	$(-2,6)$	$(-3,7)$

$$G_2 \equiv \frac{G_{2,2,j}}{\sigma_{2,2,j}}|_\nu = \frac{3}{32}(-11pq - 15qr + 5ps + rs - pt - 5rt - 3qu$$
$$+5su - tu + 7pv + 3rv + 15uv - 7qw + sw - 5tw + 11vw). \tag{6.63}$$

From the set $G_{2,2,j}$ we choose the expression $G_{2,2,0}$ as a semi-invariant, which according to Table 6.2 has the weight -1. From here, using (2.56) and (6.62), we obtain that the comitant corresponding to the quantity G_2 belongs to the type $(6, 12, 2)$.

Similarly to the previous case, we choose a comitant of the weight -1 of the differential system $s(1, 3)$ from (6.42), which contains the expressions $G_{2,2,j} + B_{2,2,j}b_2 + D_{2,2,j}d_j$ $(j = \overline{0,6})$ as a semi-invariant, and we find that it belongs to the linear space

$$S_{1,3}^{(6,12,2)}, \tag{6.64}$$

of Sibirsky algebra $S_{1,3}$.

Consider determination of G_3 of homogeneity of degree 8 with respect to the phase variables x and y in (6.62). Writing the system that consists of (6.45), (6.47), (6.49) in the matrix form

$$A_3 B_3 = C_3,$$

we obtain

$$G_3 = \frac{G_{3,i,j,k} + B_{3,i,j,k}b_i + D_{3,i,j,k}d_j + F_{3,i,j,k}f_k}{\sigma_{3,i,j,k}} \tag{6.65}$$
$$(i = \overline{0,4};\ j = \overline{0,6};\ k = \overline{0,8}).$$

Similarly to the previous case, we choose a comitant of the weight -1 of the differential system $s(1, 3)$ from (6.42), which contains the expressions $G_{3,2,j,k} + B_{3,2,j,k}b_2 + D_{3,2,j,k}d_j + F_{3,2,j,k}f_k$ $(k = \overline{0,8})$ as a semi-invariant, and we find that it belongs to the linear space

$$S_{1,3}^{(8,22,3)} \tag{6.66}$$

of Sibirsky algebra $S_{1,3}$.

Let us consider the extension of system (6.44)–(6.49), which contains the quantity G_k, and is obtained from (5.12) for differential system (6.42) and function (6.43). We write system (6.44)–(6.49) in the matrix form $A_k B_k = C_k$. We denote by m_{G_k} the number of equations and by n_{G_k} the number of unknowns of this system. Note that these numbers are written as follows:

$$m_{G_k} = \underbrace{5}_{G_1} + 7 + 9 + \dots + 2k + 3,$$

$$n_{G_k} = \underbrace{6}_{G_1} + 9 + 12 + \dots + (2k+3) + k,$$

for $k = 1, 2, 3, \dots$. From here we obtain

$$m_{G_k} = k(k+4), \tag{6.67}$$

and

$$n_{G_k} = m_{G_k} + k.$$

From these systems, it is obtained

$$G_k = \frac{G_{k,i_1,i_2,\dots,i_k} + B_{k,i_1,i_2,\dots,i_k} b_{i_1} + \dots + Z_{k,i_1,i_2,\dots,i_k} z_{i_k}}{\sigma_{k,i_1,i_2,\dots,i_k}}. \tag{6.68}$$

It is important to determine the degree of the polynomial $G_{k,i_1,i_2,\dots,i_k} + B_{k,i_1,i_2,\dots,i_k} b_{i_1} + \dots + Z_{k,i_1,i_2,\dots,i_k} z_{i_k}$ in coefficients of differential system (6.42).

Note that the degree of nonzero coefficient of the polynomial G_i $(i = \overline{1, k})$ in coefficients of differential system (6.42) in Cramer's determinant of order m_{G_k}, when the last column corresponding to the quantity G_k is replaced with the column corresponding to free members, forms the following diagram

$$G_1, G_2, G_3, \dots, G_{k-1}, G_k.$$

$$\downarrow \quad \downarrow \quad \downarrow \qquad \downarrow \quad \downarrow$$

$$2 \quad 3 \quad 4 \qquad k \quad 2$$

Then the degree of the polynomials $G_{k,i_1,i_2,\dots,i_k} + B_{k,i_1,i_2,\dots,i_k} b_{i_1} + \dots + Z_{k,i_1,i_2,\dots,i_k} z_{i_k}$ in coefficients of differential system (6.42), denoted by N_{G_k}, will be written as

$$N_{G_k} = m_{G_k} + \frac{k(k-1)}{2} + 1,$$

from where we obtain

$$N_{G_k} = \frac{1}{2}(3k^2 + 7k + 2). \tag{6.69}$$

It is the degree of homogeneity of polynomials $G_{k,i_1,i_2,...,i_k} + B_{k,i_1,i_2,...,i_k} b_{i_1} +$ $\cdots + Z_{k,i_1,i_2,...,i_k} z_{i_k}$ in coefficients of system (6.42), which is contained in a polynomial of type (δ, d_0, d_1), where δ is the degree of homogeneity of polynomial in x and y, d_0 is the degree of homogeneity of polynomial in coefficients of linear part, and d_1 is the degree of homogeneity of polynomial in coefficients of the cubic part of the system $s(1,3)$ from (6.42). Since $\delta = 2(k+1)$ and $d_1 = k$, then $d_0 = N_{G_k} - 2k$.

Based on this, we find that a comitant of the weight -1 of the differential system $s(1,3)$ from (6.42), that contains $G_{k,i_1,i_2,...,i_k} + B_{k,i_1,i_2,...,i_k} b_{i_1} + \cdots + Z_{k,i_1,i_2,...,i_k} z_{i_k}$ as a semi-invariant and corresponds to the quantity G_k for $k = 1, 2, 3, ...$, belongs to the type

$$\left(2(k+1), \frac{1}{2}(3k^2 + 5k + 2), k \right), \tag{6.70}$$

where $2(k+1)$ is the degree of homogeneity of the comitant in phase variables x, y; $\frac{1}{2}(3k^2 + 5k + 2)$ is the degree of homogeneity of comitant in the coefficients c, d, e, f of linear part; k is the degree of homogeneity of the comitant in the coefficients p, q, r, s, t, u, v, w of cubic part of the differential system $s(1,3)$ from (6.42).

Therefore, the expressions $G_{k,i_1,i_2,...,i_k} + B_{k,i_1,i_2,...,i_k} b_{i_1} + \cdots + Z_{k,i_1,i_2,...,i_k} z_{i_k}$, which determine comitants of types (6.70), corresponding to the quantity G_k $(k = 1, 2, 3, ...)$, will be called *generalized focus pseudo-quantities*, and comitants of type (6.70) for $k = 1, 2, 3, ...$ will be called *comitants which contain generalized focus pseudo-quantities* $G_{k,i_1,i_2,...,i_k} + B_{k,i_1,i_2,...,i_k} b_{i_1} + \cdots + Z_{k,i_1,i_2,...,i_k} z_{i_k}$ *as coefficients.*

Note that the spaces $S^{(2(k+1), \frac{1}{2}(3k^2+5k+2), k)}$ are generalized records of spaces (6.59), (6.64), (6.66) for $k = 1, 2, 3, ...$ of Sibirsky algebra $S_{1,3}$.

6.3 On the Upper Bound of the Number of Algebraically Independent Focus Pseudo-Quantities, that Take Part in Solving the Center and Focus Problem for the Differential System $s(1,3)$

Consider the differential system $s(1,3)$ from (6.42). In this case, according to the paper [33], using Theorem 4.4, there takes place

Theorem 6.3. *Dimension of linear space of centro-affine comitants of type* $(d) = (\delta, d_0, d_1)$ *for the differential system $s(1,3)$ from (6.42), denoted by*

$dim_{\mathbb{R}} V_{1,3}^{(d)}$, *is equal to the coefficient of the monomial* $u^{\delta} b^{d_0} d^{d_1}$ *in decomposition of generalized Hilbert series from* (4.22)–(4.24) *for Sibirsky algebra* $S_{1,3}$ *of comitants of the considered system.*

Consider the subalgebra $S_{1,3}' \subset S_{1,3}$, which we write in the form

$$S_{1,3}' = \bigoplus_{(d)} S_{1,3}^{(d')}, \tag{6.71}$$

where by $S_{1,3}^{(d')}$ the following linear spaces are denoted:

$$S_{1,3}^{(0,0,0)} = \mathbb{R},\ S_{1,3}^{(0,1,0)}, ..., S_{1,3}^{(2(k+1),\frac{1}{2}(3k^2+5k+2),k)},\ k = 1, 2, ..., \tag{6.72}$$

as well as spaces from $S_{1,3}$, which contain all kinds of their products.

Since the algebra $S_{1,3}'$ is a graded subalgebra in a finitely defined algebra $S_{1,3}$, then according to Proposition 4.1, we obtain $\varrho(S_{1,3}') \leq \varrho(S_{1,3})$. From this inequality and from the fact that according to Theorem 4.5, we have $\varrho(S_{1,3}) = 11$, according to Remark 2.3 on semi-invariants and the fact that generalized focus pseudo-quantities are coefficients of some comitants, the following is true:

Theorem 6.4. *Maximal number of algebraically independent generalized focus pseudo-quantities in the center and focus problem for differential system* (6.42) *does not exceed* 11.

According to Proposition 4.2, Observation 5.2 and equality (5.14), it follows that the maximal number of algebraically independent focus quantities L_k ($k = \overline{1,\infty}$) cannot exceed maximal number of algebraically independent generalized focus pseudo-quantities $G_{k,i_1,i_2,...,i_k} + B_{k,i_1,i_2,...,i_k} b_{i_1} + \cdots + Z_{k,i_1,i_2,...,i_k} z_{i_k}$.

Hence, according to Theorem 6.4, we have

Consequence 6.2. *Upper bound of the number of algebraically independent focus quantities that take part in solving the center and focus problem for differential system* (6.42) *does not exceed* 11.

Consider types of spaces $S_{1,3}^{(d')}$ from (6.72) for $d' = \delta' + d_0' + d_1' \leq 50$, which are obtained from expansion of the following fraction in a power series:

$$\frac{1}{(1 - b)(1 - u^4 b^5 d)(1 - u^6 b^{12} d^2)(1 - u^8 b^{22} d^3)(1 - u^{10} b^{35} d^4)}, \tag{6.73}$$

where $u^{\delta'} b^{d_0'} d^{d_1'}$ shows the type of space $S_{1,3}^{(d')}$ for $(d') = (\delta', d_0', d_1')$. In consideration of these types and generalized Hilbert series (4.22)–(4.24) of the algebra $S_{1,3}$, we can write expansion of Hilbert series of the algebra $S_{1,3}'$ for $d' = \delta' + d_0' + d_1' \leq 50$, which has the form

$$H(S'_{1,3}, u, b, d) = 1 + b + 2b^2 + 2b^3 + 3b^4 + 3b^5 + 4b^6 + 4b^7 + 5b^8 + 5b^9$$
$$+6b^{10} + 6b^{11} + 7b^{12} + 7b^{13} + 8b^{14} + 8b^{15} + 9b^{16} + 9b^{17} + 10b^{18} + 10b^{19}$$
$$+11b^{20} + 11b^{21} + 12b^{22} + 12b^{23} + 13b^{24} + 13b^{25} + 14b^{26} + 14b^{27} + 15b^{28}$$
$$+15b^{29} + 16b^{30} + 16b^{31} + 17b^{32} + 17b^{33} + 18b^{34} + 18b^{35} + 19b^{36} + 19b^{37}$$
$$+20b^{38} + 20b^{39} + 21b^{40} + 21b^{41} + 22b^{42} + 22b^{43} + 23b^{44} + 23b^{45} + 24b^{46}$$
$$+24b^{47} + 25b^{48} + 25b^{49} + 26b^{50} + (18b^5 d + 22b^6 d + 26b^7 d + 30b^8 d + 34b^9 d$$
$$+38b^{10}d + 42b^{11}d + 46b^{12}d + 50b^{13}d + 54b^{14}d + 58b^{15}d + 62b^{16}d$$
$$+66b^{17}d + 70b^{18}d + 74b^{19}d + 78b^{20}d + 82b^{21}d + 86b^{22}d + 90b^{23}d$$
$$+94b^{24}d + 98b^{25}d + 102b^{26}d + 106b^{27}d + 110b^{28}d + 114b^{29}d$$
$$+118b^{30}d + 122b^{31}d + 126b^{32}d + 130b^{33}d + 134b^{34}d + 138b^{35}d$$
$$+142b^{36}d + 146b^{37}d + 150b^{38}d + 154b^{39}d + 158b^{40}d + 162b^{41}d + 166b^{42}d$$
$$+170b^{43}d + 174b^{44}d + 178b^{45}d)u^4 + (174b^{12}d^2 + 193b^{13}d^2 + 208b^{14}d^2$$
$$+227b^{15}d^2 + 242b^{16}d^2 + 261b^{17}d^2 + 276b^{18}d^2 + 295b^{19}d^2 + 310b^{20}d^2$$
$$+329b^{21}d^2 + 344b^{22}d^2 + 363b^{23}d^2 + 378b^{24}d^2 + 397b^{25}d^2 + 412b^{26}d^2$$
$$+431b^{27}d^2 + 446b^{28}d^2 + 465b^{29}d^2 + 480b^{30}d^2 + 499b^{31}d^2$$
$$+514b^{32}d^2 + 533b^{33}d^2 + 548b^{34}d^2 + 567b^{35}d^2 + 582b^{36}d^2$$
$$+601b^{37}d^2 + 616b^{38}d^2 + 635b^{39}d^2 + 650b^{40}d^2 + 669b^{41}d^2$$
$$+684b^{42}d^2)u^6 + (136b^{10}d^2 + 152b^{11}d^2 + 172b^{12}d^2 + 188b^{13}d^2$$
$$+208b^{14}d^2 + 224b^{15}d^2 + 244b^{16}d^2 + 260b^{17}d^2 + 280b^{18}d^2$$
$$+296b^{19}d^2 + 316b^{20}d^2 + 332b^{21}d^2 + 352b^{22}d^2 + 368b^{23}d^2$$
$$+388b^{24}d^2 + 404b^{25}d^2 + 424b^{26}d^2 + 440b^{27}d^2 + 460b^{28}d^2$$
$$+476b^{29}d^2 + 496b^{30}d^2 + 512b^{31}d^2 + 532b^{32}d^2 + 548b^{33}d^2$$
$$+568b^{34}d^2 + 584b^{35}d^2 + 604b^{36}d^2 + 620b^{37}d^2 + 640b^{38}d^2$$
$$+656b^{39}d^2 + 676b^{40}d^2 + 1098b^{22}d^3 + 1155b^{23}d^3 + 1212b^{24}d^3$$
$$+1269b^{25}d^3 + 1326b^{26}d^3 + 1383b^{27}d^3 + 1440b^{28}d^3 + 1497b^{29}d^3$$
$$+1554b^{30}d^3 + 1611b^{31}d^3 + 1668b^{32}d^3 + 1725b^{33}d^3 + 1782b^{34}d^3$$
$$+1839b^{35}d^3 + 1896b^{36}d^3 + 1953b^{37}d^3 + 2010b^{38}d^3 + 2067b^{39}d^3)u^8$$
$$+(791b^{17}d^3 + 850b^{18}d^3 + 909b^{19}d^3 + 968b^{20}d^3 + 1027b^{21}d^3$$
$$+1086b^{22}d^3 + 1145b^{23}d^3 + 1204b^{24}d^3 + 1263b^{25}d^3 + 1322b^{26}d^3$$
$$+1381b^{27}d^3 + 1440b^{28}d^3 + 1499b^{29}d^3 + 1558b^{30}d^3 + 1617b^{31}d^3$$
$$+1676b^{32}d^3 + 1735b^{33}d^3 + 1794b^{34}d^3 + 1853b^{35}d^3 + 1912b^{36}d^3$$
$$+1971b^{37}d^3 + 4904b^{35}d^4 + 5056b^{36}d^4)u^{10} + (630b^{15}d^3 + 690b^{16}d^3$$
$$+750b^{17}d^3 + 810b^{18}d^3 + 870b^{19}d^3 + 930b^{20}d^3 + 990b^{21}d^3 + 1050b^{22}d^3$$
$$+1110b^{23}d^3 + 1170b^{24}d^3 + 1230b^{25}d^3 + 1290b^{26}d^3 + 1350b^{27}d^3$$

$$+1410b^{28}d^3 + 1470b^{29}d^3 + 1530b^{30}d^3 + 1590b^{31}d^3 + 1650b^{32}d^3$$
$$+1710b^{33}d^3 + 1770b^{34}d^3 + 1830b^{35}d^3 + 3142b^{24}d^4 + 3299b^{25}d^4$$
$$+3466b^{26}d^4 + 3623b^{27}d^4 + 3790b^{28}d^4 + 3947b^{29}d^4 + 4114b^{30}d^4$$
$$+4271b^{31}d^4 + 4438b^{32}d^4 + 4595b^{33}d^4 + 4762b^{34}d^4)u^{12} + (2696b^{22}d^4$$
$$+2865b^{23}d^4 + 3024b^{24}d^4 + 3193b^{25}d^4 + 3352b^{26}d^4 + 3521b^{27}d^4 \qquad (6.74)$$
$$+3680b^{28}d^4 + 3849b^{29}d^4 + 4008b^{30}d^4 + 4177b^{31}d^4 + 4336b^{32}d^4)u^{14}$$
$$+(2230b^{20}d^4 + 2390b^{21}d^4 + 2560b^{22}d^4 + 2720b^{23}d^4 + 2890b^{24}d^4$$
$$+3050b^{25}d^4 + 3220b^{26}d^4 + 3380b^{27}d^4 + 3550b^{28}d^4 + 3710b^{29}d^4$$
$$+3880b^{30}d^4 + 8817b^{29}d^5)u^{16} + 7693b^{27}d^5u^{18} + 6534b^{25}d^5u^{20} + \ldots$$

From here, an ordinary Hilbert series of the algebra $S'_{1,3}$ has the form (the first 51 terms):

$$H_{S'_{1,3}}(t) = H(S'_{1,3}, t, t, t) = 1 + t + 2t^2 + 2t^3 + 3t^4 + 3t^5 + 4t^6$$
$$+4t^7 + 5t^8 + 5t^9 + 24t^{10} + 28t^{11} + 33t^{12} + 37t^{13} + 42t^{14} + 46t^{15}$$
$$+51t^{16} + 55t^{17} + 60t^{18} + 64t^{19} + 379t^{20} + 418t^{21} + 458t^{22}$$
$$+497t^{23} + 537t^{24} + 576t^{25} + 616t^{26} + 655t^{27} + 695t^{28} + 734t^{29}$$

$$+2195t^{30} + 2353t^{31} + 2512t^{32} + 3768t^{33} + 3984t^{34} + 4199t^{35}$$
$$+4415t^{36} + 4630t^{37} + 4846t^{38} + 5061t^{39} + 13345t^{40} + 14046t^{41}$$
$$+14758t^{42} + 15459t^{43} + 16171t^{44} + 16872t^{45} + 17584t^{46} \qquad (6.75)$$
$$+18285t^{47} + 18997t^{48} + 24602t^{49} + 48510t^{50} + \ldots$$

We consider the first 51 terms in expansion of Hilbert series of the algebra $SI_{1,3}$, which according to [33] are obtained from (4.22)–(4.24) in the following way:

$$H_{SI_{1,3}}(t) = H(S_{1,3}, 0, t, t) = 1 + t + 5t^2 + 9t^3 + 24t^4 + 42t^5 + 95t^6 + 160t^7$$
$$+308t^8 + 506t^9 + 877t^{10} + 1376t^{11} + 2229t^{12} + 3358t^{13} + 5144t^{14}$$
$$+7498t^{15} + 10996t^{16} + 15545t^{17} + 22032t^{18} + 30335t^{19} + 41764t^{20}$$
$$+56226t^{21} + 75544t^{22} + 99686t^{23} + 131205t^{24} + 170114t^{25} + 219901t^{26}$$
$$+280744t^{27} + 357236t^{28} + 449800t^{29} + 564495t^{30} + 702002t^{31} + 870184t^{32}$$
$$+1070195t^{33} + 1311989t^{34} + 1597351t^{35} + 1938881t^{36} + 2339064t^{37}$$
$$+2813664t^{38} + 3366216t^{39} + 4016096t^{40} + 4768162t^{41} + 5646208t^{42}$$
$$+6656574t^{43} + 7828224t^{44} + 9169512t^{45} + 10715232t^{46} + 12476184t^{47}$$
$$+14494113t^{48} + 16782555t^{49} + 19391253t^{50} + \ldots$$
$$(6.76)$$

Since for series (6.75) and (6.76), the following inequality holds

$$H_{S'_{1,3}}(t) \le H_{SI_{1,3}}(t),$$

then in assumption that this inequality holds for remaining terms of the considered series, we obtain the inequality

$$\varrho(S'_{1,3}) \le \varrho(SI_{1,3}).$$

Note that $S'_{1,3}$ is not a subalgebra in $SI_{1,3}$.

Since from Theorem 4.5, we have $\varrho(SI_{1,3}) = 9$, then according to the last inequality, we obtain

Remark 6.2. *One of the ways to improve the upper bound of a number of algebraically independent generalized focus pseudo-quantities (as well as focus quantities) for differential system (6.42), that take part in solving the center and focus problem for a given differential system, is in the supposed inequality* $\varrho(S'_{1,3}) \le 9$.

However, on the other hand, you can easily check with (6.75), that for the first 51 terms, we have

$$H_{S'_{1,3}}(t) < \frac{1}{(1-t)^7}.$$

If we assume that this inequality is true for all terms of series $H_{S'_{1,3}}(t)$ and $(1-t)^{-7}$, then perhaps there is an improvement in the majorant assessment of the maximal number of algebraically independent focus quantities that take part in solving the center and focus problem for the differential system $s(1,3)$ from (6.42), which is expressed by the inequality $\varrho(S'_{1,3}) < 7$.

6.4 The Differential System $s(1,4)$ and Algebraically Independent Generalized Focus Pseudo-Quantities

Consider the differential system $s(1,4)$, which we write in the form

$$\begin{aligned} \dot{x} &= cx + dy + gx^4 + 4hx^3y + 6ix^2y^2 + 4jxy^3 + ky^4, \\ \dot{y} &= ex + fy + lx^4 + 4mx^3y + 6nx^2y^2 + 4oxy^3 + py^4 \end{aligned} \tag{6.77}$$

with a infinitely defined graded algebra of unimodular comitants $S_{1,4}$ [33]. For this system, we write function (5.13) in the form

$$\begin{aligned} U =\ &k_2 + a_0x^3 + 3a_1x^2y + 3a_2xy^2 + a_3y^3 + b_0x^4 + 4b_1x^3y + 6b_2x^2y^2 \\ &+ 4b_3xy^3 + b_4y^4 + c_0x^5 + 5c_1x^4y + 10c_2x^3y^2 + 10c_3x^2y^3 + 5c_4xy^4 \\ &+ c_5y^5 + d_0x^6 + 6d_1x^5y + 15d_2x^4y^2 + 20d_3x^3y^3 + 15d_4x^2y^4 + 6d_5xy^5 \\ &+ d_6y^6 + e_0x^7 + 7e_1x^6y + 21e_2x^5y^2 + 35e_3x^4y^3 + 21e_5x^2y^5 + 7e_6xy^6 \\ &+ e_7y^7 + f_0x^8 + 8f_1x^7y + 28f_2x^6y^2 + 56f_3x^5y^3 + 70f_4y^4 + 56f_5x^3y^5 \\ &+ 28f_6x^2y^6 + 8f_7xy^7 + f_8y^8 + g_0x^9 + 9g_1x^8y + 36g_2x^7y^2 + 84g_3x^6y^3 \end{aligned}$$

$$+126g_4x^5y^4 + 126g_5x^4y^5 + 84g_6x^3y^6 + 36g_7x^2y^7 + 9g_8xy^8 + g_9y^9$$
$$+h_0x^{10} + 10h_1x^9y + 45h_2x^8y^2 + 120h_3x^7y^3 + 210h_4x^6y^4 + 252h_5x^5y^5$$
$$+210h_6x^4y^6 + 120h_7x^3y^7 + 45h_8x^2y^8 + 10h_9xy^9 + h_{10}y^{10} + i_0x^{11}$$
$$+11i_1x^{10}y + 55i_2x^9y^2 + 165i_3x^8y^3 + 330i_4x^7y^4 + 462i_5x^6y^5 + 462i_6x^5y^6$$
$$+330i_7x^4y^7 + 55i_9x^2y^9 + 11i_{10}xy^{10} + i_{11}y^{11} + j_0x^{12} + 12j_1x^{11}y$$
$$+165i_8x^3y^8 + 66j_2x^{10}y^2 + 220j_3x^9y^3 + 495j_4x^8y^4 + 792j_5x^7y^5$$
$$+924j_6x^6y^6 + 792j_7x^5y^7 + 495j_8x^4y^8 + 220j_9x^3y^9 + 66j_{10}x^2y^{10}$$
$$+12j_{11}xy^{11} + j_{12}y^{12} + k_0x^{13} + 13k_1x^{12}y + 78k_2x^{11}y^2 + 286k_3x^{10}y^3$$
$$+715k_4x^9y^4 + 1287k_5x^8y^5 + 1716k_6x^7y^6 + 1716k_7x^6y^7 + 1287k_8x^5y^8$$

$$+715k_9x^4y^9 + 286k_{10}x^3y^{10} + 78k_{11}x^2y^{11} + 13k_{12}xy^{12} + k_{13}y^{13}$$
$$+l_0x^{14} + 14l_1x^{13}y + 91l_2x^{12}y^2 + 364l_3x^{11}y^3 + 1001l_4x^{10}y^4$$
$$+2002l_5x^9y^5 + 3003l_6x^8y^6 + 3432l_7x^7y^7 + 3003l_8x^6y^8 \qquad (6.78)$$
$$+2002l_9x^5y^9 + 1001l_{10}x^4y^{10} + 364l_{11}x^3y^{11} + 91l_{12}x^2y^{12}$$
$$+14l_{13}xy^{13} + l_{14}y^{14} + ...,$$

where $k_2 \neq 0$ is from (5.8), and $a_0, a_1, ..., l_{13}, l_{14}, ...$ are unknown coefficients.

Identity (5.12) along the trajectories of differential system (6.77) with function (6.78) splits into the following systems of equations (equality (5.15) is omitted):

$$x^3 : 3a_0c + 3a_1e = 0,$$
$$x^2y : 6a_1c + 3a_0d + 6a_2e + 3a_1f = 0,$$
$$xy^2 : 3a_2c + 6a_1d + 3a_3e + 6a_2f = 0, \qquad (6.79)$$
$$y^3 : 3a_2d + 3a_3f = 0;$$

$$x^4 : 4b_0c + 4b_1e - e^2G_1 = 0,$$
$$x^3y : 12b_1c + 4b_0d + 12b_2e + 4b_1f + 2ceG_1 - 2efG_1 = 0,$$
$$x^2y^2 : 12b_2c + 12b_1d + 12b_3e + 12b_2f - c^2G_1 + 2deG_1$$
$$\qquad + 2cfG_1 - f^2G_1 = 0,$$
$$xy^3 : 4b_3c + 12b_2d + 4b_4e + 12b_3f - 2cdG_1 + 2dfG_1 = 0,$$
$$y^4 : 4b_3d + 4b_4f - d^2G_1 = 0; \qquad (6.80)$$
$$x^5 : 5cc_0 + 5c_1e - 2eg + cl - fl = 0,$$
$$x^4y : 20cc_1 + 5c_0d + 20c_2e + 5c_1f + cg - fg - 8eh + 2dl$$
$$\qquad + 4cm - 4fm = 0,$$
$$x^3y^2 : 30cc_2 + 20c_1d + 30c_3e + 20c_2f + 4ch - 4fh - 12ei$$
$$\qquad + 8dm + 6cn - 6fn = 0,$$

$x^2y^3 : 20cc_3 + 30c_2d + 20c_4e + 30c_3f + 6ci - 6fi - 8ej + 12dn$
$\qquad + 4co - 4fo = 0,$

$xy^4 : 5cc_4 + 20c_3d + 5c_5e + 20c_4f + 4cj - 4fj - 2ek + 8do \qquad (6.81)$
$\qquad + cp - fp = 0,$

$y^5 : 5c_4d + 5c_5f + ck - fk + 2dp = 0;$

$x^6 : 6cd_0 + 6d_1e + 3a_0g + 3a_1l = -e^3 G_2,$

$x^5y : 6dd_0 + 30cd_1 + 30d_2e + 6d_1f + 6a_1g + 12a_0h + 6a_2l$
$\qquad + 12a_1m = 3ce^2 G_2 - 3e^2 f G_2,$

$x^4y^2 : 30dd_1 + 60cd_2 + 60d_3e + 30d_2f + 3a_2g + 24a_1h + 18a_0i$
$\qquad + 3a_3l + 24a_2m + 18a_1n = -3c^2 eG_2 + 3de^2 G_2$
$\qquad + 6cef G_2 - 3ef^2 G_2,$

$x^3y^3 : 60dd_2 + 60cd_3 + 60d_4e + 60d_3f + 12a_2h + 36a_1i + 12a_0j$
$\qquad + 12a_3m + 36a_2n + 12a_1o = c^3 G_2 - 6cdeG_2 - 3c^2 f G_2 \qquad (6.82)$
$\qquad + 6def G_2 + 3cf^2 G_2 - f^3 G_2,$

$x^2y^4 : 60dd_3 + 30cd_4 + 30d_5e + 60d_4f + 18a_2i + 24a_1j + 3a_0k$
$\qquad + 18a_3n + 24a_2o + 3a_1p = 3c^2 dG_2 - 3d^2 eG_2 - 6cdf G_2$
$\qquad + 3df^2 G_2,$

$xy^5 : 30dd_4 + 6cd_5 + 6d_6e + 30d_5f + 12a_2j + 6a_1k + 12a_3o$
$\qquad + 6a_2p = 3cd^2 G_2 - 3d^2 f G_2,$

$y^6 : 6dd_5 + 6d_6f + 3a_2k + 3a_3p = d^3 G_2;$

$x^7 : 7ce_0 + 7ee_1 + 4b_0g + 4b_1l = 0,$

$x^6y : 7de_0 + 42ce_1 + 42ee_2 + 7e_1f + 12b_1g + 16b_0h + 12b_2l$
$\qquad + 16b_1m = 0,$

$x^5y^2 : 42de_1 + 105ce_2 + 105ee_3 + 42e_2f + 12b_2g + 48b_1h$
$\qquad + 24b_0i + 12b_3l + 48b_2m + 24b_1n = 0,$

$x^4y^3 : 105de_2 + 140ce_3 + 140ee_4 + 105e_3f + 4b_3g + 48b_2h$
$\qquad + 72b_1i + 16b_0j + 4b_4l + 48b_3m + 72b_2n + 16b_1o = 0,$
$\qquad\qquad\qquad\qquad\qquad\qquad\qquad\qquad\qquad\qquad (6.83)$

$x^3y^4 : 140de_3 + 105ce_4 + 105ee_5 + 140e_4f + 16b_3h + 72b_2i$
$\qquad + 48b_1j + 4b_0k + 16b_4m + 72b_3n + 48b_2o + 4b_1p = 0,$

$x^2y^5 : 105de_4 + 42ce_5 + 42ee_6 + 105e_5f + 24b_3i + 48b_2j$
$\qquad + 12b_1k + 24b_4n + 48b_3o + 12b_2p = 0,$

$xy^6 : 42de_5 + 7ce_6 + 7ee_7 + 42e_6f + 16b_3j + 12b_2k + 16b_4o$
$\qquad + 12b_3p = 0,$

$y^7 : 7de_6 + 7e_7f + 4b_3k + 4b_4p = 0;$

$$x^8 : 8cf_0 + 8ef_1 + 5c_0g + 5c_1l = e^4G_3,$$

$$x^7y : 8df_0 + 56cf_1 + 8ff_1 + 56ef_2 + 20c_1g + 20c_0h + 20c_2l$$
$$+ 20c_1m = -4ce^3G_3 + 4e^3fG_3,$$

$$x^6y^2 : 56df_1 + 168cf_2 + 56ff_2 + 168ef_3 + 30c_2g + 80c_1h$$
$$+ 30c_0i + 30c_3l + 80c_2m + 30c_1n = 6c^2e^2G_3 - 4de^3G_3$$
$$- 12ce^2fG_3 + 6e^2f^2G_3,$$

$$x^5y^3 : 168df_2 + 280cf_3 + 168ff_3 + 280ef_4 + 20c_3g + 120c_2h$$
$$+ 120c_1i + 20c_0j + 20c_4l + 120c_3m + 120c_2n + 20c_1o =$$
$$= -4c^3eG_3 + 12cde^2G_3 + 12c^2efG_3 - 12de^2fG_3$$
$$- 12cef^2G_3 + 4ef^3G_3,$$

$$x^4y^4 : 280df_3 + 280cf_4 + 280ff_4 + 280ef_5 + 5c_4g + 80c_3h$$
$$+ 180c_2i + 80c_1j + 5c_0k + 5c_5l + 80c_4m + 180c_3n + 80c_2o$$
$$+ 5c_1p = c^4G_3 - 12c^2deG_3 + 6d^2e^2G_3 - 4c^3fG_3$$
$$+ 24cdefG_3 + 6c^2f^2G_3 - 12def^2G_3 - 4cf^3G_3 + f^4G_3,$$

$$x^3y^5 : 280df_4 + 168cf_5 + 280ff_5 + 168ef_6 + 20c_4h + 120c_3i$$
$$+ 120c_2j + 20c_1k + 20c_5m + 120c_4n + 120c_3o + 20c_2p =$$
$$= 4c^3dG_3 - 12cd^2eG_3 - 12c^2dfG_3$$
$$+ 12d^2efG_3 + 12cdf^2G_3 - 4df^3G_3,$$

$$x^2y^6 : 168df_5 + 56cf_6 + 168ff_6 + 56ef_7 + 30c_4i + 80c_3j + 30c_2k$$
$$+ 30c_5n + 80c_4o + 30c_3p = 6c^2d^2G_3 - 4d^3eG_3$$
$$- 12cd^2fG_3 + 6d^2f^2G_3,$$

$$xy^7 : 56df_6 + 8cf_7 + 56ff_7 + 8ef_8 + 20c_4j + 20c_3k + 20c_5o$$
$$+ 20c_4p = 4cd^3G_3 - 4d^3fG_3,$$

$$y^8 : 8df_7 + 8ff_8 + 5c_4k + 5c_5p = d^4G_3;$$

$$x^9 : 6d_0g + 9cg_0 + 9eg_1 + 6d_1l = 0,$$

$$x^8y : 30d_1g + 9dg_0 + 72cg_1 + 9fg_1 + 72eg_2 + 24d_0h + 30d_2l$$
$$+ 24d_1m = 0,$$

$$x^7y^2 : 60d_2g + 72dg_1 + 252cg_2 + 72fg_2 + 252eg_3 + 120d_1h$$
$$+ 36d_0i + 60d_3l + 120d_2m + 36d_1n = 0,$$

$$x^6y^3 : 60d_3g + 252dg_2 + 504cg_3 + 252fg_3 + 504eg_4 + 240d_2h$$
$$+ 180d_1i + 24d_0j + 60d_4l + 240d_3m + 180d_2n + 24d_1o = 0,$$

$$x^5y^4 : 30d_4g + 504dg_3 + 630cg_4 + 504fg_4 + 630eg_5 + 240d_3h$$
$$+ 360d_2i + 120d_1j + 6d_0k + 30d_5l + 240d_4m + 360d_3n$$
$$+ 120d_2o + 6d_1p = 0,$$

(6.84)

$x^4y^5 : 6d_5g + 630dg_4 + 504cg_5 + 630fg_5 + 504eg_6 + 120d_4h$
$\qquad + 360d_3i + 240d_2j + 30d_1k + 6d_6l + 120d_5m + 360d_4n$
$\qquad + 240d_3o + 30d_2p = 0,$

$x^3y^6 : 504dg_5 + 252cg_6 + 504fg_6 + 252eg_7 + 24d_5h + 180d_4i$
$\qquad + 240d_3j + 60d_2k + 24d_6m + 180d_5n + 240d_4o + 60d_3p = 0,$

$x^2y^7 : 252dg_6 + 72cg_7 + 252fg_7 + 72eg_8 + 36d_5i + 120d_4j$
$\qquad + 60d_3k + 36d_6n + 120d_5o + 60d_4p = 0,$

$xy^8 : 72dg_7 + 9cg_8 + 72fg_8 + 9eg_9 + 24d_5j + 30d_4k + 24d_6o$
$\qquad + 30d_5p = 0,$

$y^9 : 9dg_8 + 9fg_9 + 6d_5k + 6d_6p = 0;$

$$(6.85)$$

$x^{10} : 7e_0g + e^5G_4 + 10ch_0 + 10eh_1 + 7e_1l = 0,$

$x^9y : 42e_1g - 5ce^4G_4 + 5e^4fG_4 + 28e_0h + 10dh_0 + 90ch_1 + 10fh_1$
$\qquad + 90eh_2 + 42e_2l + 28e_1m = 0,$

$x^8y^2 : 105e_2g + 10c^2e^3G_4 - 5de^4G_4 - 20ce^3fG_4 + 10e^3f^2G_4 + 168e_1h$
$\qquad + 90dh_1 + 360ch_2 + 90fh_2 + 360eh_3 + 42e_0i + 105e_3l + 168e_2m$
$\qquad 42e_1n = 0,$

$x^7y^3 : 140e_3g - 10c^3e^2G_4 + 20cde^3G_4 + 30c^2e^2fG_4 - 20de^3fG_4$
$\qquad + 30ce^2f^2G_4 + 10e^2f^3G_4 + 420e_2h + 360dh_2 + 840ch_3 + 360fh_3$
$\qquad + 840eh_4 + 252e_1i + 28e_0j + 140e_4l + 420e_3m + 252e_2n + 28e_1o = 0,$

$x^6y^4 : 105e_4g + 5c^4eG_4 - 30c^2de^2G_4 + 10d^2e^3G_4 - 20c^3efG_4$
$\qquad + 60cde^2fG_4 + 30c^2ef^2G_4 - 30de^2f^2G_4 - 20cef^3G_4$
$\qquad + 5ef^4G_4 + 560e_3h + 840dh_3 + 1260ch_4 + 840fh_4$
$\qquad + 1260eh_5 + 630e_2i + 168e_1j + 7e_0k + 105e_5l + 560e_4m$
$\qquad + 630e_3n + 168e_2o + 7e_1p = 0,$

$x^5y^5 : 42e_5g - c^5G_4 + 20c^3deG_4 - 30cd^2e^2G_4 + 5c^4fG_4$
$\qquad - 60c^2defG_4 + 30d^2e^2fG_4 - 10c^3f^2G_4 + 60cdef^2G_4$
$\qquad + 10c^2f^3G_4 - 20def^3G_4 - 5cf^4G_4 + f^5G_4 + 420e_4h$
$\qquad + 1260dh_4 + 1260ch_5 + 1260fh_5 + 1260eh_6 + 840e_3i$
$\qquad + 420e_2j + 42e_1k + 42e_6l + 420e_5m + 840e_4n$
$\qquad + 420e_3o + 42e_2p = 0,$

$x^4y^6 : 7e_6g - 5c^4dG_4 + 30c^2d^2eG_4 - 10d^3e^2G_4 + 20c^3dfG_4$
$\qquad - 60cd^2efG_4 - 30c^2df^2G_4 + 30d^2ef^2G_4 + 20cdf^3G_4$
$\qquad - 5df^4G_4 + 168e_5h + 1260dh_5 + 840ch_6 + 1260fh_6$
$\qquad + 840eh_7 + 630e_4i + 560e_3j + 105e_2k + 7e_7l + 168e_6m$
$\qquad + 630e_5n + 560e_4o + 105e_3p = 0,$

$$x^3 y^7 : 20cd^3 eG_4 - 10c^3 d^2 G_4 + 30c^2 d^2 fG_4 - 20d^3 efG_4$$
$$- 30cd^2 f^2 G_4 + 10d^2 f^3 G_4 + 28e_6 h + 840dh_6 + 360ch_7$$
$$+ 840fh_7 + 360eh_8 + 252e_5 i + 420e_4 j + 140e_3 k + 28e_7 m$$
$$+ 252e_6 n + 420e_5 o + 140e_4 p = 0,$$

$$x^2 y^8 : 5d^4 eG_4 - 10c^2 d^3 G_4 + 20cd^3 fG_4 - 10d^3 f^2 G_4 + 360dh_7$$
$$+ +90ch_8 + 360fh_8 + 90eh_9 + 42e_6 i + 168e_5 j + 105e_4 k$$
$$+ 42e_7 n + 168e_6 o + 105e_5 p = 0,$$

$$xy^9 : 5d^4 fG_4 - 5cd^4 G_4 + 10eh_{10} + 90dh_8 + 10ch_9 + 90fh_9$$
$$+ 28e_6 j + 42e_5 k + 28e_7 o + 42e_6 p = 0,$$

$$y^{10} : 10fh_{10} - d^5 G_4 + 10dh_9 + 7e_6 k + 7e_7 p = 0;$$

(6.86)

It is evident that linear systems of equations (6.79)–(6.86) in the variables a_0, a_1, a_2, a_3, b_0, b_1,...,b_4, c_0, c_1,...,c_5, d_0, d_1,...,d_6, e_0, e_1,...,e_7, l_0, l_1,...,l_{14}, f_0, f_1,...,f_8,..., G_1, G_2, G_3, ... can be considered as a single system that can be extended by adding, after the last equation from (6.86), an infinite number of equations, obtained from the equality of coefficients $x^\alpha y^\beta$ for $\alpha + \beta > 10$ in identity (5.12).

For obtaining the quantity G_1 consider system (6.80), that contains constants $b_0, b_1, b_2, b_3, b_4, G_1$. We write the system in the matrix form

$$A_1 B_1 = C_1,$$

(6.87)

where

$$A_1 = \begin{pmatrix} 4c & 4e & 0 & 0 & 0 & -e^2 \\ 4d & 12c + 4f & 12e & 0 & 0 & 2ce - 2ef \\ 0 & 12d & 12c + 12f & 12e & 0 & 2de + 2cf - c^2 - f^2 \\ 0 & 0 & 12d & 4c + 12f & 4e & -2cd + 2df \\ 0 & 0 & 0 & 4d & 4f & -d^2 \end{pmatrix},$$

$$B_1 = \begin{pmatrix} b_0 \\ b_1 \\ b_2 \\ b_3 \\ b_4 \\ G_1 \end{pmatrix}, C_1 = \begin{pmatrix} 0 \\ 0 \\ 0 \\ 0 \\ 0 \end{pmatrix}.$$

(6.88)

Note that the elements of the first five columns of the matrix A_1 from (6.88) are linear functions in coefficients of the system $s(1,4)$ from (6.77), and nonzero elements of the sixth column (in the product (6.87) they correspond to the quantity G_1) have the degree 2 with respect to these coefficients. Since system (6.80) consists of five equations with six unknowns, one of them can be declared free. As a free variable, in system (6.80), there can be taken one of the unknowns b_i for the fixed $i = \overline{0,4}$. Then, from the system of equations (6.80), we obtain

$$G_1 = \frac{B_{1,i} b_i}{\sigma_{1,i}}$$

(6.89)

for any fixed i, equal to $0, 1, 2, 3, 4$, where $B_{1,i}, \sigma_{1,i}$ are polynomials in coefficients of system (6.77), and b_i are the undetermined coefficients of the function $U(x, y)$ from (6.78). We obtain that $b_0 = b_1 = b_2 = b_3 = b_4 = 0$ can be taken as one particular solution for system (6.80). Therefore, G_1 can be considered equal to zero.

For obtaining the quantity G_2 consider systems (6.79), (6.82). The resulting system consists of four equations (6.79) for determining a_i, $i = \overline{0, 4}$, to which seven more equations containing G_2 are added. Writing these equations in matrix form and performing similar reasoning, as in the abovementioned case, we obtain

$$G_2 = \frac{D_{2,i} d_i}{\sigma_{2,i}}, \qquad (6.90)$$

for each $i = \overline{0, 6}$. Since system (6.79) has a solution $a_0 = a_1 = a_2 = = a_3 = 0$, then we obtain that a particular solution of systems (6.79), (6.82) is $a_0 = a_1 = a_2 = a_3 = d_0 = d_1 = d_2 = d_3 = d_4 = d_5 = d_6 = 0$. Thus, G_2 can be considered equal to zero.

For obtaining the quantity G_3, use the matrix equation (see, Appendix 7)

$$A_3 B_3 = C_3, \qquad (6.91)$$

Since the dimension of the matrix A_3 is 15×16, then it is clear that we have at least one free variable. Therefore, choosing one of f_i ($i \in \{0, 1, ..., 8\}$) as a free variable and using the Cramer's rule for system (6.91), for each fixed i, we obtain

$$G_3 = \frac{G_{3,i} + F_{3,i} f_i}{\sigma_{3,i}}, \qquad (6.92)$$

where $G_{3,i}, F_{3,i}, \sigma_{3,i}$ are polynomials in coefficients of system (6.77), and f_i are undetermined coefficients of the function $U(x, y)$ from (6.78).

We are interested in the degree of polynomials $G_{3,i}$ with respect to the coefficients of the system $s(1, 4)$ from (6.77). The indicated degree coincides with the degree of the Cramer determinant Δ_{G_3}. So, the degree of $G_{3,i}$ with respect to the coefficients of the system $s(1, 4)$ from (6.77) will be $degG_3 = 16$ for all $i = \overline{0, 8}$. Taking into account the dimension of system (6.91), we obtain that $G_{3,i}$ has type $(0, 14, 2)$, i.e. $G_{3,i}$ is a homogeneous polynomial of degree 14 with respect to the coefficients of the linear part and homogeneous polynomial of degree 2 with respect to the coefficients of the nonhomogeneity of the fourth order of the system $s(1, 4)$ from (6.77). Zero in $(0, 14, 2)$ shows that the expressions $G_{3,i}$ do not contain phase variables x, y.

Since $G_{3,i}$ from (6.92) are homogeneous polynomials in coefficients of the differential system $s(1, 4)$ from (6.77), then according to the paper [43], for $i = \overline{0, 8}$, they are polynomials of isobarities with weights, respectively:

$$(7, -1), \ (6, 0), \ (5, 1), \ (4, 2), \ (3, 3), \ (2, 4), \ (1, 5), \ (0, 6), \ (-1, 7).$$

We also note that $\sigma_{3,i}$ are polynomials only in coefficients of the linear part c, d, e, f of the system $s(1, 4)$ from (6.77).

For differential system (6.77), using the formula of comitant's weight (3.3), we obtain that numerators of fractions (6.92) can be coefficients in comitants of weight -1 of type $(8, 14, 2)$. Using Lie differential operator D_3 from (6.96) for differential system (6.77), we obtain a system of eight linear nonhomogeneous partial differential equations:

$$D_3(G_{3,0} + F_{3,0}f_0) = G_{3,1} + F_{3,1}f_1, \quad D_3(G_{3,1} + F_{3,1}f_1) = -G_{3,2} - F_{3,2}f_2,$$
$$-D_3(G_{3,2} + F_{3,2}f_2) = G_{3,3} + F_{3,3}f_3, \quad D_3(G_{3,3} + F_{3,3}f_3) = -G_{3,4} - F_{3,4}f_4,$$
$$-D_3(G_{3,4} + F_{3,4}f_4) = G_{3,5} + F_{3,5}f_5, \quad D_3(G_{3,5} + F_{3,5}f_5) = -G_{3,6} - F_{3,6}f_6,$$
$$-D_3(G_{3,6} + F_{3,6}f_6) = G_{3,7} + F_{3,7}f_7, \quad D_3(G_{3,7} + F_{3,7}f_7) = -G_{3,8} - F_{3,8}f_8$$
$$(6.93)$$

with nine unknowns $f_0, f_1, ..., f_8$. According to Lemma 2.4, system (6.93) has an infinite number of solutions. Note that a particular solution for this system is $f_0 = f_1 = ... = f_8 = 0$, for which the polynomial

$$f_8'(x, y) = G_{3,0}x^8 - 8G_{3,1}x^7y - 4G_{3,2}x^6y^2 + 8G_{3,3}x^5y^3$$
$$+2G_{3,4}x^4y^4 - 8G_{3,5}x^3y^5 - 4G_{3,6}x^2y^6 + 8G_{3,7}xy^7 + G_{3,8}y^8$$
$$(6.94)$$

is a centro-affine comitant of differential system (6.77). This fact is also confirmed by Theorem 2.2 with the operators from Section 1.5 of the form

$$X_1 = x\frac{\partial}{\partial x} + D_1, \quad X_2 = y\frac{\partial}{\partial x} + D_2,$$
$$X_3 = x\frac{\partial}{\partial y} + D_3, \quad X_4 = y\frac{\partial}{\partial y} + D_4,$$
$$(6.95)$$

where

$$D_1 = d\frac{\partial}{\partial d} - e\frac{\partial}{\partial e} - 3g\frac{\partial}{\partial g} - 2h\frac{\partial}{\partial h} - i\frac{\partial}{\partial i} + k\frac{\partial}{\partial k} - 4l\frac{\partial}{\partial l} - 3m\frac{\partial}{\partial m}$$
$$-2n\frac{\partial}{\partial n} - o\frac{\partial}{\partial o},$$

$$D_2 = e\frac{\partial}{\partial c} + (f-c)\frac{\partial}{\partial d} - e\frac{\partial}{\partial f} + l\frac{\partial}{\partial g} + (m-g)\frac{\partial}{\partial h} + (n-2h)\frac{\partial}{\partial i}$$
$$+(o-3i)\frac{\partial}{\partial j} + (p-4j)\frac{\partial}{\partial k} - l\frac{\partial}{\partial m} - 2m\frac{\partial}{\partial n} - 3n\frac{\partial}{\partial o} - 4o\frac{\partial}{\partial p},$$

$$D_3 = -d\frac{\partial}{\partial c} + (c-f)\frac{\partial}{\partial e} + d\frac{\partial}{\partial f} - 4h\frac{\partial}{\partial g} - 3i\frac{\partial}{\partial h} - 2j\frac{\partial}{\partial i} - k\frac{\partial}{\partial j}$$
$$+(g-4m)\frac{\partial}{\partial l} + (h-3n)\frac{\partial}{\partial m} + (i-2o)\frac{\partial}{\partial n} + (j-p)\frac{\partial}{\partial o} + k\frac{\partial}{\partial p},$$

$$D_4 = -d\frac{\partial}{\partial d} + e\frac{\partial}{\partial e} - h\frac{\partial}{\partial h} - 2i\frac{\partial}{\partial i} - 3j\frac{\partial}{\partial j} - 4k\frac{\partial}{\partial k} + l\frac{\partial}{\partial l}$$
$$-n\frac{\partial}{\partial n} - 2o\frac{\partial}{\partial o} - 3p\frac{\partial}{\partial p},$$
$$(6.96)$$

for differential system (6.77), for which the following equalities hold:

$$X_1(f_8') = X_4(f_8') = f_8', \; X_2(f_8') = X_3(f_8') = 0.$$

Differential system (6.93) has an infinite number of solutions $f_0, f_1, ..., f_8$, which define centro-affine comitants of the type $(8, 14, 2)$, which are written as

$$
\begin{aligned}
f_8''(x,y) = & (G_{3,0} + F_{3,0}f_0)x^8 - 8(G_{3,1} + F_{3,1}f_1)x^7 y - 4(G_{3,2} \\
& + F_{3,2}f_2)x^6 y^2 + 8(G_{3,3} + F_{3,3}f_3)x^5 y^3 + 2(G_{3,4} + F_{3,4}f_4)x^4 y^4 \\
& - 8(G_{3,5} + F_{3,5}f_5)x^3 y^5 - 4(G_{3,6} + F_{3,6}f_6)x^2 y^6 + 8(G_{3,7} \\
& + F_{3,7}f_7)xy^7 + (G_{3,8} + F_{3,8}f_8)y^8.
\end{aligned}
\tag{6.97}
$$

According to the abovementioned, comitant (6.97) belongs to the linear space $S_{1,4}^{(8,14,2)}$.

Note that comitant (6.94) on the variety $\mathcal{V}$ from (5.11) for the differential system $s(1,4)$ from (6.77) has the form

$$f_8'(x,y)|_{\mathcal{V}} = L_3(x^2 + y^2)^4 \; (G_3|_{\mathcal{V}} = L_3), \tag{6.98}$$

where

$$
\begin{aligned}
L_3 = 648(& 7gh + 18hi + 3gj + 18ij + 3hk + 7jk - 7gl - 3il - 8hm - 7lm \\
& -3gn + 3kn - 18mn + 8jo - 3lo - 18no + 3ip + 7kp - 3mp - 7op)
\end{aligned}
$$

is the first nonzero Lyapunov quantity of differential system (6.77) on an invariant variety $\mathcal{V}$.

Consider the extension of system (6.79)–(6.86), which contains the quantity G_{3k}, which is obtained from identity (5.12) for differential system (6.77) and function (6.78). System (6.79)–(6.86) is written in matrix form $A_{3k}B_{3k} = C_{3k}$. We denote by $m_{G_{3k}}$ the number of equations, and by $n_{G_{3k}}$ the number of unknowns of this system.

Note that this number is written as

$$m_{G_{3k}} = \underbrace{6}_{G_3} + \underbrace{9 + 12}_{G_6} + \underbrace{15 + 18}_{G_9} + 21 + \cdots + \underbrace{6k + 3(2k+1)}_{G_{3k}},$$

for $k = 1, 2, 3, \ldots$. From here, we obtain

$$m_{G_{3k}} = 6k^2 + 9k. \tag{6.99}$$

From these systems, we obtain

$$G_{3k} = \frac{G_{3k,i_1,i_2,...,i_k} + B_{3k,i_1,i_2,...,i_k}b_{i_1} + \cdots + Z_{3k,i_1,i_2,...,i_k}z_{i_k}}{\sigma_{3k,i_1,i_2,...,i_k}}. \tag{6.100}$$

It is important to determine the degree of the polynomial $G_{3k,i_1,i_2,...,i_k} + B_{3k,i_1,i_2,...,i_k}b_{i_1} + \cdots + Z_{3k,i_1,i_2,...,i_k}z_{i_k}$ in coefficients of differential system (6.77).

Note that the degree of nonzero polynomial coefficient of G_{3i} $(i = \overline{1, k})$ in coefficients of differential system (6.77) in Cramer's determinant of the order $m_{G_{3k}}$, when the coefficients in G_{3k} are replaced with free members of the considered system, forms the following diagram:

$$G_3, \ G_6, \ G_9, \ ..., \ G_{3(k-1)}, \ G_{3k}.$$

$$\downarrow \quad \downarrow \quad \downarrow \qquad \downarrow \qquad \downarrow$$

$$4 \quad 7 \quad 10 \quad 3(k-1)+1 \ 2$$

Then the degree of the polynomials $G_{3k,i_1,i_2,...,i_k} + B_{3k,i_1,i_2,...,i_k} b_{i_1} + \cdots$ $+Z_{3k,i_1,i_2,...,i_k} z_{i_k}$ with respect to the coefficients of differential system (6.77), denoted by $N_{G_{3k}}$, will be written as

$$N_{G_{3k}} = m_{G_{3k}} - k + \left[\frac{1}{2}(3k^2 - 3k + 2) + k \right],$$

from where we obtain

$$N_{G_{3k}} = \frac{1}{2}(15k^2 + 15k + 2). \tag{6.101}$$

It is the degree of homogeneity of polynomials $G_{3k,i_1,i_2,...,i_k} + B_{3k,i_1,i_2,...,i_k} b_{i_1} + \cdots + Z_{3k,i_1,i_2,...,i_k} z_{i_k}$ with respect to the coefficients of system (6.77), which are polynomials of type

$$(0, d_0, d_1), \tag{6.102}$$

where d_0 is the degree of homogeneity of the polynomial with respect to the coefficients of the linear part, d_1 is the degree of homogeneity of the polynomial with respect to the coefficients of the nonhomogeneity of the fourth-order of the system $s(1,4)$ from (6.77). Since $d_1 = 2k$, then $d_0 = N_{G_{3k}} - 2k$. Therefore, according to formula (3.3), we obtain that $\delta = 2(3k + 1)$, when the weight $g = -1$.

Based on this, we find that a comitant of the weight -1 of the system $s(1,4)$ from (6.77), which contains $G_{3k,i_1,i_2,...,i_k} + B_{3k,i_1,i_2,...,i_k} b_{i_1} + \cdots + Z_{3k,i_1,i_2,...,i_k} z_{i_k}$ corresponding to the quantity G_{3k} for $k = 1, 2, 3, ...$ $(G_n = 0$ if $n \neq 3k)$ as a semi-invariant (coefficient at the highest degree of x), belongs to the type

$$\left(2(3k + 1), \frac{1}{2}(15k^2 + 11k + 2), 2k \right), \tag{6.103}$$

where $2(3k+1)$ is the degree of homogeneity of the comitant in phase variables x, y; $\frac{1}{2}(15k^2 + 11k + 2)$ is the degree of homogeneity of the comitant in the coefficients of linear part; $2k$ is the degree of homogeneity of the comitant in the coefficients of the nonhomogeneity of the fourth order of the system $s(1,4)$ from (6.77).

Consider the differential system $s(1,4)$ from (6.77). In this case, according to the paper [33] and using Theorem 4.7, the following is true

Theorem 6.5. *Dimension of linear space of centro-affine comitants of type* $(d) = (\delta, d_0, d_1)$ *for the differential system* $s(1,4)$ *from* (6.77), *denoted by* $dim_{\mathbb{R}} V_{1,4}^{(d)}$, *is equal to the coefficient of the monomial* $u^{\delta} b^{d_0} e^{d_1}$ *in decomposition of generalized Hilbert series from* (4.31)–(4.32) *for Sibirsky algebra* $S_{1,4}$ *of comitants of the considered system.*

Consider the subalgebra $S_{1,4}' \subset S_{1,4}$, which we write in the form

$$S_{1,4}' = \bigoplus_{(d)} S_{1,4}^{(d')}, \tag{6.104}$$

where by $S_{1,4}^{(d')}$ the following linear spaces are denoted:

$$S_{1,4}^{(0,0,0)} = \mathbb{R}, \; S_{1,4}^{(0,1,0)}, \; ..., \; S_{1,4}^{2(3k+1),\frac{1}{2}(15k^2+11k+2),2k}, \; k = 1, 2, ..., \tag{6.105}$$

as well as spaces from $S_{1,4}$, which contain all kinds of their products.

Since the algebra $S_{1,4}'$ is a graded subalgebra in a finitely defined algebra $S_{1,4}$, then according to Proposition 4.1, we obtain $\varrho(S_{1,4}') \leq \varrho(S_{1,4})$. From this inequality and Theorem 4.9, we have $\varrho(S_{1,4}) = 13$. According to Remark 5.1 on semi-invariants and the fact that generalized focus pseudo-quantities are coefficients of some comitants, there takes place

Theorem 6.6. *Maximal number of algebraically independent generalized focus pseudo-quantities in the center and focus problem for differential system* (6.77) *does not exceed 13.*

From Proposition 4.2 and Remark 5.1, it follows that the maximal number of algebraically independent focus quantities L_k $(k = \overline{1, \infty})$ cannot exceed the maximal number of algebraically independent generalized focus pseudo-quantities $G_{3k,i_1,i_2,...,i_k} + B_{3k,i_1,i_2,...,i_k} b_{i_1} + \cdots + Z_{3k,i_1,i_2,...,i_k} z_{i_k}$. There from, using Theorem 6.6, we obtain

Consequence 6.3. *Maximal number of algebraically independent focus quantities that take part in solving the center and focus problem for system* (6.77) *does not exceed 13.*

The generalized generating function of the space $V_{1,4}$, consisting of direct sum of spaces (6.104), can be written as

$$\Phi(V_{1,4}, u, b, e) = dim_{\mathbb{R}} S_{1,4}^{(0,0,0)} + dim_{\mathbb{R}} S_{1,4}^{(0,1,0)} b$$

$$+ \sum_{k=1}^{\infty} dim_{\mathbb{R}} S_{1,4}^{(2(k+1),\frac{1}{2}(15k^2+11k+2),2k)} u^{2(k+1)} b^{\frac{1}{2}(15k^2+11k+2)} e^{2k}. \tag{6.106}$$

Using computer, expand the Hilbert series $H(S_{1,4}, u, b, e)$ in a power series. Then for (6.106), we have

$$\Phi(V_{1,4}, u, b, e) = 1 + b + 153u^4b^{14}e^2 + 4589u^6b^{42}e^4 + 49632u^8b^{85}e^6 + \ldots$$

$$+ C_{2(k+1),\frac{1}{2}(15k^2+11k+2),2k} u^{2(k+1)} b^{\frac{1}{2}(15k^2+11k+2)} e^{2k} + \ldots,$$

$$(6.107)$$

where $C_{2(k+1),\frac{1}{2}(15k^2+11k+2),2k}$ is an undetermined coefficient.

Using this generalized generating function and the Hilbert series $H(S_{1,4}, u, b, e)$ from Theorem 4.7, the first terms up to $u^8b^{85}e^6$ in the Hilbert series $H(S'_{1,4}, u, b, e)$ were obtained:

$$H(S'_{1,4}, u, b, e) = 1 + b + 2b^2 + 2b^3 + 3b^4 + 3b^5 + 4b^6 + 4b^7 + 5b^8 + 5b^9$$
$$+6b^{10} + 6b^{11} + 7b^{12} + 7b^{13} + 8b^{14} + 8b^{15} + 9b^{16} + 9b^{17} + 10b^{18} + 10b^{19}$$
$$+11b^{20} + 11b^{21} + 12b^{22} + 12b^{23} + 13b^{24} + 13b^{25} + 14b^{26} + 14b^{27} + 15b^{28}$$
$$+15b^{29} + 16b^{30} + 16b^{31} + 17b^{32} + 17b^{33} + 18b^{34} + 18b^{35} + 19b^{36} + 19b^{37}$$
$$+20b^{38} + 20b^{39} + 21b^{40} + 21b^{41} + 22b^{42} + 22b^{43} + 23b^{44} + 23b^{45} + 24b^{46}$$
$$+24b^{47} + 25b^{48} + 25b^{49} + 26b^{50} + 26b^{51} + 27b^{52} + 27b^{53} + 28b^{54} + 28b^{55}$$
$$+29b^{56} + 29b^{57} + 30b^{58} + 30b^{59} + 31b^{60} + 31b^{61} + 32b^{62} + 32b^{63} + 33b^{64}$$
$$+33b^{65} + 34b^{66} + 34b^{67} + 35b^{68} + 35b^{69} + 36b^{70} + 36b^{71} + 37b^{72} + 37b^{73}$$
$$+38b^{74} + 38b^{75} + 39b^{76} + 39b^{77} + 40b^{78} + 40b^{79} + 41b^{80} + 41b^{81} + 42b^{82}$$
$$+42b^{83} + 43b^{84} + 43b^{85} + 242b^{14}e^2u^4 + 264b^{15}e^2u^4 + 281b^{16}e^2u^4$$
$$+303b^{17}e^2u^4 + 320b^{18}e^2u^4 + 342b^{19}e^2u^4 + 359b^{20}e^2u^4 + 381b^{21}e^2u^4$$
$$+398b^{22}e^2u^4 + 420b^{23}e^2u^4 + 437b^{24}e^2u^4 + 459b^{25}e^2u^4 + 476b^{26}e^2u^4$$
$$+498b^{27}e^2u^4 + 515b^{28}e^2u^4 + 537b^{29}e^2u^4 + 554b^{30}e^2u^4 + 576b^{31}e^2u^4$$
$$+593b^{32}e^2u^4 + 615b^{33}e^2u^4 + 632b^{34}e^2u^4 + 654b^{35}e^2u^4 + 671b^{36}e^2u^4$$
$$+693b^{37}e^2u^4 + 710b^{38}e^2u^4 + 732b^{39}e^2u^4 + 749b^{40}e^2u^4 + 771b^{41}e^2u^4$$
$$+788b^{42}e^2u^4 + 810b^{43}e^2u^4 + 827b^{44}e^2u^4 + 849b^{45}e^2u^4 + 866b^{46}e^2u^4$$
$$+888b^{47}e^2u^4 + 905b^{48}e^2u^4 + 927b^{49}e^2u^4 + 944b^{50}e^2u^4 + 966b^{51}e^2u^4$$
$$+983b^{52}e^2u^4 + 1005b^{53}e^2u^4 + 1022b^{54}e^2u^4 + 1044b^{55}e^2u^4 + 1061b^{56}e^2u^4$$
$$+1083b^{57}e^2u^4 + 1100b^{58}e^2u^4 + 1122b^{59}e^2u^4 + 1139b^{60}e^2u^4 + 1161b^{61}e^2u^4$$
$$+1178b^{62}e^2u^4 + 1200b^{63}e^2u^4 + 1217b^{64}e^2u^4 + 1239b^{65}e^2u^4 + 1256b^{66}e^2u^4$$
$$+1278b^{67}e^2u^4 + 1295b^{68}e^2u^4 + 1317b^{69}e^2u^4 + 1334b^{70}e^2u^4 + 1356b^{71}e^2u^4$$
$$+1373b^{72}e^2u^4 + 1395b^{73}e^2u^4 + 1412b^{74}e^2u^4 + 1434b^{75}e^2u^4 + 1451b^{76}e^2u^4$$
$$+1473b^{77}e^2u^4 + 1490b^{78}e^2u^4 + 1512b^{79}e^2u^4 + 1529b^{80}e^2u^4 + 1551b^{81}e^2u^4$$
$$+1568b^{82}e^2u^4 + 1590b^{83}e^2u^4 + 1607b^{84}e^2u^4 + 1629b^{85}e^2u^4 + 9591b^{42}e^4u^6$$
$$+9845b^{43}e^4u^6 + 10084b^{44}e^4u^6 + 10338b^{45}e^4u^6 + 10577b^{46}e^4u^6$$

$$+10831b^{47}e^4u^6 + 11070b^{48}e^4u^6 + 11324b^{49}e^4u^6 + 11563b^{50}e^4u^6$$

$$+11817b^{51}e^4u^6 + 12056b^{52}e^4u^6 + 12310b^{53}e^4u^6 + 12549b^{54}e^4u^6$$

$$+12803b^{55}e^4u^6 + 13042b^{56}e^4u^6 + 13296b^{57}e^4u^6 + 13535b^{58}e^4u^6$$

$$+13789b^{59}e^4u^6 + 14028b^{60}e^4u^6 + 14282b^{61}e^4u^6 + 14521b^{62}e^4u^6$$

$$+14775b^{63}e^4u^6 + 15014b^{64}e^4u^6 + 15268b^{65}e^4u^6 + 15507b^{66}e^4u^6$$

$$+15761b^{67}e^4u^6 + 16000b^{68}e^4u^6 + 16254b^{69}e^4u^6 + 16493b^{70}e^4u^6$$

$$+16747b^{71}e^4u^6 + 16986b^{72}e^4u^6 + 17240b^{73}e^4u^6 + 17479b^{74}e^4u^6$$

$$+17733b^{75}e^4u^6 + 17972b^{76}e^4u^6 + 18226b^{77}e^4u^6 + 18465b^{78}e^4u^6$$

$$+18719b^{79}e^4u^6 + 18958b^{80}e^4u^6 + 19212b^{81}e^4u^6 + 19451b^{82}e^4u^6$$

$$+19705b^{83}e^4u^6 + 19944b^{84}e^4u^6 + 20198b^{85}e^4u^6 + 7110b^{28}e^4u^8$$

$$+7393b^{29}e^4u^8 + 7691b^{30}e^4u^8 + 7974b^{31}e^4u^8 + 8272b^{32}e^4u^8$$

$$+8555b^{33}e^4u^8 + 8853b^{34}e^4u^8 + 9136b^{35}e^4u^8 + 9434b^{36}e^4u^8$$

$$+9717b^{37}e^4u^8 + 10015b^{38}e^4u^8 + 10298b^{39}e^4u^8 + 10596b^{40}e^4u^8$$

$$+10879b^{41}e^4u^8 + 11177b^{42}e^4u^8 + 11460b^{43}e^4u^8 + 11758b^{44}e^4u^8$$

$$+12041b^{45}e^4u^8 + 12339b^{46}e^4u^8 + 12622b^{47}e^4u^8 + 12920b^{48}e^4u^8$$

$$+13203b^{49}e^4u^8 + 13501b^{50}e^4u^8 + 13784b^{51}e^4u^8 + 14082b^{52}e^4u^8$$

$$+14365b^{53}e^4u^8 + 14663b^{54}e^4u^8 + 14946b^{55}e^4u^8 + 15244b^{56}e^4u^8$$

$$+15527b^{57}e^4u^8 + 15825b^{58}e^4u^8 + 16108b^{59}e^4u^8 + 16406b^{60}e^4u^8$$

$$+16689b^{61}e^4u^8 + 16987b^{62}e^4u^8 + 17270b^{63}e^4u^8 + 17568b^{64}e^4u^8$$

$$+17851b^{65}e^4u^8 + 18149b^{66}e^4u^8 + 18432b^{67}e^4u^8 + 18730b^{68}e^4u^8$$

$$+19013b^{69}e^4u^8 + 19311b^{70}e^4u^8 + 19594b^{71}e^4u^8 + 19892b^{72}e^4u^8$$

$$+20175b^{73}e^4u^8 + 20473b^{74}e^4u^8 + 20756b^{75}e^4u^8 + 21054b^{76}e^4u^8$$

$$+21337b^{77}e^4u^8 + 21635b^{78}e^4u^8 + 21918b^{79}e^4u^8 + 22216b^{80}e^4u^8$$

$$+22499b^{81}e^4u^8 + 22797b^{82}c^4u^8 + 23080b^{83}e^4u^8 + 23378b^{84}e^4u^8$$

$$+23661b^{85}e^4u^8 + 137561b^{85}e^6u^8 + \dots.$$

Hence, the ordinary Hilbert series $H_{S'_{1,4}}(t)$ of the algebra $S'_{1,4}$ will have the form (the first 100 terms):

$$H_{S'_{1,4}}(t) = 1 + t + 2t^2 + 2t^3 + 3t^4 + 3t^5 + 4t^6 + 4t^7 + 5t^8 + 5t^9 + 6t^{10}$$

$$+6t^{11} + 7t^{12} + 7t^{13} + 8t^{14} + 8t^{15} + 9t^{16} + 9t^{17} + 10t^{18} + 10t^{19} + 253t^{20}$$

$$+275t^{21} + 293t^{22} + 315t^{23} + 333t^{24} + 355t^{25} + 373t^{26} + 395t^{27} + 413t^{28}$$

$$+435t^{29} + 453t^{30} + 475t^{31} + 493t^{32} + 515t^{33} + 533t^{34} + 555t^{35} + 573t^{36}$$

$$+595t^{37} + 613t^{38} + 635t^{39} + 7763t^{40} + 8068t^{41} + 8384t^{42} + 8689t^{43}$$

$$+9005t^{44} + 9310t^{45} + 9626t^{46} + 9931t^{47} + 10247t^{48} + 10552t^{49}$$

$+10868t^{50} + 11173t^{51} + 21080t^{52} + 21639t^{53} + 22194t^{54} + 22753t^{55}$

$+23308t^{56} + 23867t^{57} + 24422t^{58} + 24981t^{59} + 25536t^{60} + 26095t^{61}$

$+26650t^{62} + 27209t^{63} + 27764t^{64} + 28323t^{65} + 28878t^{66} + 29437t^{67}$

$+29992t^{68} + 30551t^{69} + 31106t^{70} + 31665t^{71} + 32220t^{72} + 32779t^{73}$

$+33334t^{74} + 33893t^{75} + 34448t^{76} + 35007t^{77} + 35562t^{78} + 36121t^{79}$ (6.108)

$+36676t^{80} + 37235t^{81} + 37790t^{82} + 38349t^{83} + 38904t^{84} + 39463t^{85}$

$+39974t^{86} + 40533t^{87} + 41087t^{88} + 41646t^{89} + 42200t^{90} + 42759t^{91}$

$+41667t^{92} + 42204t^{93} + 42741t^{94} + 43278t^{95} + 23378t^{96} + 23661t^{97}$

$$+137561t^{99} + \dots .$$

We consider the first 100 terms in the expansion of Hilbert series of the algebra $SI_{1,4}$, which according to [33] is obtained from (4.31)–(4.32) in the following way:

$$H_{SI_{1,4}}(t) = H(S_{1,4}, 0, t, t) = 1 + t + 2t^2 + 5t^3 + 14t^4 + 26t^5 + 57t^6$$

$$+119t^7 + 248t^8 + 461t^9 + 864t^{10} + 1547t^{11} + 2737t^{12} + 4601t^{13}$$

$$+7662t^{14} + 12383t^{15} + 19768t^{16} + 30664t^{17} + 47066t^{18} + 70770t^{19}$$

$$+105300t^{20} + 153783t^{21} + 222506t^{22} + 317223t^{23} + 448337t^{24}$$

$$+625302t^{25} + 865296t^{26} + 1184226t^{27} + 1609007t^{28} + 2164498t^{29}$$

$$+2892657t^{30} + 3832653t^{31} + 5047384t^{32} + 6595561t^{33} + 8570829t^{34}$$

$$+11061230t^{35} + 14202137t^{36} + 18120878t^{37} + 23011677t^{38} + 29058179t^{39}$$

$$+36532673t^{40} + 45692819t^{41} + 56917559t^{42} + 70566839t^{43}$$

$$+87158250t^{44} + 107183955t^{45} + 131345992t^{46} + 160313871t^{47}$$

$$+195025339t^{48} + 236375592t^{49} + 285608968t^{50} + 343916178t^{51}$$

$$+412927382t^{52} + 494204023t^{53} + 589868721t^{54} + 701958526t^{55}$$

$$+833207339t^{56} + 986241378t^{57} + 1164562071t^{58}$$

$$+1371538331t^{59} + 1611612886t^{60} + 1889064095t^{61}$$

$$+2209499727t^{62} + 2578327522t^{63} + 3002568564t^{64}$$

$$+3488999055t^{65} + 4046367551t^{66} + 4683127424t^{67}$$

$$+5410102263t^{68} + 6237758509t^{69} + 7179427892t^{70}$$

$$+8248015477t^{71} + 9459839180t^{72} + 10830705810t^{73}$$

$$+12380506870t^{74} + 14128528398t^{75} + 16098882290t^{76}$$

$$+18314961754t^{77} + 20805884383t^{78} + 23599922669t^{79}$$

$$+26732062272t^{80} + 30236294034t^{81} + 34154507484t^{82}$$

$$+38527430455t^{83} + 43404991223t^{84} + 48835760568t^{85}$$

$$+54879055119t^{86} + 61592616546t^{87} + 69046625211t^{88}$$

$$+77309433488t^{89} + 86463824763t^{90} + 96590473934t^{91}$$
$$+107786664234t^{92} + 120147246938t^{93} + 133786223574t^{94}$$
$$+148814805866t^{95} + 165366127962t^{96} + 183570124286t^{97} \qquad (6.109)$$
$$+203581864473t^{98} + 225552766408t^{99} + \dots.$$

Since for series (6.108) and (6.109), the following inequality holds

$$H_{S'_{1,4}}(t) \le H_{SI_{1,4}}(t),$$

then in assumption that this inequality holds for remaining terms of the considered series, we obtain the inequality

$$\varrho(S'_{1,4}) \le \varrho(SI_{1,4}).$$

Note that $S'_{1,4}$ is not a subalgebra in $SI_{1,4}$. Since from Theorem 4.9, we have $\varrho(SI_{1,4}) = 11$, then according to the last inequality, we obtain that the following hypothesis can be true

Hypothesis 6.1. *Maximal number of algebraically independent generalized focus pseudo-quantities (as well as focus quantities) for differential system (6.77), which take part in solving the center and focus problem for given differential system, may not exceed 11.*

6.5 On the Upper Bound of the Number of Algebraically Independent Focus Pseudo-Quantities for the Differential System $s(1,5)$

Consider the differential system $s(1,5)$, which we write in the form

$$\dot{x} = cx + dy + gx^5 + 5hx^4y + 10kx^3y^2 + 10lx^2y^3 + 5mxy^4 + ny^5,$$
$$\dot{y} = ex + fy + px^5 + 5qx^4y + 10rx^3y^2 + 10sx^2y^3 + 5uxy^4 + vy^5 \qquad (6.110)$$

with a finitely defined graded algebra of unimodular comitants $S_{1,5}$ [33,34]. For this system, we write function (5.13) in the form

$$U = k_2 + a_0x^3 + 3a_1x^2y + 3a_2xy^2 + a_3y^3 + b_0x^4 + 4b_1x^3y + 6b_2x^2y^2$$
$$+4b_3xy^3 + b_4y^4 + c_0x^5 + 5c_1x^4y + 10c_2x^3y^2 + 10c_3x^2y^3 + 5c_4xy^4$$
$$+c_5y^5 + d_0x^6 + 6d_1x^5y + 15d_2x^4y^2 + 20d_3x^3y^3 + 15d_4x^2y^4 + 6d_5xy^5$$
$$+d_6y^6 + e_0x^7 + 7e_1x^6y + 21e_2x^5y^2 + 35e_3x^4y^3 + 21e_5x^2y^5 + 7e_6xy^6$$
$$+e_7y^7 + f_0x^8 + 8f_1x^7y + 28f_2x^6y^2 + 56f_3x^5y^3 + 70f_4y^4$$
$$+56f_5x^3y^5 + 28f_6x^2y^6 + 8f_7xy^7 + f_8y^8 + \dots, \qquad (6.111)$$

where $k_2 \not\equiv 0$ is from (5.8), and $a_0, a_1, ..., f_7, f_8, ...$ are unknown coefficients.

Identity (5.12) along the trajectories of differential system (6.110) with function (6.111) splits into the following systems of equations (equality (5.15) is omitted):

$$
\begin{aligned}
x^3 &: 3a_0c + 3a_1e = 0,\\
x^2y &: 6a_1c + 3a_0d + 6a_2e + 3a_1f = 0,\\
xy^2 &: 3a_2c + 6a_1d + 3a_3e + 6a_2f = 0,\\
y^3 &: 3a_2d + 3a_3f = 0;
\end{aligned}
\tag{6.112}
$$

$$
\begin{aligned}
x^4 &: 4b_0c + 4b_1e - e^2G_1 = 0,\\
x^3y &: 12b_1c + 4b_0d + 12b_2e + 4b_1f + 2ceG_1 - 2efG_1 = 0,\\
x^2y^2 &: 12b_2c + 12b_1d + 12b_3e + 12b_2f - c^2G_1 + 2deG_1\\
&\quad + 2cfG_1 - f^2G_1 = 0,\\
xy^3 &: 4b_3c + 12b_2d + 4b_4e + 12b_3f - 2cdG_1 + 2dfG_1 = 0,\\
y^4 &: 4b_3d + 4b_4f - d^2G_1 = 0;
\end{aligned}
\tag{6.113}
$$

$$
\begin{aligned}
x^5 &: 5cc_0 + 5c_1e = 0,\\
x^4y &: 20cc_1 + 5c_0d + 20c_2e + 5c_1f = 0,\\
x^3y^2 &: 30cc_2 + 20c_1d + 30c_3e + 20c_2f = 0,\\
x^2y^3 &: 20cc_3 + 30c_2d + 20c_4e + 30c_3f = 0,\\
xy^4 &: 5cc_4 + 20c_3d + 5c_5e + 20c_4f = 0,\\
y^5 &: 5c_4d + 5c_5f = 0;\\
x^6 &: 6cd_0 + 6d_1e - 2eg + e^3G_2 + cp - fp = 0,\\
x^5y &: 6dd_0 + 30cd_1 + 30d_2e + 6d_1f + cg - fg - 3ce^2G_2\\
&\quad + 3e^2fG_2 - 10eh + 2dp + 5cq - 5fq = 0,\\
x^4y^2 &: 330dd_1 + 60cd_2 + 60d_3e + 30d_2f + 3c^2eG_2 - 3de^2G_2\\
&\quad - 6cefG_2 + 3ef^2G_2 + 5ch - 5fh - 20ek + 10dq + 10cr\\
&\quad - 10fr = 0,\\
x^3y^3 &: 60dd_2 + 60cd_3 + 60d_4e + 60d_3f - c^3G_2 + 6cdeG_2\\
&\quad + 3c^2fG_2 - 6defG_2 - 3cf^2G_2 + f^3G_2 + 10ck - 10fk\\
&\quad - 20el + 20dr + 10cs - 10fs = 0,\\
x^2y^4 &: 660dd_3 + 30cd_4 + 30d_5e + 60d_4f - 3c^2dG_2 + 3d^2eG_2\\
&\quad + 6cdfG_2 - 3df^2G_2 + 10cl - 10fl - 10em + 20ds + 5cu\\
&\quad - 5fu = 0,
\end{aligned}
\tag{6.114}
$$

$$xy^5 : 30dd_4 + 6cd_5 + 6d_6e + 30d_5f - 3cd^2G_2 + 3d^2fG_2 + 5cm$$
$$- 5fm - 2en + 10du + cv - fv = 0, \qquad (6.115)$$
$$y^6 : 6dd_5 + 6d_6f - d^3G_2 + cn - fn + 2dv = 0;$$

$$x^7 : 7ce_0 + 7ee_1 + 3a_0g + 3a_1p = 0,$$
$$x^6y : 7de_0 + 42ce_1 + 42ee_2 + 7e_1f + 6a_1g + 15a_0h + 6a_2p$$
$$+ 15a_1q = 0,$$
$$x^5y^2 : 42de_1 + 105ce_2 + 105ee_3 + 42e_2f + 3a_2g + 30a_1h + 30a_0k$$
$$+ 3a_3p + 30a_2q + 30a_1r = 0,$$
$$x^4y^3 : 105de_2 + 140ce_3 + 140ee_4 + 105e_3f + 15a_2h + 60a_1k + 30a_0l$$
$$+ 15a_3q + 60a_2r + 30a_1s = 0,$$
$$x^3y^4 : 140de_3 + 105ce_4 + 105ee_5 + 140e_4f + 30a_2k + 60a_1l + 15a_0m$$
$$+ 30a_3r + 60a_2s + 15a_1u = 0, \qquad (6.116)$$
$$x^2y^5 : 105de_4 + 42ce_5 + 42ee_6 + 105e_5f + 30a_2l + 30a_1m$$
$$+ 3a_0n + 30a_3s + 30a_2u + 3a_1v = 0,$$
$$xy^6 : 42de_5 + 7ce_6 + 7ee_7 + 42e_6f + 15a_2m + 6a_1n + 15a_3u$$
$$+ 6a_2v = 0,$$
$$y^7 : 7de_6 + 7e_7f + 3a_2n + 3a_3v = 0;$$

$$x^8 : 8cf_0 + 8ef_1 + 4b_0g + 4b_1p - e^4G_3 = 0,$$
$$x^7y : 8df_0 + 56cf_1 + 8ff_1 + 56ef_2 + 12b_1g + 20b_0h + 12b_2p$$
$$+ 20b_1q + 4ce^3G_3 - 4e^3fG_3 = 0,$$
$$x^6y^2 : 56df_1 + 168cf_2 + 56ff_2 + 168ef_3 + 12b_2g + 60b_1h + 40b_0k$$
$$+ 12b_3p + 60b_2q + 40b_1r - 6c^2e^2G_3 + 4de^3G_3 + 12ce^2fG_3$$
$$- 6e^2f^2G_3 = 0,$$
$$x^5y^3 : 168df_2 + 280cf_3 + 168ff_3 + 280ef_4 + 4b_3g + 60b_2h$$
$$+ 120b_1k + 40b_0l + 4b_4p + 60b_3q + 120b_2r + 40b_1s$$
$$+ 4c^3eG_3 - 12cde^2G_3 - 12c^2efG_3 + 12de^2fG_3$$
$$+ 12cef^2G_3 - 4ef^3G_3 = 0,$$
$$x^4y^4 : 280df_3 + 280cf_4 + 280ff_4 + 280ef_5 + 20b_3h + 120b_2k$$
$$+ 120b_1l + 20b_0m + 20b_4q + 120b_3r + 120b_2s + 20b_1u$$
$$- c^4G_3 + 12c^2deG_3 - 6d^2e^2G_3 + 4c^3fG_3 - 24cdefG_3$$
$$- 6c^2f^2G_3 + 12def^2G_3 + 4cf^3G_3 - f^4G_3 = 0,$$
$$x^3y^5 : 280df_4 + 168cf_5 + 280ff_5 + 168ef_6 + 40b_3k + 120b_2l$$
$$+ 60b_1m + 4b_0n + 40b_4r + 120b_3s + 60b_2u + 4b_1v$$

$$- 4c^3 dG_3 + 12cd^2 eG_3 + 12c^2 df G_3 - 12d^2 ef G_3$$
$$- 12cdf^2 G_3 + 4df^3 G_3 = 0,$$

$$x^2 y^6 : 168df_5 + 56cf_6 + 168ff_6 + 56ef_7 + 40b_3 l + 60b_2 m$$
$$+ 12b_1 n + 40b_4 s + 60b_3 u + 12b_2 v - 6c^2 d^2 G_3 + 4d^3 eG_3$$
$$+ 12cd^2 f G_3 - 6d^2 f^2 G_3 = 0,$$

$$xy^7 : 56df_6 + 8cf_7 + 56ff_7 + 8ef_8 + 20b_3 m + 12b_2 n + 20b_4 u$$
$$+ 12b_3 v - 4cd^3 G_3 + 4d^3 f G_3 = 0,$$

$$y^8 : 8df_7 + 8ff_8 + 4b_3 n + 4b_4 v - d^4 G_3 = 0.$$

(6.117)

It is evident that linear systems of equations (6.112)–(6.117) with respect to the variables a_0, a_1, a_2, a_3, b_0, b_1,...,b_4, c_0, c_1,...,c_5, d_0, d_1,...,d_6, e_0, e_1,...,e_7, f_0, f_1,...,f_8,..., G_1, G_2, G_3,... can be considered as a single system, which can be extended by adding, after the last equation from (6.117), an infinite number of equations, obtained from the equality of coefficients $x^\alpha y^\beta$ for $\alpha + \beta > 8$ in identity (5.12).

For obtaining the quantity G_1, we write system (6.113) in the matrix form

$$A_1 B_1 = C_1,$$

(6.118)

where

$$A_1 = \begin{pmatrix} 4c & 4e & 0 & 0 & 0 & -e^2 \\ 4d & 12c + 4f & 12e & 0 & 0 & 2ce - 2ef \\ 0 & 12d & 12c + 12f & 12e & 0 & 2de + 2cf - c^2 - f^2 \\ 0 & 0 & 12d & 4c + 12f & 4e & -2cd + 2df \\ 0 & 0 & 0 & 4d & 4f & -d^2 \end{pmatrix},$$

$$B_1 = \begin{pmatrix} b_0 \\ b_1 \\ b_2 \\ b_3 \\ b_4 \\ G_1 \end{pmatrix}, \quad C_1 = \begin{pmatrix} 0 \\ 0 \\ 0 \\ 0 \\ 0 \end{pmatrix}.$$

(6.119)

Since the dimension of the matrix A_1 is 5×6, then it is clear that we have at least one free variable. Therefore, choosing one of b_i ($i \in \{0, , 1, ..., 4\}$) as a free variable, using the Cramer's rule for system (6.118), we obtain

$$G_1 = \frac{B_{1,i} b_i}{\sigma_{1,i}},$$

(6.120)

for any fixed i, equal to $0, 1, 2, 3, 4$, where $B_{1,i}, \sigma_{1,i}$ are polynomials in coefficients of system (6.110), and b_i are undetermined coefficients of the function $U(x, y)$ from (6.111). We obtain that $b_0 = b_1 = b_2 = b_3 = b_4 = 0$ can be taken as one particular solution of system (6.113). This means that G_1 can be considered equal to zero.

System (6.115) consists of seven equations. Writing these equations in the matrix form $A_2 B_2 = C_2$ (see, Appendix 8) and performing similar reasoning as in the abovementioned case, we obtain

$$G_2 = \frac{G_{2,i} + D_{2,i} d_i}{\sigma_{2,i}} \tag{6.121}$$

for each $i = \overline{0,6}$.

We are interested in the degree of polynomials $G_{2,i}$ with respect to the coefficients of the system $s(1,5)$ from (6.110). The indicated degree coincides with the degree of the Cramer determinant Δ_{G_2}. So the degree of $G_{2,i}$ with respect to the coefficients of the system $s(1,5)$ from (6.110) will be $deg G_{2,i} = 8$ for all $i = \overline{0,6}$. Considering the dimension of the system, we obtain that $G_{2,i}$ has type $(0,7,1)$, i.e. G_2 is a homogeneous polynomial of degree 7 with respect to the coefficients of the linear part and homogeneous of degree 1 with respect to the coefficients of the nonhomogeneity of the fifth order of the system $s(1,5)$ from (6.110). Zero in $(0,7,1)$ shows that expressions $G_{2,i}$ does not contain phase variables x, y.

Note that in addition to these results obtained from the analysis of system (6.115), computer calculations were carried out and explicit forms of polynomials $G_{2,i}$ from (6.121) were determined for each fixed $i = \overline{0,6}$ (see, Appendix 9).

It was determined that $G_{2,i}$ $(i = \overline{0,6})$ are homogeneous polynomials with respect to the coefficients of the system $s(1,5)$ from (6.110), and at the same time, for $i = 0, 1, 2, 3, 4, 5, 6$, they are polynomials of isobarities with weights, respectively

$$(5,-1), \ (4,0), \ (3,1), \ (2,2), \ (1,3), \ (0,4), \ (-1,5). \tag{6.122}$$

We also note that $\sigma_{2,i}$ are polynomials only on the coefficients c, d, e, f of the linear part of the system $s(1,5)$ from (6.110).

In the future, we will need an explicit form of the operators $X_1, ..., X_4$ of Lie algebra L_4 for system (6.116), the expressions for which are obtained from §5:

$$X_1 = x\frac{\partial}{\partial x} + D_1, \ X_2 = y\frac{\partial}{\partial x} + D_2, \ X_3 = x\frac{\partial}{\partial y} + D_3, \ X_4 = y\frac{\partial}{\partial y} + D_4, \tag{6.123}$$

where

$$D_1 = d\frac{\partial}{\partial d} - e\frac{\partial}{\partial e} - 4g\frac{\partial}{\partial g} - 3h\frac{\partial}{\partial h} - 2k\frac{\partial}{\partial k} - l\frac{\partial}{\partial l} + n\frac{\partial}{\partial n} - 5p\frac{\partial}{\partial p}$$

$$-4q\frac{\partial}{\partial q} - 3r\frac{\partial}{\partial r} - 2s\frac{\partial}{\partial s} - u\frac{\partial}{\partial u},$$

$$D_2 = e\frac{\partial}{\partial c} + (f - c)\frac{\partial}{\partial d} - e\frac{\partial}{\partial f} + p\frac{\partial}{\partial g} + (q - g)\frac{\partial}{\partial h} + (r - 2h)\frac{\partial}{\partial k}$$

$$+(s - 3k)\frac{\partial}{\partial l} + (u - 4l)\frac{\partial}{\partial m} + (v - 5m)\frac{\partial}{\partial n} - p\frac{\partial}{\partial q} - 2q\frac{\partial}{\partial r} - 3r\frac{\partial}{\partial s}$$

$$-4s\frac{\partial}{\partial u} - 5u\frac{\partial}{\partial v},$$

$$D_3 = -d\frac{\partial}{\partial c} + (c - f)\frac{\partial}{\partial e} + d\frac{\partial}{\partial f} - 5h\frac{\partial}{\partial g} - 4k\frac{\partial}{\partial h} - 3l\frac{\partial}{\partial k}$$

$$-2m\frac{\partial}{\partial l} - n\frac{\partial}{\partial m} + (g - 5q)\frac{\partial}{\partial p} + (h - 4r)\frac{\partial}{\partial q} + (k - 3s)\frac{\partial}{\partial r}$$

$$+(l - 2u)\frac{\partial}{\partial s} + (m - v)\frac{\partial}{\partial u} + n\frac{\partial}{\partial v},$$

$$D_4 = -d\frac{\partial}{\partial d} + e\frac{\partial}{\partial e} - h\frac{\partial}{\partial h} - 2k\frac{\partial}{\partial k} - 3l\frac{\partial}{\partial l} - 4m\frac{\partial}{\partial m}$$

$$-5n\frac{\partial}{\partial n} + p\frac{\partial}{\partial p} - r\frac{\partial}{\partial r} - 2s\frac{\partial}{\partial s} - 3u\frac{\partial}{\partial u} - 4v\frac{\partial}{\partial v}. \tag{6.124}$$

Using formula of comitant's weight (3.3), for differential system (6.110), we obtain that numerators of fractions (6.121) can be coefficients in comitants of the weight -1 of type $(6, 7, 1)$. Using Lie differential operator D_3 from (6.124) for differential system (6.110), we obtain the system of six linear nonhomogeneous partial differential equations:

$$\begin{aligned}
D_3(G_{2,0} + D_{2,0}f_0) &= G_{2,1} + D_{2,1}d_1, \\
D_3(G_{2,1} + D_{2,1}d_1) &= -G_{2,2} - D_{2,2}d_2, \\
-D_3(G_{2,2} + D_{2,2}d_2) &= G_{2,3} + D_{2,3}d_3, \\
D_3(G_{2,3} + D_{2,3}d_3) &= -G_{2,4} - D_{2,4}d_4, \\
-D_3(G_{2,4} + D_{2,4}d_4) &= G_{2,5} + D_{2,5}d_5, \\
D_3(G_{2,5} + D_{2,5}d_5) &= -G_{2,6} - D_{2,6}d_6
\end{aligned} \tag{6.125}$$

with seven unknowns $d_0, d_1, ..., d_6$. According to Lemma 2.4, system (6.125) has an infinite number of solutions. Note that a particular solution of this system is $d_0 = d_1 = ... = d_6 = 0$, for which the polynomial

$$f_6'(x, y) = G_{2,0}x^6 - 6G_{2,1}x^5y - 3G_{2,2}x^4y^2 + 2G_{2,3}x^3y^3$$

$$+3G_{2,4}x^2y^4 - 6G_{2,5}xy^5 - G_{2,6}y^6 \tag{6.126}$$

is a centro-affine comitant of differential system (6.110). This fact is also confirmed by Theorem 2.2 with the operators $X_1 - X_4$ from (6.123), (6.124) for differential system (6.110), for which

$$X_1(f_6') = X_4(f_6') = f_6', \ X_2(f_6') = X_3(f_6') = 0.$$

It is obvious that differential system (6.125) has an infinite number of solutions $d_0, d_1, ..., d_6$, which define centro-affine comitants of type $(6, 7, 1)$, which are written as

$$f_6''(x, y) = (G_{2,0} + D_{2,0}d_0)x^6 - 6(G_{2,1} + D_{2,1}d_1)x^5y - 3(G_{2,2}$$
$$+ D_{2,2}d_2)x^4y^2 + 2(G_{2,3} + D_{2,3}d_3)x^3y^3 + 3(G_{2,4} + D_{2,4}d_4)x^2y^4 \quad (6.127)$$
$$- 6(G_{2,5} + D_{2,5}d_5)xy^5 - (G_{2,6} + D_{2,6}d_6)y^6.$$

According to the abovementioned, comitant (6.126) belongs to the linear space $S_{1,5}^{(6,7,1)}$.

Note that comitant (6.126) on the variety $\mathcal{V}$ from (5.9) for differential system (6.110) has the form

$$f_6'(x, y)|_\mathcal{V} = L_2(x^2 + y^2)^3 \quad (G_2|_\mathcal{V} = L_2), \quad (6.128)$$

where

$$L_2 = 20(g + 2k + m + q + 2s + v),$$

is the first nonzero Lyapunov quantity of differential system (6.110) on an invariant variety $\mathcal{V}$.

Consider the extension of system (6.112)–(6.117), which is obtained from identity (5.12) for differential system (6.110) and function (6.111), that contains the quantity G_{2k}, which we write in the matrix form $A_{2k}B_{2k} = C_{2k}$. We denote by $m_{G_{2k}}$ the number of equations, and by $n_{G_{2k}}$ the number of unknowns of this system.

Note that this number is written as:

$$m_{G_{2k}} = \underbrace{\underbrace{\underbrace{2 \cdot 2 + 3}_{G_2} + 2 \cdot 4 + 3}_{G_4} + 2 \cdot 6 + 3}_{G_6} + ... + 2 \cdot 2k + 3,$$
$$\underbrace{}_{G_{2k}}$$

for $k = 1, 2, 3, \ldots$. There from we obtain

$$m_{G_{2k}} = 2k^2 + 5k. \quad (6.129)$$

From these systems, it is obtained

$$G_{2k} = \frac{G_{2k,i_1,i_2,...,i_k} + B_{2k,i_1,i_2,...,i_k}b_{i_1} + \cdots + Z_{2k,i_1,i_2,...,i_k}z_{i_k}}{\sigma_{2k,i_1,i_2,...,i_k}}. \quad (6.130)$$

It is important to determine the degree of the polynomial $G_{2k,i_1,i_2,\ldots,i_k} + +B_{2k,i_1,i_2,\ldots,i_k} b_{i_1} + \cdots + Z_{2k,i_1,i_2,\ldots,i_k} z_{i_k}$ with respect to the coefficients of the differential system (6.110).

Note that the degree of nonzero coefficient of the polynomial G_{2i} $(i = \overline{1, k})$ with respect to the coefficients of system (6.110) in Cramer's determinant of the order $m_{G_{2k}}$, when the coefficients in G_{2k} are replaced with the free members of the considered system, forms the following diagram:

$$G_2, \; G_4, \; G_6, \; \ldots \,, G_{2(k-1)}, \qquad G_{2k}.$$

$$\downarrow \quad \downarrow \quad \downarrow \qquad \downarrow \qquad\quad \downarrow$$

$$7 \quad 18 \quad 33 \quad k(2k+1) - 3 \quad 1$$

Then the degree of the polynomials $G_{2k,i_1,i_2,\ldots,i_k} + B_{2k,i_1,i_2,\ldots,i_k} b_{i_1} + \cdots + Z_{2k,i_1,i_2,\ldots,i_k} z_{i_k}$ with respect to the coefficients of system (6.110), denoted by $N_{G_{2k}}$, will be written as

$$N_{G_{2k}} = m_{G_{2k}} + 2\frac{k(k-1)}{2} + 1,$$

from where we obtain

$$N_{G_{2k}} = 3k^2 + 4k + 1. \tag{6.131}$$

It is the degree of homogeneity of polynomials $G_{2k,i_1,i_2,\ldots,i_k} + B_{2k,i_1,i_2,\ldots,i_k} b_{i_1} + \cdots + Z_{2k,i_1,i_2,\ldots,i_k} z_{i_k}$ with respect to the coefficients of system (6.110), which are polynomials of type

$$(0, d_0, d_1), \tag{6.132}$$

where d_0 is the degree of homogeneity of the polynomial with respect to the coefficients of the linear part, d_1 is the degree of homogeneity of the polynomial with respect to the coefficients of the nonhomogeneity of the fifth order of the system $s(1, 5)$ from (6.110). Since $\delta = 2(2k+1)$ and $d_1 = k$, then $d_0 = N_{G_k} - k$.

Based on this, we find that a comitant of weight -1 of the system $s(1, 5)$ from (6.110), which contains $G_{2k,i_1,i_2,\ldots,i_k} + +B_{2k,i_1,i_2,\ldots,i_k} b_{i_1} + \cdots + Z_{2k,i_1,i_2,\ldots,i_k} z_{i_k}$ corresponding to the quantity G_{2k} for $k = 1, 2, 3, \ldots$ ($G_n = 0$ if $n \neq 2k$) as a semi-invariant, has type

$$\left(2(2k + 1), 3k^2 + 3k + 1, k\right), \tag{6.133}$$

where $2(2k+1)$ is the degree of homogeneity of the comitant in phase variables x, y; $3k^2 + 3k + 1$ is the degree of homogeneity of the comitant in the coefficients of linear part; and k is the degree of homogeneity of the comitant in the coefficients of the nonhomogeneity of the fifth order of the system $s(1, 5)$ from (6.110).

Consider the differential system $s(1, 5)$ from (6.110). In this case, according to the paper [33] using Theorem 4.10 the following is true:

Theorem 6.7. *Dimension of linear space of centro-affine comitants of type* $(d) = (\delta, d_0, d_1)$ *for the differential system* $s(1,5)$ *from* (6.110), *denoted by* $dim_{\mathbb{R}} V_{1,5}^{(d)}$, *is equal to the coefficient of the monomial* $u^{\delta} b^{d_0} f^{d_1}$ *in decomposition of generalized Hilbert series from* (4.42)–(4.43) *for Sibirsky algebra* $S_{1,5}$ *of comitants of the considered system.*

Consider the subalgebra $S_{1,5}' \subset S_{1,5}$, which we write in the form

$$S_{1,5}' = \bigoplus_{(d)} S_{1,5}^{(d')}, \tag{6.134}$$

where by $S_{1,5}^{(d')}$ the linear spaces are denoted:

$$S_{1,5}^{(0,0,0)} = \mathbb{R}, \quad S_{1,5}^{(0,1,0)}, ..., S_{1,5}^{2(2k+1),3k^2+3k+1,k}, \quad k = 1, 2, ..., \tag{6.135}$$

as well as spaces from $S_{1,5}$, which contain all kinds of their products.

Since algebra $S_{1,5}'$ is a graded subalgebra in a finitely defined algebra $S_{1,5}$, then according to Proposition 4.1, we obtain $\varrho(S_{1,5}') \leq \varrho(S_{1,5})$. From this inequality and from the fact that $\varrho(S_{1,5}) = 15$ (see Theorem 4.12), according to Remark 2.3 on semi-invariants and the fact that generalized focus pseudo-quantities are coefficients of some comitants, we have that the following takes place:

Theorem 6.8. *Maximal number of algebraically independent generalized focus pseudo-quantities in the center and focus problem for differential system* (6.110) *does not exceed 15.*

According to Proposition 4.2, Remark 5.1, and equality (5.14), it follows that the maximal number of algebraically independent focus quantities L_k $(k = \overline{1, \infty})$ cannot exceed maximal number of algebraically independent generalized focus pseudo-quantities $G_{2k,i_1,i_2,...,i_k} + B_{2k,i_1,i_2,...,i_k} b_{i_1} + \cdots + Z_{2k,i_1,i_2,...,i_k} z_{i_k}$.

There from using Theorem 6.8, we obtain

Consequence 6.4. *Maximal number of algebraically independent focus quantities that take part in solving the center and focus problem for differential system* (6.110) *does not exceed 15.*

The generalized generating function of space (6.134) can be written as:

$$\Phi(V_{1,5}, u, b, f) = dim_{\mathbb{R}} S_{1,5}^{(0,0,0)} + dim_{\mathbb{R}} S_{1,5}^{(0,1,0)} b$$

$$+ \sum_{k=1}^{\infty} dim_{\mathbb{R}} S_{1,5}^{(2(2k+1),3k^2+3k+1,k)} u^{2(2k+1)} b^{3k^2+3k+1} f^k. \tag{6.136}$$

Using computer, expand the Hilbert series $H(S_{1,5}, u, b, f)$ from (24.7)–(24.9) in a power series. Then for (6.136), we have

$$\Phi(V_{1,5}, u, b, f) = 1 + b + 33u^6 b^7 f + 585u^{10} b^{19} f^2 + 5616u^{14} b^{37} f^3 + ...$$

$$+ C_{2(2k+1),3k^2+3k+1,k} u^{2(2k+1)} b^{(3k^2+3k+1)} f^k + ..., \tag{6.137}$$

where $C_{2(k+1),\frac{1}{2}(15k^2+11k+2),2k}$ is an undetermined coefficient.

Using this generalized generating function and the Hilbert series $H(S_{1,5}, u, b, f)$ from Theorem 6.7, there were obtained the first terms up to $u^{22}b^{91}f^5$ $(\delta + d_0 + d_1 \leq 118)$ in the Hilbert series $H(S'_{1,5}, u, b, f)$:

$$H(S'_{1,5}, u, b, f) = 1 + b + 2b^2 + 2b^3 + 3b^4 + 3b^5 + 4b^6 + 4b^7 + 5b^8 + 5b^9$$
$$+6b^{10} + 6b^{11} + 7b^{12} + 7b^{13} + 8b^{14} + 8b^{15} + 9b^{16} + 9b^{17} + 10b^{18} + 10b^{19}$$
$$+11b^{20} + 11b^{21} + 12b^{22} + 12b^{23} + 13b^{24} + 13b^{25} + 14b^{26} + 14b^{27} + 15b^{28}$$
$$+15b^{29} + 16b^{30} + 16b^{31} + 17b^{32} + 17b^{33} + 18b^{34} + 18b^{35} + 19b^{36} + 19b^{37}$$
$$+20b^{38} + 20b^{39} + 21b^{40} + 21b^{41} + 22b^{42} + 22b^{43} + 23b^{44} + 23b^{45}$$
$$+24b^{46} + \ldots + u^{22}(46666b^{78}f^4 + 47358b^{79}f^4 + 48029b^{80}f^4 + 48721b^{81}f^4$$
$$+49392b^{82}f^4 + 50084b^{83}f^4 + 50755b^{84}f^4 + 51447b^{85}f^4 + 52118b^{86}f^4$$
$$+52810b^{87}f^4 + 53481b^{88}f^4 + 54173b^{89}f^4 + 54844b^{90}f^4 + 55536b^{91}f^4$$
$$+176322b^{91}f^5) + \ldots$$

Hence, the ordinary Hilbert series $H_{S'_{1,5}}(t)$ of the algebra $S'_{1,5}$ will have the form (the first 119 terms):

$$H_{S'_{1,5}}(t) = 1 + t + 2t^2 + 2t^3 + 3t^4 + 3t^5 + 4t^6 + 4t^7 + 5t^8 + 5t^9 + 6t^{10}$$
$$+6t^{11} + 7t^{12} + 7t^{13} + 41t^{14} + 47t^{15} + 54t^{16} + 60t^{17} + 67t^{18} + 73t^{19}$$
$$+80t^{20} + 86t^{21} + 93t^{22} + 99t^{23} + 106t^{24} + 112t^{25} + 119t^{26} + 125t^{27}$$
$$+504t^{28} + 546t^{29} + 595t^{30} + 1222t^{31} + 1306t^{32} + 1389t^{33} + 1473t^{34}$$
$$+1556t^{35} + 1640t^{36} + 1723t^{37} + 1807t^{38} + 1890t^{39} + 1974t^{40} + 2057t^{41}$$
$$+4598t^{42} + 4863t^{43} + 5129t^{44} + 8915t^{45} + 9362t^{46} + 9808t^{47}$$
$$+10255t^{48} + 10701t^{49} + 11148t^{50} + 11594t^{51} + 12041t^{52}$$
$$+12487t^{53} + 18550t^{54} + 19175t^{55} + 19801t^{56} + 20426t^{57}$$
$$+21052t^{58} + 37686t^{59} + 38983t^{60} + 40300t^{61} + 61616t^{62} + 63602t^{63}$$
$$+65589t^{64} + 67575t^{65} + 69562t^{66} + 71548t^{67} + 73535t^{68} + 75521t^{69}$$
$$+77508t^{70} + 79494t^{71} + 81481t^{72} + 83467t^{73} + 85454t^{74} + 87440t^{75}$$
$$+89427t^{76} + 91413t^{77} + 93400t^{78} + 95386t^{79} + 97373t^{80} + 99359t^{81}$$
$$+101346t^{82} + 139347t^{83} + 141998t^{84} + 144669t^{85} + 147320t^{86}$$
$$+149991t^{87} + 152642t^{88} + 155313t^{89} + 157964t^{90} + 160635t^{91}$$
$$+163286t^{92} + 165957t^{93} + 168608t^{94} + 171279t^{95} + 173930t^{96}$$
$$+176601t^{97} + 179252t^{98} + 181380t^{99} + 184025t^{100} + 186690t^{101}$$
$$+189335t^{102} + 192000t^{103} + 191289t^{104} + 193913t^{105} + 193109t^{106}$$
$$+195697t^{107} + 198265t^{108} + 185392t^{109} + 187781t^{110} + 174723t^{111}$$

$$+176931t^{112} + 163780t^{113} + 108892t^{114} + 110253t^{115}$$
$$+54903t^{116} + 55595t^{117} + 176382t^{118} + \dots \tag{6.138}$$

We consider the first 119 terms in the expansion of Hilbert series of the algebra $SI_{1,5}$ from (4.42)–(4.43). Replacing $z = t$, we obtain

$$H_{SI_{1,5}}(t) = H(S_{1,5}, 0, t, t) = 1 + t + 4t^2 + 8t^3 + 26t^4 + 53t^5 + 146t^6$$
$$+305t^7 + 704t^8 + 1417t^9 + 2920t^{10} + 5533t^{11} + 10500t^{12} + 18825t^{13}$$
$$+33444t^{14} + 57120t^{15} + 96303t^{16} + 157599t^{17} + 254508t^{18} + 401472t^{19}$$
$$+625182t^{20} + 955251t^{21} + 1442076t^{22} + 2142840t^{23} + 3149178t^{24}$$
$$+4566267t^{25} + 6554694t^{26} + 9300484t^{27} + 13076140t^{28} + 18198949t^{29}$$
$$+25118690t^{30} + 34359893t^{31} + 46645739t^{32} + 62820314t^{33} + 84019460t^{34}$$
$$+111568250t^{35} + 147213784t^{36} + 192990661t^{37} + 251534302t^{38}$$
$$+325907859t^{39} + 420016674t^{40} + 538389135t^{41} + 686719824t^{42}$$
$$+871593216t^{43} + 1101188574t^{44} + 1384936842t^{45} + 1734423882t^{46}$$
$$+2162969685t^{47} + 2686776843t^{48} + 3324416523t^{49} + 4098277602t^{50}$$
$$+5033946165t^{51} + 6162015960t^{52} + 7517347113t^{53} + 9141313732t^{54}$$
$$+11080921339t^{55} + 13391579524t^{56} + 16136061599t^{57}$$
$$+19387898270t^{58} + 23230161917t^{59} + 27759598166t^{60}$$
$$+33085209860t^{61} + 39333260630t^{62} + 46645639450t^{63}$$
$$+55185881485t^{64} + 65137293814t^{65} + 76710167634t^{66}$$
$$+90139636710t^{67} + 105694278048t^{68} + 123673683567t^{69}$$
$$+144418662324t^{70} + 168308506209t^{71} + 195773044560t^{72}$$
$$+227289574704t^{73} + 263397050880t^{74} + 304692662715t^{75}$$
$$+351848461524t^{76} + 405607571979t^{77} + 466803608430t^{78}$$
$$+536356493511t^{79} + 615295057113t^{80} + 704752453114t^{81}$$
$$+805992362230t^{82} + 920404001941t^{83} + 1049532417158t^{84}$$
$$+1195073066570t^{85} + 1358906736914t^{86} + 1543093711865t^{87}$$
$$+1749913882256t^{88} + 1981860511040t^{89} + 2241686191348t^{90}$$
$$+2532396208270t^{91} + 2857301061364t^{92} + 3220009454859t^{93}$$
$$+3624488148738t^{94} + 1981860511040t^{89} + 2241686191348t^{90}$$
$$+2532396208270t^{91} + 2857301061364t^{92} + 3220009454859t^{93}$$
$$+3624488148738t^{94} + 4075054619802t^{95} + 4576445103411t^{96}$$
$$+5133806955846t^{97} + 5752775830170t^{98} + 6439467809532t^{99}$$
$$+7200566719746t^{100} + 8043316119915t^{101} + 8975617875264t^{102}$$

$$+10006024079634t^{103} + 11143848111366t^{104} + 12399156613041t^{105}$$
$$+13782894327786t^{106} + 15306876284184t^{107} + 16983927856168t^{108}$$
$$+18827877291745t^{109} + 20853712561492t^{110} + 23077574432714t^{111}$$
$$+25516931762351t^{112} + 28190575349780t^{113} + 31118813506166t^{114} \quad (6.139)$$
$$+34323466911800t^{115} + 37828086439832t^{116} + 41657949323224t^{117}$$
$$+45840301322554t^{118} + \dots$$

Since for series (6.138) and (6.139), the following inequality holds:

$$H_{S'_{1,5}}(t) \le H_{SI_{1,5}}(t),$$

then in assumption that this inequality holds for remaining terms of the considered series, we obtain the inequality

$$\varrho(S'_{1,5}) \le \varrho(SI_{1,5}).$$

Note that $S'_{1,5}$ is not a subalgebra in $SI_{1,5}$. Since from Theorem 4.12, we have $\varrho(SI_{1,5}) = 13$, then according to the last inequality, we obtain that the following can be true:

Hypothesis 6.2. *Maximal number of algebraically independent generalized focus pseudo-quantities (as well as focus quantities) for differential system (6.110) that take part in solving the center and focus problem for the given differential system may not exceed 13.*

6.6 Comitants that Have Generalized Focus Pseudo-Quantities of the System $s(1, 2, 3)$ as Coefficients, and Their Sibirsky Graded Algebra

Consider the differential system $s(1, 2, 3)$, which we write in the form

$$\dot{x} = cx + dy + gx^2 + 2hxy + kx^2 + px^3 + 3qx^2y + 3rxy^2 + sy^3,$$
$$\dot{y} = ex + fy + lx^2 + 2mxy + ny^2 + tx^3 + 3ux^2y + 3vxy^2 + wy^3 \quad (6.140)$$

with a finitely defined graded algebra of unimodular comitants $S_{1,2,3}$. For this system, we write function (5.13) in the form

$$U = k_2 + a_0x^3 + 3a_1x^2y + 3a_2xy^2 + a_3y^3 + b_0x^4 + 4b_1x^3y + 6b_2x^2y^2$$
$$+4b_3xy^3 + b_4y^4 + c_0x^5 + 5c_1x^4y + 10c_2x^3y^2 + 10c_3x^2y^3 + 5c_4xy^4$$
$$+c_5y^5 + d_0x^6 + 6d_1x^5y + 15d_2x^4y^2 + 20d_3x^3y^3 + 15d_4x^2y^4 + 6d_5xy^5$$

$$+d_6y^6 + e_0x^7 + 7e_1x^6y + 21e_2x^5y^2 + 35e_3x^4y^3 + 21e_5x^2y^5 + 7e_6xy^6$$
$$+e_7y^7 + f_0x^8 + 8f_1x^7y + 28f_2x^6y^2 + 56f_3x^5y^3 + 70f_4y^4$$
$$+56f_5x^3y^5 + 28f_6x^2y^6 + 8f_7xy^7 + f_8y^8 + \dots,$$

$$(6.141)$$

where $k_2 \not\equiv 0$ is from (5.8), and $a_0, a_1, \dots, f_7, f_8, \dots$ are unknown coefficients.

Identity (5.12) along the trajectories of differential system (6.140) with function (6.141) splits into the following systems of equations with respect to the variables $a_0, a_1, \dots, f_7, f_8, G_1, G_2, G_3, \dots$ (equality (5.15) is omitted). For obtaining the quantity G_1, we write equations, in which the identity (5.12) decomposes in the case of differential system (6.140) into the matrix form

$$\widetilde{A}_1\widetilde{B}_1 = \widetilde{C}_1, \tag{6.142}$$

where

$$\widetilde{A}_1 = \begin{pmatrix} 3c & 3e & 0 & 0 & 0 & 0 & 0 \\ 3d & 6c+3f & 6e & 0 & 0 & 0 & 0 \\ 0 & 6d & 3c+6f & 3e & 0 & 0 & 0 \\ 0 & 0 & 3d & 3f & 0 & 0 & 0 \\ 3g & 3l & 0 & 0 & 4c & 4e & 0 \\ 6h & 6g+6m & 6l & 0 & 4d & 12c+4f & 12e \\ 3k & 12h+3n & 3g+12m & 3l & 0 & 12d & 12c+12f \\ 0 & 6k & 6h+6n & 6m & 0 & 0 & 12d \\ 0 & 0 & 3k & 3n & 0 & 0 & 0 \end{pmatrix}$$

$$\begin{pmatrix} 0 & 0 & 0 \\ 0 & 0 & 0 \\ 0 & 0 & 0 \\ 0 & 0 & 0 \\ 0 & 0 & -e^2 \\ 0 & 0 & 2ce-2ef \\ 12e & 0 & -c^2+2de+2cf-f^2 \\ 4c+12f & 4e & -2cd+2df \\ 4d & 4f & -d^2 \end{pmatrix},$$

$$\widetilde{B}_1 = \begin{pmatrix} a_0 \\ a_1 \\ a_2 \\ a_3 \\ b_0 \\ b_1 \\ b_2 \\ b_3 \\ b_4 \\ G_1 \end{pmatrix}, \quad \widetilde{C}_1 = \begin{pmatrix} 2eg - cl + fl \\ -cg + fg + 4eh - 2dl - 2cm + 2fm \\ -2ch + 2fh + 2ek - 4dm - cn + fn \\ -ck + fk - 2dn \\ 2ep - ct + ft \\ -cp + fp + 6eq - 2dt - 3cu + 3fu \\ -3cq + 3fq + 6er - 6du - 3cv + 3fv \\ -3cr + 3fr + 2es - 6dv - cw + fw \\ -cs + fs - 2dw \end{pmatrix}.$$

$$(6.143)$$

For each fixed $i \in \{0, 1, ..., 4\}$ using Cramer's rule from system (6.142), we obtain

$$G_1 = \frac{\widetilde{G}_{1,i} + \widetilde{B}_{1,i} b_i}{\widetilde{\sigma}_{1,i}}, \qquad (6.144)$$

where $\widetilde{G}_{1,i}, \widetilde{B}_{1,i}, \widetilde{\sigma}_{1,i}$ are polynomials in coefficients of differential system (6.140), and b_i are undetermined coefficients of the function $U(x, y)$ from (6.141).

Studying matrix equation (6.142) for differential system (6.140), we find that the focus pseudo-quantity $\widetilde{G}_{1,i}$ for any fixed i from (6.144) can be written as

$$\widetilde{G}_{1,i} = \widetilde{G}'_{1,i} + \widetilde{G}''_{1,i}, \ (i = 0, 1, 2, 3, 4), \qquad (6.145)$$

where $\widetilde{G}'_{1,i}$ ($\widetilde{G}''_{1,i}$ respectively) are homogeneous polynomials of degree 8 (9 respectively) with respect to the coefficients of the linear part and homogeneous of degree 2 with respect to the coefficients of the quadratic part (of degree 1 with respect to the coefficients of the cubic part, respectively) of differential system (6.140).

From (1.40)–(1.41), (1.44)–(1.45), (1.47)–(1.48), (1.50)–(1.51) operators of Lie algebra L_4 for differential system (6.140) are obtained:

$$\mathcal{X}_1 = x\frac{\partial}{\partial x} + d\frac{\partial}{\partial d} - e\frac{\partial}{\partial e} - g\frac{\partial}{\partial g} + k\frac{\partial}{\partial k} - 2l\frac{\partial}{\partial l} - m\frac{\partial}{\partial m} - 2p\frac{\partial}{\partial p} - q\frac{\partial}{\partial q}$$

$$+ s\frac{\partial}{\partial s} - 3t\frac{\partial}{\partial t} - 2u\frac{\partial}{\partial u} - v\frac{\partial}{\partial v},$$

$$\mathcal{X}_2 = y\frac{\partial}{\partial x} + e\frac{\partial}{\partial c} + (f - c)\frac{\partial}{\partial d} - e\frac{\partial}{\partial f} + l\frac{\partial}{\partial g} + (m - g)\frac{\partial}{\partial h} + (n$$

$$- 2h)\frac{\partial}{\partial k} - l\frac{\partial}{\partial m} - 2m\frac{\partial}{\partial n} + t\frac{\partial}{\partial p} + (u - p)\frac{\partial}{\partial q} + (v - 2q)\frac{\partial}{\partial r} + (w$$

$$- 3r)\frac{\partial}{\partial s} - t\frac{\partial}{\partial u} - 2u\frac{\partial}{\partial v} - 3v\frac{\partial}{\partial w},$$

$$\mathcal{X}_3 = x\frac{\partial}{\partial y} - d\frac{\partial}{\partial c} + (c - f)\frac{\partial}{\partial e} + d\frac{\partial}{\partial f} - 2h\frac{\partial}{\partial g} - k\frac{\partial}{\partial h} + (g - 2m)\frac{\partial}{\partial l}$$

$$+ (h - n)\frac{\partial}{\partial m} + k\frac{\partial}{\partial n} - 3q\frac{\partial}{\partial p} - 2r\frac{\partial}{\partial q} - s\frac{\partial}{\partial r} + (p - 3u)\frac{\partial}{\partial t} + (q$$

$$- 2v)\frac{\partial}{\partial u} + (r - w)\frac{\partial}{\partial v} + s\frac{\partial}{\partial w},$$

$$\mathcal{X}_4 = y\frac{\partial}{\partial y} - d\frac{\partial}{\partial d} + e\frac{\partial}{\partial e} - h\frac{\partial}{\partial h} - 2k\frac{\partial}{\partial k} + l\frac{\partial}{\partial l} - n\frac{\partial}{\partial n} - q\frac{\partial}{\partial q} - 2r\frac{\partial}{\partial r}$$

$$- 3s\frac{\partial}{\partial s} + t\frac{\partial}{\partial t} - v\frac{\partial}{\partial v} - 2w\frac{\partial}{\partial w}.$$

Applying these operators to expressions from (6.145), we find the equalities

$$\mathcal{X}_1(\widetilde{f}'_4) = \mathcal{X}_4(\widetilde{f}'_4) = \widetilde{f}'_4, \ \mathcal{X}_2(\widetilde{f}'_4) = \mathcal{X}_3(\widetilde{f}'_4) = 0,$$

$$\mathcal{X}_1(\widetilde{f}''_4) = \mathcal{X}_4(\widetilde{f}''_4) = \widetilde{f}''_4, \ \mathcal{X}_2(\widetilde{f}''_4) = \mathcal{X}_3(\widetilde{f}''_4) = 0,$$

where

$$\widetilde{f}'_4(x,y) = \widetilde{G}'_{1,0}x^4 - 4\widetilde{G}'_{1,1}x^3y + 2\widetilde{G}'_{1,2}x^2y^2 + 4\widetilde{G}'_{1,3}xy^3 - \widetilde{G}'_{1,4}y^4,$$
$$\widetilde{f}''_4(x,y) = \widetilde{G}''_{1,0}x^4 - 4\widetilde{G}''_{1,1}x^3y + 2\widetilde{G}''_{1,2}x^2y^2 + 4\widetilde{G}''_{1,3}xy^3 - \widetilde{G}''_{1,4}y^4 \tag{6.146}$$

are comitants of weight -1 of differential system (6.140), and $\widetilde{G}'_{1,i}$, $\widetilde{G}''_{1,i}$ are from (6.145).

According to the abovementioned and (3.1), comitants (6.146) belong to the linear spaces

$$S_{1,2,3}^{(4,8,2,0)}, \ S_{1,2,3}^{(4,9,0,1)}, \tag{6.147}$$

which are components of graded Sibirsky algebra of comitants $S_{1,2,3}$ for differential system (6.140).

According to (6.144), for example, for $b_i = 0$ $(i = \overline{0,4})$ on the variety $\mathcal{V}$ from (5.11) for (6.145), (6.146), we find that between the first focus quantity L_1 of differential system (6.140) and comitants (6.146), the following equality holds:

$$\left[\widetilde{f}'_4(x,y) + \widetilde{f}''_4(x,y)\right]\big|_{\mathcal{V}} = 8L_1(x^2 + y^2)^2 \ (G_1|_{\mathcal{V}} = 8L_1),$$

where

$$L_1 = \frac{1}{4}\left\{2[g(l-h) - k(h+n) + m(l+n)] - 3[p+r+u+w]\right\}.$$

If you exclude the numerical constant and consider the notation of coefficients of this differential system, then the last expression coincides with the focus quantity of this system from [10, p. 25].

For obtaining the quantity G_2 for differential system (6.140) from identity (5.12), similarly, we obtain the following matrix equation (see Appendix 10):

$$\widetilde{A}_2\widetilde{B}_2 = \widetilde{C}_2. \tag{6.148}$$

For any fixed $i \in \{0,1,...,4\}$, $j \in \{0,1,2,...,6\}$, we find the expression

$$G_2 = \frac{\widetilde{G}_{2,i,j} + \widetilde{B}_{2,i,j}b_i + \widetilde{D}_{2,i,j}d_j}{\widetilde{\sigma}_{2,i,j}}. \tag{6.149}$$

Studying matrix equation (6.148), we find that focus pseudo-quantity from (6.149) can be written as a homogeneity of degree 24 and presented as

$$\widetilde{G}_{2,i,j} = \widetilde{G}'_{2,i,j} + \widetilde{G}''_{2,i,j} + \widetilde{G}'''_{2,i,j}, \tag{6.150}$$

where $\widetilde{G}'_{2,i,j}$, $\widetilde{G}''_{2,i,j}$ and $\widetilde{G}'''_{2,i,j}$ are homogeneities of a similar type from (3.1), i.e. of the form $(0, d_0, d_1, d_2)$. Hence, we have the formulas $(0, 20, 4, 0)$, $(0, 21, 2, 1)$ and $(0, 22, 0, 2)$, respectively. Note that on the variety $\mathcal{V}$ from (5.11) for differential system (6.140) quantities $\widetilde{G}_{2,2,j}$ $(j = \overline{0,6})$ have the form

$$\widetilde{G}_{2,2,j}|_{\mathcal{V}} = 2304L_2 \ (j = 0,2,4,6), \ \widetilde{G}_{2,2,j}|_{\mathcal{V}} = 0, \ (j = 1,3,5).$$

On the other hand, the second focus quantity L_2 of differential system (6.140) can be written using an expression from (6.150) as follows:

$$24L_2 = \widetilde{G}'_{2,2,j}|_\nu + \widetilde{G}''_{2,2,j}|_\nu + \widetilde{G}'''_{2,2,j}|_\nu \ (j = 0, 2, 4, 6),$$

where

$$\widetilde{G}'_{2,2,j}|_\nu = 4(62g^3h - 2gh^3 + 95g^2hk - 2h^3k + 38ghk^2 + 5hk^3 - 62g^3l$$
$$+27gh^2l - 39g^2kl + 29h^2kl - 15gk^2l - 8ghl^2 + 15hkl^2 - 5gl^3 + 53g^2hm$$
$$+66ghkm + 13hk^2m - 127g^2lm - 6h^2lm - 68gklm - 15k^2lm - 13hl^2m$$
$$-5l^3m + 6ghm^2 + 6hkm^2 - 63glm^2 - 29klm^2 + 2lm^3 + 6g^3n + 61gh^2n$$
$$+72g^2kn + 63h^2kn + 33gk^2n + 5k^3n - 10ghln + 68hkln - 33gl^2n$$
$$+15kl^2n - 72g^2mn - 6h^2mn + 10gkmn + 8k^2mn - 66hlmn - 38l^2mn$$
$$-61gm^2n - 27km^2n + 2m^3n + 72ghn^2 + 127hkn^2 - 72gln^2 + 39kln^2$$
$$-53hmn^2 - 95lmn^2 - 6gn^3 + 62kn^3 - 62mn^3),$$

$$\widetilde{G}''_{2,2,j}|_\nu = -2(186g^2p + 10h^2p + 117gkp + 45k^2p + 59hlp + 15l^2p$$
$$+159gmp + 75kmp + 18m^2p + 143hnp + 89lnp + 196n^2p - 69ghq$$
$$-57hkq + 69glq + 12klq + 9lmq + 60gnq + 3knq + 21mnq + 168g^2r$$
$$-6h^2r + 69gkr + 15k^2r + 87hlr + 45l^2r + 123gmr + 39kmr + 18m^2r$$
$$+171hnr + 129lnr + 222n^2r - 13ghs - 17hks - 15gls - 16hms - 15lms$$
$$-16gns - 17kns - 19mns - 19ght - 15hkt - 17glt - 16hmt - 17lmt$$
$$-16gnt - 15knt - 13mnt + 222g^2u + 18h^2u + 129gku + 45k^2u + 39hlu$$
$$+15l^2u + 171gmu + 87kmu - 6m^2u + 123hnu + 69lnu + 168n^2u + 21ghv$$
$$+9hkv + 3glv + 12klv - 57lmv + 60gnv + 69knv - 69mnv + 196g^2w$$
$$+18h^2w + 89gkw + 15k^2w + 75hlw + 45l^2w + 143gmw + 59kmw$$
$$+10m^2w + 159hnw + 117lnw + 186n^2w),$$

$$\widetilde{G}'''_{2,2,j}|_\nu = -9(11pq + 15qr - 5ps - rs + pt + 5rt + 3qu - 5su + tu - 7pv$$
$$-3rv - 15uv + 7qw - sw + 5tw - 11vw).$$

Choose a comitant of weight -1 of the differential system $s(1, 2, 3)$ from (6.140), which contains the expression $\widetilde{G}_{2,i,j} + \widetilde{B}_{2,i,j}b_i + \widetilde{D}_{2,i,j}d_j$ as semi-invariant. According to decomposition (6.150) and the types arising from it, we find that this comitant is the sum of the comitants, which belong to the linear spaces

$$S_{1,2,3}^{(6,20,4,0)}, \ S_{1,2,3}^{(6,21,2,1)}, \ S_{1,2,3}^{(6,22,0,2)}. \tag{6.151}$$

Using the same process and the matrix equation

$$\widetilde{A}_3\widetilde{B}_3 = \widetilde{C}_3$$

for any fixed $i \in \{0, 1, ..., 4\}$, $j \in \{0, 1, ..., 6\}$, $k \in \{0, 1, ..., 8\}$ we obtain

$$G_3 = \frac{\widetilde{G}_{3,i,j,k} + \widetilde{B}_{3,i,j,k} b_i + \widetilde{D}_{3,i,j,k} d_j + \widetilde{F}_{3,i,j,k} f_j}{\widetilde{\sigma}_{3,i,j,k}}. \tag{6.152}$$

Similarly to the previous case, we find that the focus pseudo-quantity $\widetilde{G}_{3,i,j,k}$ is decomposed into a sum of four members of the same degree 43 with respect to the coefficients of differential system (6.140), which according to (3.1) will be written as $(0, d_0, d_1, d_2)$ and belong to the types $(0, 37, 6, 0)$, $(0, 38, 4, 1)$, $(0, 39, 2, 2)$ and $(0, 40, 0, 3)$. In this case, it follows that a comitant of the weight -1, which has one of the numerators of expression (6.152) as a semi-invariant, consists of the sum of comitants of system (6.140), which belong to the linear spaces

$$S_{1,2,3}^{(8,37,6,0)}, \quad S_{1,2,3}^{(8,38,4,1)}, \quad S_{1,2,3}^{(8,39,2,2)}, \quad S_{1,2,3}^{(8,40,0,3)}. \tag{6.153}$$

Following this process, we obtain a series of linear spaces (6.147), (6.151), (6.153) etc. of the comitants of system (6.140). Note that the corresponding generalized focus pseudo-quantities G_k of the given system are the sum of the coefficients of these comitants.

Similarly, it is not difficult to derive a general formula of comitants that have generalized focus pseudo-quantities corresponding to the G_k as coefficients, which decompose into sum of comitants of differential system (6.140) of the corresponding types

$$\left(2(k+1), \frac{1}{2}(5k^2 + 9k + 2) + i, 2(k-i), i \right) \quad (i = \overline{0, k}).$$

Consider the subalgebra $S'_{1,2,3} \subset S_{1,2,3}$, which we write in the form

$$S'_{1,2,3} = \bigoplus_{(d)} S_{1,2,3}^{(d')}, \tag{6.154}$$

where by $S_{1,2,3}^{(d')}$ the linear spaces are denoted:

$$S_{1,2,3}^{(0,0,0,0)} = \mathbb{R}, \quad S_{1,2,3}^{(0,1,0,0)}, \ldots, S_{1,2,3}^{(2(k+1),\frac{1}{2}(5k^2+9k+2)+i,2(k-i),i)}$$
$$(i = \overline{0, k}, k = 1, 2, ...) \tag{6.155}$$

as well as spaces from $S_{1,2,3}$, which contain all kinds of their products. Since the algebra $S'_{1,2,3}$ is a graded subalgebra in a finitely defined algebra $S_{1,2,3}$, then according to Proposition 4.1, we obtain $\varrho(S'_{1,2,3}) \le \varrho(S_{1,2,3})$. From this inequality and from the fact that using formula (2.54), in which for differential system (6.140) we have $m_0 = 1$, $m_1 = 2$, $m_2 = 3$, we obtain $\varrho(S_{1,2,3}) = 17$.

Then according to Remark 2.3 on semi-invariants and the fact that generalized focus pseudo-quantities are coefficients of some comitants of the given differential system, we obtain that the following is true:

Theorem 6.9. *Maximal number of algebraically independent generalized focus pseudo-quantities in the center and focus problem for differential system (6.140) does not exceed 17.*

According to Proposition 4.2, Observation 5.2, and equality (5.14), it follows that maximal number of algebraically independent focus quantities L_k $(k = \overline{1, \infty})$ cannot exceed maximal number of algebraically independent generalized focus pseudo-quantities. There from using Theorem 6.10, we obtain

Consequence 6.5. *Maximal number of algebraically independent focus quantities that take part in solving the center and focus problem for differential system (6.140) does not exceed 17.*

6.7   On the Upper Bound of the Number of Algebraically Independent Focus Quantities that Take Part in Solving the Center and Focus Problem for the Differential System $s(1, m_1, ..., m_\ell)$

Consider the differential system $s(1, m_1, ..., m_\ell)$:

$$\frac{dx}{dt} = \sum_{i=0}^{\ell} P_{m_i}(x, y), \quad \frac{dy}{dt} = \sum_{i=0}^{\ell} Q_{m_i}(x, y), \tag{6.156}$$

where P_{m_i} and Q_{m_i} are homogeneous polynomials of degree $m_i \geq 1$ with respect to x and y, and $m_0 = 1$. According to Section 5.1, the problem is to determine a majorant estimate of the number λ of algebraically independent elements from (5.1), (5.2), i.e. for any nontrivial polynomial with respect to the variables $L_{i_1}, L_{i_2}, ..., L_{i_\lambda}$ the inequality $F(L_{i_1}, L_{i_2}, ..., L_{i_\lambda}) \neq 0$, $(\lambda \leq \omega)$ holds.

Focus quantities of differential system (6.156) form an infinite sequence of polynomials from coefficients of this system (5.1).

It is known from the paper [33], that differential system $s(1, m_1, ..., m_\ell)$ admits the group $GL(2, \mathbb{R})$, to which the reductive Lie algebra L_4 corresponds, that consists of operators (1.40)–(1.41), (1.44)–(1.45), (1.47)–(1.48), (1.50)–(1.51) of linear representation of this group in the space of phase variables and coefficients of this system. This algebra generates a graded Sibirsky algebra of invariant polynomials with respect to the unimodular group $SL(2, \mathbb{R}) \subset GL(2, \mathbb{R})$, which we write in the form

$$S_{1,m_1,...,m_\ell} = \sum_{(d)} S^{(d)}_{1,m_1,...,m_\ell}, \qquad (6.157)$$

where (d) is called a type of the space $S^{(d)}_{1,m_1,...,m_\ell}$ and has the form (3.1), which is a finite-dimensional linear space of invariant polynomials (homogeneous comitants, invariants of degree δ with respect to the phase variables x, y and of degree d_i with respect to the coefficients of the polynomial P_{m_i} and Q_{m_i} of system (6.156)).

Similarly to the considered examples in this chapter, it can be shown that to each focus quantity L_k $(k = \overline{1, \infty})$ one can associate finite-dimensional linear spaces of invariant polynomials (unimodular comitants [33])

$$S^{(d^{(k)})}_{1,m_1,...,m_\ell} \quad (k = 1, 2, ...), \qquad (6.158)$$

where

$$(d^{(k)}) = (\delta^{(k)}, d^{(k)}_0, d^{(k)}_1, ..., d^{(k)}_\ell) \qquad (6.159)$$

is a type of a space which contains comitants that have focus pseudo-quantities of the differential system $s(1, m_1, ..., m_\ell)$ as coefficients.

Existence of spaces of comitants (6.158) is argued by Lemma 2.4. These spaces are characterized by the fact that they contain at least one homogeneous polynomial with respect to x, y, in which the coefficients are generalized focus pseudo-quantities, which are characterized by the fact that on the invariant variety $\mathcal{V}$ from (5.11) their semi-invariants, except for a numerical constant, go to the corresponding focus quantity L_k. Thus, we obtain the sequence of spaces $\mathbb{R} = S^{(0,0,...,0)}_{1,m_1,...,m_\ell}, S^{(0,1,...,0)}_{1,m_1,...,m_\ell}, ..., S^{(\delta^{(k)},d^{(k)}_0,d^{(k)}_1,...,d^{(k)}_\ell)}_{1,m_1,...,m_\ell}, ...$ from $S_{1,m_1,...,m_\ell}$. Using them, we form a graded subalgebra $S'_{1,m_1,...,m_\ell}$, which satisfies the inclusion

$$S'_{1,m_1,...,m_\ell} \subset S_{1,m_1,...,m_\ell},$$

where according to Proposition 4.1, it follows that between their Krull dimensions, the following inequality holds:

$$\varrho(S'_{1,m_1,...,m_\ell}) \le \varrho(S_{1,m_1,...,m_\ell}). \qquad (6.160)$$

From formula (2.54) and Remark 2.2, we obtain

$$\varrho(S_{1,m_1,...,m_\ell}) = 2\left(\sum_{i=1}^{\ell} m_i + \ell\right) + 3. \qquad (6.161)$$

Similar to the examples considered in Sections 6.1–6.6 and using (6.160) and (6.161), it can be shown that the following is true

Lemma 6.1. *Maximal number of algebraically independent generalized focus pseudo-quantities in the center and focus problem for differential system*

(6.156) *does not exceed the number from* (6.161), *i.e. does not exceed maximal number of all possible nonzero coefficients of the right-hand sides of this system minus one.*

Considering that generalized focus pseudo-quantities, being semi-invariants in comitants that are contained in all homogeneous spaces of the algebra $S'_{1,m_1,...,m_\ell}$ of the system $s(1, m_1, ..., m_\ell)$ from (1.1)–(1.2) on the variety $\mathcal{V}$ from (5.9) or, what is the same, from (5.11), are being transformed up to a numerical factor in the focus quantities L_1, L_2,..., L_k,... of this system, using Lemma 6.1, we obtain that there takes place

Theorem 6.10. *Maximal number of algebraically independent focus quantities of differential system* (6.156), *that take part in solving the center and focus problem, does not exceed the number from* (6.161), *i.e., does not exceed maximal number of all possible nonzero coefficients of the right-hand sides of this system minus one.*

Recall that for the differential systems $s(1,2)$ and $s(1,3)$ the number of essential conditions of center is $\omega = 3$ and 5, respectively, and for the differential system $s(1,2,3)$, according to one hypothesis, it is $\omega \leq 13$.

From Theorem 6.10, we have that maximal number of algebraically independent focus quantities of the differential system $s(1,2)$ does not exceed 9, for the differential system $s(1,3)$ it does not exceed 11, and for the differential system $s(1,2,3)$ it does not exceed 17.

These arguments and Proposition 4.2 with the variety $\mathcal{V}$ from (5.9), which is equivalent to (5.11), and the previously defined algebra $S'_{1,m_1,...,m_\ell}$, allow us to conclude that the following can be true

Hypothesis 6.3. *Number of essential conditions of center ω* (5.2), *that solve the center and focus problem for differential system* (6.156), *which at the origin has a singular point of the second group, does not exceed the number from* (6.161), *i.e., does not exceed maximal number of all possible nonzero coefficients of the right-hand sides of this system minus one.*

Observation 6.1. *Equality* (6.161) *shows that the quantity ϱ is equal to maximal number of all possible nonzero coefficients of the right-hand sides of system* (6.156) *minus one.*

These results were first reported in the paper [38].

Note that in Section 3.3 from Sibirsky's monograph [38], it is determined that an estimate of the lower limit of number of essential conditions of center when differential system (1.1) contains all nonhomogeneities of degree from 1 to q ($q > 1$). In other words, for the differential system $s(\Gamma)$, where $\Gamma = \{1, 2, ..., q\}$, there takes place

Theorem 6.11. [38]. *Number of essential conditions of center is not less than $q^2 - q$ for even q and than $q^2 - q - 1$ for odd q.*

Adapting the result from Theorem 6.11 for the differential system $s(1, 2, ..., q)$, we find

$$\varrho(S_{1,2,...,q}) = 2\left(\sum_{i=2}^{q} i + q - 1\right) + 3 = q^2 + 3q - 1.$$

We obtain that between the estimates from Theorems 6.10 and 6.11 [38] there are relations

$$\varrho(S_{1,2,...,q}) = q^2 + 3q - 1 = \underbrace{[q^2 - q] + 4q - 1}_{q=2k} \text{ or } \underbrace{[q^2 - q - 1] + 4q}_{q=2k+1}.$$

So we get that for the differential system $s(1, 2, ..., q)$ the estimation from Theorem 6.10 is greater than the estimation from Theorem 6.11 [38] by $4q - 1$ if q is even and by $4q$ if q is odd.

6.8 Comments to Chapter Six

The results of this chapter allowed us to approach to an estimation for the upper bound of the number of algebraically independent focus quantities, which are involved in solving the center and focus problem for any differential system of the form (1.1)–(1.2), the problem formulated more than 130 years ago by the French mathematician Poincaré. These results made it possible to propose a valid hypothesis on the upper bound of the number of essential focus quantities, which are involved in solving the considered problem for any differential system of the form (1.1)–(1.2).

In these studies, an important role was played by Hilbert series for Sibirsky algebras of invariants and comitants and also by Lie algebra of representation of centro-affine group in the space of phase variables and coefficients of systems of the form (1.1)–(1.2).

The main result is that the number of algebraically independent focus quantities for the differential system $s(1, m_1, ..., m_\ell)$ of the form (6.156), which has at the origin a singular point of center or focus type does not exceed the number $\varrho = 2(\sum_{i=1}^{\ell} m_i + \ell) + 3$, which is the Krull dimension of the Sibirsky graded algebra of comitants $S_{1,m_1,...,m_\ell}$ of system (6.156).

This is the solution of the generalized center and focus problem for the system $s(1, m_1, ..., m_\ell)$ of the form (6.156).

7

On the Upper Bound of the Number of Functionally Independent Focus Quantities that Take Part in Solving the Center and Focus Problem for Lyapunov System

7.1 Lie Operators of Representation of the Rotation Group $SO(2, \mathbb{R})$ in the Space of Coefficients of Lyapunov System (5.6)

Consider the center and focus problem for classical system of differential equations (5.6), which has the general form

$$\dot{x} = y + \sum_{i=1}^{\ell} P_{m_i}(x, y), \ \dot{y} = -x + \sum_{i=1}^{\ell} Q_{m_i}(x, y) \ (\ell < \infty), \qquad (7.1)$$

where

$$P_{m_i}(x, y) = \sum_{k=0}^{m_i} \binom{m_i}{k} \overset{i1}{a_k} x^{m_i-k} y^k,$$

$$Q_{m_i}(x, y) = \sum_{k=0}^{m_i} \binom{m_i}{k} \overset{i2}{a_k} x^{m_i-k} y^k \ (m_i \in \Gamma; \ i = \overline{1, \ell}). \qquad (7.2)$$

The set $\Gamma = \{m_i\}_{i=1}^{\ell}$ is a finite set of different natural numbers. Recall that we will call system (7.1) as a *Lyapunov system* [20] and denote it by $s\mathcal{L}(1, m_1, ..., m_\ell)$.

We study the action of the rotation group $SO(2, \mathbb{R})$, given by formulas

$$\overline{x} = x \cos \varphi + y \sin \varphi, \ \overline{y} = -x \sin \varphi + y \cos \varphi \ (0 \le \varphi < \pi), \qquad (7.3)$$

on system (7.1)–(7.2).

Due to transformations (7.3) in system (7.1)–(7.2), we obtain

$$\dot{\overline{x}} = \overline{y} + \sum_{i=1}^{\ell} P_{m_i}(\overline{x}, \overline{y}), \ \dot{\overline{y}} = -\overline{x} + \sum_{i=1}^{\ell} Q_{m_i}(\overline{x}, \overline{y}) \ (\ell < \infty), \qquad (7.4)$$

DOI: 10.1201/9781003193074-8

where $\overline{x}$, $\overline{y}$ have the form (7.3), and the coefficients $\overset{i}{b}{}_k^j$ ($i = \overline{1,\ell}$; $j = 1,2$; $k = \overline{1,m_i}$) in the polynomials

$$P_{m_i}(\overline{x},\overline{y}) = \sum_{k=0}^{m_i} \binom{m_i}{k} \overset{i}{b}{}_k^1 \overline{x}^{m_i-k}\overline{y}^k,$$

$$Q_{m_i}(\overline{x},\overline{y}) = \sum_{k=0}^{m_i} \binom{m_i}{k} \overset{i}{b}{}_k^2 \overline{x}^{m_i-k}\overline{y}^k \ (m_i \in \Gamma; \ i = \overline{1,\ell}) \tag{7.5}$$

will be written in the form

$$\overset{i}{b}{}_k^1 = \overset{i}{a}{}_k^1 \cos^{m_i+1}\varphi + [(m_i - k)\overset{i}{a}{}_{k+1}^1 - k\overset{i}{a}{}_{k-1}^1 + \overset{i}{a}{}_k^2]\cos^{m_i}\varphi \sin\varphi$$
$$+ o(\sin\varphi),$$

$$\overset{i}{b}{}_k^2 = \overset{i}{a}{}_k^2 \cos^{m_i+1}\varphi + [-\overset{i}{a}{}_k^1 + (m_i - k)\overset{i}{a}{}_{k+1}^2 - k\overset{i}{a}{}_{k-1}^2]\cos^{m_i}\varphi \sin\varphi \tag{7.6}$$
$$+ o(\sin\varphi).$$

Note that $o(\sin\varphi)$ is a linear function of coefficients of system (7.1)–(7.2) and contains $\sin\varphi$ of degree not less than two in each term.

From (7.3) and (7.6), according to Sections 1.2 and 1.4, it follows that a linear representation of the group $SO(2,\mathbb{R})$ is a one-parameter group depending on the parameter φ, whose value in $\varphi = 0$ corresponds to the identity transformation

$$\overline{x} = x, \ \overline{y} = y, \ \overset{i}{b}{}_k^j = \overset{i}{a}{}_k^j \ (i = \overline{1,\ell}; \ j = 1,2; \ k = \overline{1,m_i}).$$

With this, according to formula (1.35), we obtain

$$\xi^1(x,y) = y, \ \xi^2 = -x,$$

$$\overset{i}{\eta}{}_k^1 = (m_i - k)\overset{i}{a}{}_{k+1}^1 - k\overset{i}{a}{}_{k-1}^1 + \overset{i}{a}{}_k^2 \ (i = \overline{1,\ell}; \ k = \overline{0,m_i}),$$

$$\overset{i}{\eta}{}_k^2 = -\overset{i}{a}{}_k^1 + (m_i - k)\overset{i}{a}{}_{k+1}^2 - k\overset{i}{a}{}_{k-1}^2 \ (i = \overline{1,\ell}; \ k = \overline{0,m_i}).$$

Substituting these equalities in (1.38)–(1.39), we find that there takes place

Theorem 7.1. *Lie operator of representation of the group $SO(2,\mathbb{R})$ in the space $E^{N+2}(x,y,A)$ of system (7.1)–(7.2) has the form*

$$X = y\frac{\partial}{\partial x} - x\frac{\partial}{\partial y} + D, \tag{7.7}$$

where

$$D = \sum_{i=1}^{\ell}\sum_{k=0}^{m_i}\left\{ [(m_i - k)\overset{i}{a}{}_{k+1}^1 - k\overset{i}{a}{}_{k-1}^1 + \overset{i}{a}{}_k^2]\frac{\partial}{\partial \overset{i}{a}{}_k^1} \right.$$

$$\left. + [-\overset{i}{a}{}_k^1 + (m_i - k)\overset{i}{a}{}_{k+1}^2 - k\overset{i}{a}{}_{k-1}^2]\frac{\partial}{\partial \overset{i}{a}{}_k^2} \right\}. \tag{7.8}$$

By A we denote a totality of coefficients of nonlinearities of the right parts of system (7.1)–(7.2).

Consequence 7.1. *Lie operator of representation of the group $SO(2, \mathbb{R})$ in the space of coefficients $E^N(A)$ of system (7.1)–(7.2) has the form (7.8).*

Remark 7.1. *Theorem 7.1 and Consequence 7.1 are known from the papers [33,34] for the system $s(1, m_1, ..., m_\ell)$ of general form.*

Remark 7.2. *Using defining equations (1.66), it can be verified that operators (7.7)–(7.8) are admitted by the Lyapunov system $s\mathcal{L}(1, m_1, ..., m_\ell)$ from (7.1)–(7.2).*

7.2 Comitants of System (7.1) − (7.2) for the Rotation Group and Concept of Functional Basis

Definition 7.1. *The polynomial $F(x, y, A)$ in coefficients of system (7.1)–(7.2) and the phase variables x, y is called algebraic comitant of this system under the rotation group $SO(2, \mathbb{R})$ from (7.3), if for all admissible A, x, y and φ the following identity holds*

$$F(\overline{x}, \overline{y}, B) = F(x, y, A), \tag{7.9}$$

where A (B) is a totality of coefficients of system (7.1)–(7.2) ((7.4)–(7.5)), and (x, y), $(\overline{x}, \overline{y})$ are the phase variables of the same systems.
If the comitant F of system (7.1)–(7.2) does not depend on the phase variables x, y, then, according to [38], it is called invariant of the indicated system under the rotation group.

Definition 7.2. *A totality of algebraic comitants of system (7.1)–(7.2) with respect to the rotation group*

$$\{F_\alpha(x, y, A), \ \alpha \in \mathbb{N}^+\} \tag{7.10}$$

is called a functional basis of comitants of indicated system under the group $SO(2, \mathbb{R})$ if any comitant $F(x, y, A)$ of considered system under the rotation group can be represented as unambiguous function of comitants (7.10).
From the papers [33,34] it follows that there takes place

Theorem 7.2. *In order that the polynomial $F(x, y, A)$ to be a comitant of system (7.1)–(7.2) under the rotation group (7.3), i.e., to satisfy equality (7.9), it is necessary and sufficient for it to satisfy the equation*

$$X(F) = 0, \tag{7.11}$$

where X is a Lie operator from (7.7)–(7.8).

Using Theorem 7.2, it is easy to verify that there takes place

Consequence 7.2. *In order that the polynomial $I(A)$ to be an invariant of system (7.1)–(7.2) under the rotation group (7.3), it is necessary and sufficient for it to satisfy the equation*

$$D(I) = 0, \tag{7.12}$$

where D is a Lie operator from (7.8).

Remark 7.3. *From (7.11) and (7.12), it follows that comitants and invariants of the Lyapunov system $s\mathcal{L}(1, m_1, ..., m_\ell)$ (7.1)–(7.2) are solutions of homogeneous first-order partial differential equations with Lie operators (7.7) and (7.8). Note that the number of variables involved in operators (7.7) and (7.8) is equal, respectively, to*

$$N = 2\left(\sum_{i=1}^{\ell} m_i + \ell + 1\right) \tag{7.13}$$

and

$$N_1 = 2\left(\sum_{i=1}^{\ell} m_i + \ell\right). \tag{7.14}$$

From the general theory of equations of types (7.11) and (7.12) (see, e.g.,[27]) it is known that *maximal number* of functionally independent solutions is equal to $N - 1$ and $N_1 - 1$, respectively.

Example 7.1. Consider differential system (5.16) with quadratic nonlinearities written in a tensor notation.

It is known from the paper [37] that generators of Sibirsky algebra for system (5.16) are the following invariants and comitants:

$$I_1 = a_\alpha^\alpha, \ I_2 = a_\beta^\alpha a_\alpha^\beta, \ I_3 = a_p^\alpha a_{\alpha q}^\beta a_{\beta\gamma}^\gamma \varepsilon^{pq}, \ I_4 = a_p^\alpha a_{\beta q}^\beta a_{\alpha\gamma}^\gamma \varepsilon^{pq},$$

$$I_5 = a_p^\alpha a_{\gamma q}^\beta a_{\alpha\beta}^\gamma \varepsilon^{pq}, \ I_6 = a_p^\alpha a_\gamma^\beta a_{\alpha q}^\gamma a_{\beta\delta}^\delta \varepsilon^{pq},$$

$$I_7 = a_{pr}^\alpha a_{\alpha q}^\beta a_{\beta s}^\gamma a_{\gamma\delta}^\delta \varepsilon^{pq} \varepsilon^{rs}, \ I_8 = a_{pr}^\alpha a_{\alpha q}^\beta a_{\delta s}^\gamma a_{\beta\gamma}^\delta \varepsilon^{pq} \varepsilon^{rs},$$

$$I_9 = a_{pr}^\alpha a_{\beta q}^\beta a_{\gamma s}^\gamma a_{\alpha\delta}^\delta \varepsilon^{pq} \varepsilon^{rs}, \ I_{10} = a_p^\alpha a_\delta^\beta a_\nu^\gamma a_{\alpha q}^\delta a_{\beta\gamma}^\nu \varepsilon^{pq},$$

$$I_{11} = a_p^\alpha a_{qr}^\beta a_{\beta s}^\gamma a_{\alpha\gamma}^\delta a_{\delta\mu}^\mu \varepsilon^{pq} \varepsilon^{rs}, I_{12} = a_p^\alpha a_{qr}^\beta a_{\beta s}^\gamma a_{\alpha\delta}^\delta a_{\gamma\mu}^\mu \varepsilon^{pq} \varepsilon^{rs}, \tag{7.15}$$

$$I_{13} = a_p^\alpha a_{qr}^\beta a_{\gamma s}^\gamma a_{\alpha\beta}^\delta a_{\delta\mu}^\mu \varepsilon^{pq} \varepsilon^{rs}, I_{14} = a_p^\alpha a_r^\beta a_{\alpha q}^\gamma a_{\beta s}^\delta a_{\gamma\delta}^\mu a_{\mu\nu}^\nu \varepsilon^{pq} \varepsilon^{rs},$$

$$I_{15} = a_{pr}^\alpha a_{qk}^\beta a_{\alpha s}^\gamma a_{\delta l}^\delta a_{\beta\gamma}^\mu a_{\mu\nu}^\nu \varepsilon^{pq} \varepsilon^{rs} \varepsilon^{kl},$$

$$I_{16} = a_p^\alpha a_r^\beta a_\delta^\gamma a_{\alpha q}^\delta a_{\beta s}^\mu a_{\gamma\tau}^\nu a_{\mu\nu}^\tau \varepsilon^{pq} \varepsilon^{rs}.$$

$$K_1 = a_{\alpha\beta}^{\alpha} x^{\beta}, \quad K_2 = a_{\alpha}^{p} x^{\alpha} x^{q} \varepsilon_{pq}, \quad K_3 = a_{\beta}^{\alpha} a_{\alpha\gamma}^{\beta} x^{\gamma}, \quad K_4 = a_{\gamma}^{\alpha} a_{\alpha\beta}^{\beta} x^{\gamma},$$

$$K_5 = a_{\alpha\beta}^{p} x^{\alpha} x^{\beta} x^{q} \varepsilon_{pq}, \quad K_6 = a_{\alpha\beta}^{\alpha} a_{\gamma\delta}^{\beta} x^{\gamma} x^{\delta}, \quad K_7 = a_{\beta\gamma}^{\alpha} a_{\alpha\delta}^{\beta} x^{\gamma} x^{\delta},$$

$$K_8 = a_{\gamma}^{\alpha} a_{\delta}^{\beta} a_{\alpha\beta}^{\gamma} x^{\delta}, \quad K_9 = a_{\alpha p}^{\alpha} a_{\gamma q}^{\beta} a_{\beta\delta}^{\gamma} x^{\delta} \varepsilon^{pq}, \quad K_{10} = a_{\alpha p}^{\alpha} a_{\delta q}^{\beta} a_{\beta\gamma}^{\gamma} x^{\delta} \varepsilon^{pq},$$

$$K_{11} = a_{\alpha}^{p} a_{\beta\gamma}^{\alpha} x^{\beta} x^{\gamma} x^{q} \varepsilon_{pq}, \quad K_{12} = a_{\beta}^{\alpha} a_{\alpha\gamma}^{\beta} a_{\delta\mu}^{\gamma} x^{\delta} x^{\mu},$$

$$K_{13} = a_{\gamma}^{\alpha} a_{\alpha\beta}^{\beta} a_{\delta\mu}^{\gamma} x^{\delta} x^{\mu}, \quad K_{14} = a_{p}^{\alpha} a_{\alpha q}^{\beta} a_{\beta\delta}^{\gamma} a_{\gamma\mu}^{\delta} x^{\mu} \varepsilon^{pq},$$

$$K_{15} = a_{p}^{\alpha} a_{\alpha q}^{\beta} a_{\beta\mu}^{\gamma} a_{\gamma\delta}^{\delta} x^{\mu} \varepsilon^{pq}, \quad K_{16} = a_{p}^{\alpha} a_{\beta q}^{\beta} a_{\alpha\mu}^{\gamma} a_{\gamma\delta}^{\delta} x^{\mu} \varepsilon^{pq},$$

$$K_{17} = a_{\beta\nu}^{\alpha} a_{\alpha\gamma}^{\beta} a_{\delta\mu}^{\gamma} x^{\delta} x^{\mu} x^{\nu}, \quad K_{18} = a_{\mu p}^{\alpha} a_{\alpha q}^{\beta} a_{\beta\nu}^{\gamma} a_{\gamma\delta}^{\delta} x^{\mu} x^{\nu} \varepsilon^{pq},$$

$$K_{19} = a_{p}^{\alpha} a_{\nu}^{\beta} a_{\alpha q}^{\gamma} a_{\beta\mu}^{\delta} a_{\gamma\delta}^{\mu} x^{\nu} \varepsilon^{pq}, \quad K_{20} = a_{pr}^{\alpha} a_{\nu q}^{\beta} a_{\alpha s}^{\gamma} a_{\beta\gamma}^{\delta} a_{\delta\mu}^{\mu} x^{\nu} \varepsilon^{pq} \varepsilon^{rs}.$$

$$(7.16)$$

We write system (5.16) in the form (7.1)–(7.2). According to notation (5.18), it will take the form

$$\dot{x} = y + gx^2 + 2hxy + ky^2,$$
$$\dot{y} = -x + \ell x^2 + 2mxy + ny^2,$$
$$(7.17)$$

for which Lie operators (7.7)–(7.8) will have the form

$$X = y\frac{\partial}{\partial x} - x\frac{\partial}{\partial y} + D,$$
$$(7.18)$$

where

$$D = (2h + \ell)\frac{\partial}{\partial g} + (-g + k + m)\frac{\partial}{\partial h} + (2h + n)\frac{\partial}{\partial k}$$
$$+ (-g + 2m)\frac{\partial}{\partial \ell} + (-h - \ell + n)\frac{\partial}{\partial m} - (k + 2m)\frac{\partial}{\partial n}.$$
$$(7.19)$$

Then, taking into account notation (5.18) for invariants and comitants (7.15)–(7.16), obtained for system (7.17), we have

$$D(I_j) = 0 \, (j = \overline{1, 16}), \quad X(K_i) = 0 \, (i = \overline{1, 20}).$$
$$(7.20)$$

Using equalities (7.20) and Theorem 7.2, it is proved the following

Lemma 7.1. *Centro-affine invariants and comitants* (7.15)–(7.16) *of system* (5.16) *are invariants and comitants of the rotation group* $SO(2, \mathbb{R})$ *for system* (7.17).

The reciprocal of Lemma 7.1 is not always true.

7.3 General Formulas that Interconnect Coefficients of Comitants of the Lyapunov System $s\mathcal{L}(1, m_1, ..., m_\ell)$ Among Themselves with Respect to the Rotation Group

We write the comitant K of the Lyapunov system (7.1)–(7.2) with respect to the rotation group in the form

$$
\begin{aligned}
K =\; & A_0 x^m + A_1 x^{m-1}y + A_2 x^{m-2}y^2 + A_3 x^{m-3}y^3 + A_4 x^{m-4}y^4 \\
& + A_5 x^{m-5}y^5 + A_6 x^{m-6}y^6 + ... + A_{m-7}x^7 y^{m-7} + A_{m-6}x^6 y^{m-6} \\
& + A_{m-5}x^5 y^{m-5} + A_{m-4}x^4 y^{m-4} + A_{m-3}x^3 y^{m-3} + A_{m-2}x^2 y^{m-2} \\
& + A_{m-1}x^1 y^{m-1} + A_m y^m,
\end{aligned}
\tag{7.21}
$$

where A_i $(i = \overline{1, m})$ are polynomials in coefficients of system (7.1)–(7.2).

We consider equality (7.11) taking into account the form of Lie operator (7.7). Using (7.21) and Lie operator (7.7), we obtain

$$
X(K) = y\frac{\partial K}{\partial x} - x\frac{\partial K}{\partial y} + D(K).
\tag{7.22}
$$

Then the terms on the right side (7.22), considering (7.21), are written in the form

$$
\begin{aligned}
y\frac{\partial K}{\partial x} =\; & mA_0 x^{m-1} + (m-1)A_1 x^{m-2}y + (m-2)A_2 x^{m-3}y^2 \\
& + (m-3)A_3 x^{m-4}y^3 + (m-4)A_4 x^{m-5}y^4 + (m-5)A_5 x^{m-6}y^5 \\
& + (m-6)A_6 x^{m-7}y^6 + ... + 7A_{m-7}x^6 y^{m-6} + 6A_{m-6}x^5 y^{m-5} \\
& + 5A_{m-5}x^4 y^{m-4} + 4A_{m-4}x^3 y^{m-3} + 3A_{m-3}x^2 y^{m-2} \\
& + 2A_{m-2}xy^{m-1} + A_{m-1}y^m, \\[2mm]
-x\frac{\partial K}{\partial y} =\; & -A_1 x^m - 2A_2 x^{m-1}y - 3A_3 x^{m-2}y^2 - 4A_4 x^{m-3}y^3 \\
& - 5A_5 x^{m-4}y^4 - 6A_6 x^{m-5}y^5 - ... - (m-7)A_{m-7}x^8 y^{m-8} \\
& - (m-6)A_{m-6}x^7 y^{m-7} - (m-5)A_{m-5}x^6 y^{m-6} \\
& - (m-4)A_{m-4}x^5 y^{m-5} - (m-3)A_{m-3}x^4 y^{m-4} \\
& - (m-2)A_{m-2}x^3 y^{m-3} - (m-1)A_{m-1}x^2 y^{m-2} - mA_m xy^{m-1}, \\[2mm]
D(K) =\; & D(A_0)x^m + D(A_1)x^{m-1}y + D(A_2)x^{m-2}y^2 \\
& + D(A_3)x^{m-3}y^3 + D(A_4)x^{m-4}y^4 + D(A_5)x^{m-5}y^5 \\
& + D(A_6)x^{m-6}y^6 + ... + D(A_{m-7})x^7 y^{m-7} + D(A_{m-6})x^6 y^{m-6} \\
& + D(A_{m-5})x^5 y^{m-5} + D(A_{m-4})x^4 y^{m-4} + D(A_{m-3})x^3 y^{m-3} \\
& + D(A_{m-2})x^2 y^{m-2} + D(A_{m-1})x^1 y^{m-1} + D(A_m)y^m.
\end{aligned}
\tag{7.23}
$$

Considering relation (7.11) and the last expressions from $X(K) = 0$, we have the following equalities:

$$x^m : D(A_0) - A_1 = 0,$$
$$x^{m-1}y : mA_0 + 2A_2 + D(A_1) = 0,$$
$$x^{m-2}y^2 : (m-1)A_1 - 3A_3 + D(A_2) = 0,$$
$$x^{m-3}y^3 : (m-2)A_2 - 4A_4 + D(A_3) = 0,$$
$$x^{m-4}y^4 : (m-3)A_3 - 5A_5 + D(A_4) = 0,$$
$$x^{m-5}y^5 : (m-4)A_4 - 6A_6 + D(A_5) = 0,$$
$$x^{m-6}y^6 : (m-5)A_5 - 7A_7 + D(A_6) = 0,$$

$$x^{m-7}y^7 : (m-6)A_6 - 8A_8 + D(A_7) = 0,$$

$$\dots\dots\dots\dots\dots\dots\dots\dots\dots\dots\dots\dots\dots\dots$$

$$x^6 y^{m-6} : 7A_{m-7} - (m-5)A_{m-5} + D(A_{m-6}) = 0,$$
$$x^5 y^{m-5} : 6A_{m-6} - (m-4)A_{m-4} + D(A_{m-5}) = 0,$$
$$x^4 y^{m-4} : 5A_{m-5} - (m-3)A_{m-3} + D(A_{m-4}) = 0,$$
$$x^3 y^{m-3} : 4A_{m-4} - (m-2)A_{m-2} + D(A_{m-3}) = 0,$$
$$x^2 y^{m-2} : 3A_{m-3} - (m-1)A_{m-1} + D(A_{m-2}) = 0,$$
$$xy^{m-1} : 2A_{m-2} - (m)A_m + D(A_{m-1}) = 0,$$
$$y^m : A_{m-1} + D(A_m) = 0.$$

From this and Theorem 7.2, we obtain that there holds

Theorem 7.3. *Polynomial* (7.21) *is a comitant of the Lyapunov system* $s\mathcal{L}(1, m_1, ..., m_\ell)$ *from* (7.1) *with respect to the rotation group* $SO(2, \mathbb{R})$ *if and only if its coefficients satisfy the equalities*

$$A_1 = D(A_0), \quad D(A_m) = -A_{m-1},$$
$$A_k = \frac{1}{k}[D(A_{k-1}) + (m-k+2)A_{k-2}] \ (k = \overline{2, m}),$$

(7.24)

where A_i $(i = \overline{0, m})$ *are taken from* (7.21), *and* D *is from* (7.8).

For example, we can verify the validity of Theorem 7.3 on comitants (7.16) with notation (5.18) for system (7.17) with operators (7.19).

Consequence 7.3. *If the coefficient* A_0 *at the highest degree of* x *of the comitant* K *from* (7.21) *of the Lyapunov system* (7.1)–(7.2) *with respect to the rotation group* $SO(2, \mathbb{R})$ *and its degree with respect to* x *and* y *are known, then the remaining coefficients can be constructed by formulas* (7.24).

By analogy with the comitants of the group of centro-affine transformations for system (1.1)–(1.2), we will call the coefficient A_0 of the comitant K

from (7.21) of the Lyapunov system (7.1)–(7.2) with respect to the rotation group $SO(2, \mathbb{R})$ as *a semi-invariant*.

It can be easily verified that there takes place

Observation 7.1. *If the comitants $K_1, K_2, ..., K_r$ of the Lyapunov system (7.1)–(7.2) with respect to the group $SO(2, \mathbb{R})$ are functionally independent, then its semi-invariants can also be functionally independent.*

There from it follows

Remark 7.4. *The number of functionally independent semi-invariants for the functionally independent comitants $K_1, K_2, ..., K_r$ of the Lyapunov system (7.1)–(7.2) with respect to the group $SO(2, \mathbb{R})$ does not exceed r.*

7.4 On the Invariance of Focus Quantities in the Center and Focus Problem with Respect to the Rotation Group

In the paper [38, p. 84], it is shown that there takes place

Remark 7.5. *The conditions of existence of center for the Lyapunov system $s\mathcal{L}(1, m_1, ..., m_\ell)$ from (7.1)–(7.2) are invariants of this system with respect to the rotation group $SO(2, \mathbb{R})$.*

It is known that focus quantities for the Lyapunov system $s\mathcal{L}(1, m_1, ..., m_\ell)$ from (5.6) are ambiguously constructed since equations (5.5) with the Lyapunov function (5.4) contain arbitrary constants, which can take different values.

We will show on some examples that focus quantities in the center and focus problem for the Lyapunov system $s\mathcal{L}(1, m_1, ..., m_\ell)$ from (7.1)–(7.2) can be *invariants* and *semi-invariants* of this system with respect to the rotation group $SO(2, \mathbb{R})$. For this, we use the results obtained in Section 7.3.

Consider the Lyapunov system $s\mathcal{L}(1, 2)$ from (7.17) and Lie operator (7.18)–(7.19), admitted by this system.

In the paper of A. P. Sadovsky [35, p. 110], three focus quantities are given that solve the center and focus problem for the system $s\mathcal{L}(1, 2)$ from (7.17) and have the following form:

$$L_1 S = \frac{1}{2}(-gh - hk + gl + lm - kn + mn),$$

$$L_2 S = \frac{1}{24}(62g^3h - 2gh^3 + 119g^2hk - 2h^3k + 62ghk^2 + 5hk^3$$

$$-62g^3l + 27gh^2l - 63g^2kl + 29h^2kl - 15gk^2l - 8ghl^2 + 15hkl^2$$

$$-5gl^3 + 101g^2hm + 138ghkm + 37hk^2m - 175g^2lm - 6h^2lm$$

$$-116gklm - 15k^2lm - 13hl^2m - 5l^3m + 54ghm^2 + 54hkm^2$$
$$-159glm^2 - 53klm^2 - 46lm^3 + 6g^3n + 37gh^2n + 72g^2kn + 39h^2kn$$
$$+57gk^2n + 5k^3n + 14ghln + 68hkln - 33gl^2n + 15kl^2n - 72g^2mn$$
$$-6h^2mn + 34gkmn + 32k^2mn - 42hlmn - 38l^2mn - 109gm^2n$$
$$-3km^2n - 46m^3n + 48ghn^2 + 79hkn^2 - 48gln^2 + 39kln^2$$
$$-29hmn^2 - 71lmn^2 - 6gn^3 + 38kn^3 - 38mn^3),$$

$$L_3S = \frac{1}{2304}(-44725g^5h + 11142g^3h^3 - 88gh^5 - 124537g^4hk$$
$$+30842g^2h^3k - 88h^5k - 121728g^3hk^2 + 27186gh^3k^2 - 45492g^2hk^3$$
$$+7486h^3k^3 - 2651ghk^4 + 925hk^5 + 44725g^5l - 51066g^3h^2l$$
$$-1216gh^4l + 84372g^4kl - 95044g^2h^2kl - 1704h^4kl + 53320g^3k^2l$$
$$-44602gh^2k^2l + 10096g^2k^3l - 2880h^2k^3l - 465gk^4l + 28368g^3hl^2$$
$$-3362gh^3l^2 + 7708g^2hkl^2 - 5802h^3kl^2 - 13582ghk^2l^2 - 3650hk^3l^2$$
$$+2436g^3l^3 + 1858gh^2l^3 + 9332g^2kl^3 - 1700h^2kl^3 + 3650gk^2l^3$$
$$-875ghl^4 + 465hkl^4 - 925gl^5 - 157320g^4hm + 7528g^2h^3m$$
$$-356424g^3hkm + 19904gh^3km - 264096g^2hk^2m + 12376h^3k^2m$$
$$-67288ghk^3m - 2296hk^4m + 211165g^4lm - 72038g^2h^2lm$$
$$+888h^4lm + 310816g^3klm - 109100gh^2klm + 144112g^2k^2lm$$
$$-26630h^2k^2lm + 17812gk^3lm - 465k^4lm + 67344g^2hl^2m$$
$$+2776h^3l^2m + 12024ghkl^2m - 7552hk^2l^2m + 14808g^2l^3m$$
$$+978h^2l^3m + 17844gkl^3m + 3650k^2l^3m - 1800hl^4m - 925l^5m$$
$$-214106g^3hm^2 - 4560gh^3m^2 - 214106g^2hkm^2 - 4560h^3km^2$$
$$-29106ghk^2m^2 - 29106hk^3m^2 + 389610g^3lm^2 - 16864gh^2lm^2$$
$$+249820g^2klm^2 - 32896h^2klm^2 + 131634gk^2lm^2 + 7716k^3lm^2$$
$$+60798ghl^2m^2 + 9126hkl^2m^2 + 15478gl^3m^2 + 8512kl^3m^2$$
$$-131528g^2hm^3 - 188448ghkm^3 - 56920hk^2m^3 + 350410g^2lm^3$$
$$+7632h^2lm^3 + 83832gklm^3 + 40842k^2lm^3 + 15912hl^2m^3$$
$$+3106l^3m^3 - 29432ghm^4 - 29432hkm^4 + 152800glm^4$$
$$+60456klm^4 + 25560lm^5 - 4560g^5n - 49528g^3h^2n$$
$$-2168gh^4n - 56129g^4kn - 71382g^2h^2kn - 2656h^4kn$$
$$-86332g^3k^2n - 8356gh^2k^2n - 42376g^2k^3n + 11242h^2k^3n$$
$$-3576gk^4n + 925k^5n - 8288g^3hln - 21396gh^3ln$$
$$-107608g^2hkln - 28148h^3kln - 78028ghk^2ln - 6580hk^3ln$$

$$+41832g^3l^2n - 3496gh^2l^2n + 29628g^2kl^2n - 28690h^2kl^2n$$
$$-7552gk^2l^2n - 3650k^3l^2n + 7132ghl^3n - 5780hkl^3n$$
$$-520gl^4n + 465kl^4n + 44753g^4mn - 87182g^2h^2mn$$
$$+888h^4mn - 78864g^3kmn - 108972gh^2kmn - 155168g^2k^2mn$$
$$-11358h^2k^2mn - 60956gk^3mn - 3221k^4mn + 51096g^2hlmn$$
$$+9632h^3lmn - 124544ghklmn - 50104hk^2lmn + 144264g^2l^2mn$$
$$+16334h^2l^2mn + 63884gkl^2mn - 1522k^2l^2mn + 5272hl^3mn$$
$$-1445l^4mn + 173116g^3m^2n - 43264gh^2m^2n + 17830g^2km^2n$$
$$-59296h^2km^2n + 82532gk^2m^2n - 25890k^3m^2n + 121900ghlm^2n$$
$$-26836hklm^2n + 146916gl^2m^2n + 39066kl^2m^2n + 222962g^2m^3n$$
$$+7632h^2m^3n - 103272gkm^3n - 18814k^2m^3n + 53184hlm^3n$$
$$+38574l^2m^3n + 125304gm^4n + 32960km^4n + 25560m^5n$$
$$-56942g^3hn^2 - 23378gh^3n^2 - 141270g^2hkn^2 - 27690h^3kn^2$$
$$-60328ghk^2n^2 + 6856hk^3n^2 + 48734g^3ln^2 - 38598gh^2ln^2$$
$$-11808g^2kln^2 - 81368h^2kln^2 - 38236gk^2ln^2 - 3700k^3ln^2$$
$$+22272ghl^2n^2 - 43344hkl^2n^2 + 12584gl^3n^2 - 4080kl^3n^2$$
$$-57664g^2hmn^2 + 6856h^3mn^2 - 184664ghkmn^2$$
$$-49232hk^2mn^2 + 202966g^2lmn^2 + 36790h^2lmn^2$$
$$+13848gklmn^2 - 28284k^2lmn^2 + 43488hl^2mn^2$$
$$+11604l^3mn^2 + 28654ghm^2n^2 - 68410hkm^2n^2$$
$$+243030glm^2n^2 + 34596klm^2n^2 + 37272hm^3n^2 + 75942lm^3n^2$$
$$-288g^3n^3 - 43772gh^2n^3 - 48542g^2kn^3 - 64906h^2kn^3$$
$$-30696gk^2n^3 + 3100k^3n^3 + 4176ghln^3 - 85408hkln^3$$
$$+31676gl^2n^3 - 20456kl^2n^3 + 55598g^2mn^3 + 21434h^2mn^3$$
$$-58400gkmn^3 - 31408k^2mn^3 + 69384hlmn^3 + 39200l^2mn^3$$
$$+100760gm^2n^3 - 6790km^2n^3 + 40474m^3n^3 - 18481ghn^4$$
$$-56701hkn^4 + 25249gln^4 - 30484kln^4 + 32968hmn^4$$
$$+43993lmn^4 + 3408gn^5 - 16917kn^5 + 16917mn^5).$$

In L_iS, the letter S emphasizes that L_iS are focus quantities of A. P. Sadovsky.

Remark 7.6. *Note that for focus quantities* (7.25) *using Lie operator* (7.19), *we find*

$$D(L_1S) = 0, \ D(L_2S) \neq 0, \ D(L_3S) \neq 0,$$

whence it follows that L_1S is an invariant of the system $s\mathcal{L}(1,2)$ under the rotation group $SO(2,\mathbb{R})$, and L_2S, L_3S are not.

In view of Theorems 7.2 and 7.3 and operators (7.18)–(7.19), we find that to the focus quantities L_2S and L_3S from (7.25), the following comitants of the rotation group correspond:

$$K(L_2S) = L_2Sx^4 + A_1x^3y + A_2x^2y^2 + A_3xy^3 + A_4y^4, \qquad (7.25)$$

where

$$A_1 = D(L_2S), \; A_2 = \frac{1}{2}[D(A_1) + 4L_2S],$$

$$A_3 = \frac{1}{3}[D(A_3) + 2A_2], \; A_4 = \frac{1}{4}[D(A_3) + 2A_2], \qquad (7.26)$$

and

$$K(L_3S) = L_3Sx^{12} + A_1x^{11}y + A_2x^{10}y^2 + A_3x^9y^3 + A_4x^8y^4$$
$$+A_5x^7y^5 + A_6x^6y^6 + A_7x^5y^7 + A_8x^4y^8 + A_9x^3y^9 + A_{10}x^2y^{10} \qquad (7.27)$$
$$+A_{11}xy^{11} + A_{12}y^{12},$$

for

$$A_1 = D(L_3S), \; A_2 = \frac{1}{2}[D(A_1) + 12L_3S], \; A_3 = \frac{1}{3}[D(A_2) + 11A_1],$$

$$A_4 = \frac{1}{4}[D(A_3) + 10A_2], \; A_5 = \frac{1}{5}[D(A_4) + 9A_3],$$

$$A_6 = \frac{1}{6}[D(A_5) + 8A_4], \; A_7 = \frac{1}{7}[D(A_6) + 7A_5],$$

$$A_8 = \frac{1}{8}[D(A_7) + 6A_6], \; A_9 = \frac{1}{9}[D(A_8) + 5A_7], \qquad (7.28)$$

$$A_{10} = \frac{1}{10}[D(A_9) + 4A_8], \; A_{11} = \frac{1}{11}[D(A_{10}) + 3A_9],$$

$$A_{12} = \frac{1}{12}[D(A_{11}) + 2A_{10}].$$

Using Lie operator (7.18)–(7.19) from (7.26)–(7.27) and (7.28)–(7.29) we find

$$X[K(L_2S)] = X[K(L_3S)] = 0. \qquad (7.29)$$

Therefore, from Remark 7.6 and equalities (7.30), we obtain that the following holds:

Lemma 7.2. *The focus quantity L_1S is an invariant, but L_2S and L_3S are semi-invariants of the system $s\mathcal{L}(1,2)$ from (7.17) with respect to the rotation group $SO(2,\mathbb{R})$.*

Remark 7.7. *The other three focus quantities that solve the center and focus problem for the system $s\mathcal{L}(1,2)$ from (7.17) were introduced to us by Professor Iu. F. Calin, built by another method. They have the form*

$$L_1C = L_1S = \frac{1}{2}(-gh - hk + gl + lm - kn + mn),$$

$$L_2C = \frac{1}{24}(53g^3h + 16gh^3 + 95g^2hk + 16h^3k + 47ghk^2 + 5hk^3$$

$$-53g^3l + 18gh^2l - 48g^2kl + 38h^2kl - 15gk^2l - 17ghl^2 + 15hkl^2$$

$$-5gl^3 + 62g^2hm + 84ghkm + 22hk^2m - 127g^2lm - 24h^2lm - 86gklm$$

$$-15k^2lm - 22hl^2m - 5l^3m + 24ghm^2 + 24hkm^2 - 90glm^2 - 38klm^2$$

$$-16lm^3 + 6g^3n + 70gh^2n + 63g^2kn + 90h^2kn + 42gk^2n + 5k^3n$$

$$-10ghln + 86hkln - 42gl^2n + 15kl^2n - 63g^2mn - 24h^2mn + 10gkmn$$

$$+17k^2mn - 84hlmn - 47l^2mn - 70gm^2n - 18km^2n - 16m^3n + 63ghn^2$$

$$+127hkn^2 - 63gln^2 + 48kln^2 - 62hmn^2 - 95lmn^2 - 6gn^3$$

$$+53kn^3 - 53mn^3),$$

$$L_3C = \frac{1}{2304}(-31393g^5h - 17022g^3h^3 - 1144gh^5 - 77985g^4hk$$

$$-27330g^2h^3k - 1144h^5k - 67264g^3hk^2 - 8666gh^3k^2 - 21292g^2hk^3$$

$$+1642h^3k^3 + 305ghk^4 + 925hk^5 + 31393g^5l - 27138g^3h^2l - 10960gh^4l$$

$$+50720g^4kl - 70216g^2h^2kl - 13080h^4kl + 30636g^3k^2l - 41898gh^2k^2l$$

$$+5940g^2k^3l - 5844h^2k^3l - 465gk^4l + 35076g^3hl^2 - 1706gh^3l^2$$

$$+16672g^2hkl^2 - 14946h^3kl^2 - 6030ghk^2l^2 - 3650hk^3l^2 + 828g^3l^3$$

$$+7270gh^2l^3 + 6664g^2kl^3 - 3560h^2kl^3 + 3650gk^2l^3 + 265ghl^4 + 465hkl^4$$

$$-925gl^5 - 79756g^4hm - 30792g^2h^3m - 164120g^3hkm - 37792gh^3km$$

$$-106536g^2hk^2m - 7000h^3k^2m - 21512ghk^3m + 660hk^4m + 119405g^4lm$$

$$+4582g^2h^2lm + 3096h^4lm + 157960g^3klm - 48148gh^2klm$$

$$+72848g^2k^2lm - 19738h^2k^2lm + 9500gk^3lm - 465k^4lm + 74424g^2hl^2m$$

$$+12424h^3l^2m + 31600ghkl^2m + 6844g^2l^3m + 7530h^2l^3m + 12508gkl^3m$$

$$+3650k^2l^3m - 660hl^4m - 925l^5m - 74114g^3hm^2 - 15376gh^3m^2$$

$$-127550g^2hkm^2 - 15376h^3km^2 - 60966ghk^2m^2 - 7530hk^3m^2$$

$$+168022g^3lm^2 + 41568gh^2lm^2 + 177140g^2klm^2 + 57158gk^2lm^2$$

$$+3560k^3lm^2 + 52258ghl^2m^2 + 19738hkl^2m^2 + 4374gl^3m^2 + 5844kl^3m^2$$

$$= 29432g^2hm^3 - 41856ghkm^3 - 12424hk^2m^3 + 104770g^2lm^3$$

$$+15376h^2lm^3 + 82980gklm^3 + 14946k^2lm^3 + 7000hl^2m^3 - 1642l^3m^3$$

$$-3096ghm^4 - 3096hkm^4 + 25904glm^4 + 13080klm^4 + 1144lm^5$$

$$-4128g^5n - 89564g^3h^2n - 23784gh^4n - 41357g^4kn - 183390g^2h^2kn$$

$$-25904h^4kn - 51912g^3k^2n - 91176gh^2k^2n - 21132g^2k^3n - 4374h^2k^3n$$

$$-620gk^4n + 925k^5n + 30248g^3hln - 43620gh^3ln - 64504g^2hkln$$

$$-82980h^3kln - 67900ghk^2ln - 12508hk^3ln + 43332g^3l^2n + 17324gh^2l^2n$$

$$+30812g^2kl^2n - 57158h^2kl^2n - 3650k^3l^2n + 21092ghl^3n - 9500hkl^3n$$

$$+620gl^4n + 465kl^4n + 32573g^4mn - 89970g^2h^2mn + 3096h^4mn$$
$$-32600g^3kmn - 175220gh^2kmn - 61672g^2k^2mn - 52258h^2k^2mn$$
$$-21092gk^3mn - 265k^4mn + 162352g^2hlmn + 41856h^3lmn$$
$$-31600hk^2lmn + 119932g^2l^2mn + 60966h^2l^2mn + 67900gkl^2mn$$
$$+6030k^2l^2mn + 21512hl^3mn - 305l^4mn + 93296g^3m^2n + 38762g^2km^2n$$
$$-41568h^2km^2n - 17324gk^2m^2n - 7270k^3m^2n + 175220ghlm^2n$$
$$+48148hklm^2n + 91176gl^2m^2n + 41898kl^2m^2n + 78650g^2m^3n$$
$$+15376h^2m^3n + 43620gkm^3n + 1706k^2m^3n + 37792hlm^3n$$
$$+8666l^2m^3n + 23784gm^4n + 10960km^4n + 1144m^5n$$
$$-68550g^3hn^2 - 78650gh^3n^2 - 204974g^2hkn^2 - 104770h^3kn^2$$
$$-119932ghk^2n^2 - 6844hk^3n^2 + 59766g^3ln^2 - 38762gh^2ln^2$$
$$-177140h^2kln^2 - 30812gk^2ln^2 - 6664k^3ln^2 + 61672ghl^2n^2$$
$$-72848hkl^2n^2 + 21132gl^3n^2 - 5940kl^3n^2 + 29432h^3mn^2$$
$$-162352ghkmn^2 - 74424hk^2mn^2 + 204974g^2lmn^2$$
$$+127550h^2lmn^2 + 64504gklmn^2 - 16672k^2lmn^2$$
$$+106536hl^2mn^2 + 21292l^3mn^2 + 89970ghm^2n^2$$
$$-4582hkm^2n^2 + 183390glm^2n^2 + 70216klm^2n^2$$
$$+30792hm^3n^2 + 27330lm^3n^2 - 93296gh^2n^3$$
$$-59766g^2kn^3 - 168022h^2kn^3 - 43332gk^2n^3 - 828k^3n^3$$
$$+32600ghln^3 - 157960hkln^3 + 51912gl^2n^3 - 30636kl^2n^3$$
$$+68550g^2mn^3 + 74114h^2mn^3 - 30248gkmn^3 - 35076k^2mn^3$$
$$+164120hlmn^3 + 67264l^2mn^3 + 89564gm^2n^3 + 27138km^2n^3$$

$$+17022m^3n^3 - 32573ghn^4 - 119405hkn^4 + 41357gln^4$$
$$-50720kln^4 + 79756hmn^4 + 77985lmn^4 + 4128gn^5 \tag{7.30}$$
$$-31393kn^5 + 31393mn^5).$$

In L_iC, the letter C emphasizes that L_iC are focus quantities of Iu. F. Calin. Using the operator D from (7.19) and (7.31), we find

$$D(L_1C) = D(L_2C) = D(L_3C) = 0.$$

Therefore, from Remark 7.3, we have that focus quantities (7.31) are invariants of the system $s\mathcal{L}(1,2)$ from (7.17) with respect to the rotation group $SO(2,\mathbb{R})$.

In a similar way that was used for focus quantities of the system $s\mathcal{L}(1,2)$ one can find sequences of focus quantities for the systems $s\mathcal{L}(1,3)$, $s\mathcal{L}(1,4)$, $s\mathcal{L}(1,2,3)$ etc., which are invariants and semi-invariants of these systems relative to the rotation group $SO(2,\mathbb{R})$.

7.5 On the Upper Bound of the Number of Algebraically Independent Focus Quantities that Take Part in Solving the Center and Focus Problem for the Lyapunov System $s\mathcal{L}(1, m_1, ..., m_\ell)$

Consider the set of comitants $\{F(x, y, A)\}$ of the Lyapunov system $s\mathcal{L}(1, m_1, ..., m_\ell)$ of type (7.9)–(7.10) with respect to the rotation group $SO(2, \mathbb{R})$. It follows from Theorem 7.2 that equality (7.11) holds for them, where X is a Lie operator of linear representation of the rotation group $SO(2, \mathbb{R})$ in the space $E^{N+2}(x, y, A)$ of the system $s\mathcal{L}(1, m_1, ..., m_\ell)$. Then, according to Remark 7.3, we have that maximal number of functionally independent solutions of equation (7.11) is equal to

$$\mu = 2 \left(\sum_{i=1}^{\ell} m_i + \ell \right) + 1, \tag{7.31}$$

that corresponds to the number of coefficients in nonlinear part of the system $s\mathcal{L}(1, m_1, ..., m_\ell)$ plus one. Since the focal quantities in the center and focus problem for the Lyapunov system $s\mathcal{L}(1, m_1, ..., m_\ell)$ from (7.9)–(7.10), in general, are semi-invariants of this system with respect to the rotation group $SO(2, \mathbb{R})$, then according to Remark 7.4, the number θ of functionally independent focus quantities for this system does not exceed the number μ from (7.32).

Thus, we obtain

Theorem 7.4. *The number θ of functionally independent focus quantities in the center and focus problem for the Lyapunov system $s\mathcal{L}(1, m_1, ..., m_\ell)$ does not exceed the number μ from (7.32), i.e., the following inequality holds*

$$\theta \leq \mu = 2 \left(\sum_{i=1}^{\ell} m_i + \ell \right) + 1. \tag{7.32}$$

Hence, we have

Remark 7.8. *The number θ of functionally independent focus quantities from (44.2) in the center and focus problem for the Lyapunov system $s\mathcal{L}(1, m_1, ..., m_\ell)$ from (7.9)–(7.10) does not exceed maximal number of all possible nonzero coefficients of this system minus one.*

Since the number of focus quantities that can solve the center and focus problem for the system $s\mathcal{L}(1, m_1, ..., m_\ell)$ could be considered functionally independent, then it makes sense to assume that the following is true:

Hypothesis 7.1. *The number of essential focus quantities that solve the center and focus problem for the Lyapunov system $s\mathcal{L}(1, m_1, ..., m_\ell)$ from (7.9)–(7.10) does not exceed maximal number of all possible nonzero coefficients of this system minus one.*

7.6 Comments to Chapter Seven

As it was mentioned earlier in Section 5.2, the system $s(1, m_1, ..., m_\ell)$ is written in the form

$$\dot{x} = cx + dy + \sum_{i=1}^{\ell} P_{m_i}(x, y), \ \dot{y} = ex + fy + \sum_{i=1}^{\ell} Q_{m_i}(x, y), \qquad (7.33)$$

and the Lyapunov system $s\mathcal{L}(1, m_1, ..., m_\ell)$ has the form

$$\dot{x} = y + \sum_{i=1}^{\ell} P_{m_i}(x, y), \ \dot{y} = -x + \sum_{i=1}^{\ell} Q_{m_i}(x, y), \qquad (7.34)$$

where P_{m_i} and Q_{m_i} are homogeneous polynomials of degree m_i with respect to phase variables x and y. The set $\{1, m_1, ..., m_\ell\}$ consists of a finite number of distinct positive integers.

In *Hypothesis* 6.3, it was noted that the number of the essential center conditions that solve the center and focus problem for system (7.34) with a singular point of center or focus type at the origin does not exceed the number

$$\varrho = 2\left(\sum_{i=1}^{\ell} m_i + \ell\right) + 3, \qquad (7.35)$$

but *Hypothesis* 7.1 claims that the same number of the essential center conditions for system (7.35) is not greater than the number

$$\mu = 2\left(\sum_{i=1}^{\ell} m_i + \ell\right) + 1. \qquad (7.36)$$

Since the system $s(1, m_1, ..., m_\ell)$ from (7.34), in the case of a singular point of center or focus type at the origin, can be reduced to the Lyapunov form $s\mathcal{L}(1, m_1, ..., m_\ell)$ from (7.35) by a centro-affine transformation and time rescaling, then (7.36) implies that the number of the essential center conditions for system $s(1, m_1, ..., m_\ell)$ and for system $s\mathcal{L}(1, m_1, ..., m_\ell)$ cannot exceed μ.

Hence, we can claim the following

Main Hypothesis 7.1. *The number of essential focus quantities ω that solves the center and focus problem for system (7.34) with a singular point of center or focus type at the origin does not exceed the number μ from (7.36), i.e. does not exceed maximal number of all possible nonzero coefficients of this system minus one.*

This means an improvement by two units of the upper bound of the number of essential focus quantities ω, which are involved in solving the center and focus problem for system (7.33).

The main result of this book can be formulated concisely as follows: Let
$$N = 2 \sum_{i=0}^{\ell} (m_i + 1)$$ *be the maximal possible number of nonzero coefficients of system* (7.33), *where* $m_0 = 1$. *Then the number of algebraically independent focus quantities from* (5.1) *does not exceed* $N - 1$, *which is the Krull dimension of Sibirsky algebra of comitants for system* (7.33). *It is also shown that this number could be reduced to* $N - 3$, *which is the Krull dimension of Sibirsky algebra of invariants for the mentioned system. It is assumed that the number of essential focus quantities* ω *from* (5.2) *does not exceed* $N - 1$ *and it can be improved up to* $N - 3$, *and their construction will begin with the first algebraically independent nonzero focus quantities obtained consecutively up to the mentioned estimations.*

Appendix 1

The Polynomials $R_k(b,e)$, that Define $N_{1,4}(u,b,e)$

$R_0(b,e) = 1 - e^2 + 4e^4 + e^6 + 18e^8 + 11e^{10} + 35e^{12} + 13e^{14} + 35e^{16}$
$+11e^{18} + 18e^{20} + e^{22} + 4e^{24} - e^{26} + e^{28} + b(e^2 + 5e^4 + 13e^6 + 26e^8$
$+29e^{10} + 40e^{12} + 19e^{14} + 36e^{16} - 5e^{18} + 6e^{20} - 15e^{22} + e^{24} - 5e^{26}$
$+2e^{28} - 2e^{30}) + b^2(e^2 + 8e^4 + 16e^6 + 26e^8 + 27e^{10} + 20e^{12} + 12e^{14}$
$-11e^{16} - 29e^{18} - 31e^{20} - 22e^{22} - 11e^{24} - 4e^{26} - 2e^{28} - e^{30} + e^{32})$
$+b^3(e^2 + 10e^4 + 10e^6 + 24e^8 - 5e^{10} + 7e^{12} - 64e^{14} - 49e^{16} - 107e^{18}$
$-55e^{20} - 58e^{22} - 10e^{24} - 10e^{26} + 3e^{28} + e^{30}) + b^4(e^2 + 6e^4 + 9e^6$
$+10e^8 - 7e^{10} - 29e^{12} - 87e^{14} - 75e^{16} - 117e^{18} - 29e^{20} - 30e^{22}$
$+26e^{24} + 2e^{26} + 17e^{28} - 3e^{30} + 4e^{32}) + b^5(5e^4 + 3e^6 + 10e^8 - 21e^{10}$
$-38e^{12} - 82e^{14} - 76e^{16} - 72e^{18} + e^{20} + 20e^{22} + 44e^{24} + 32e^{26} + 17e^{28}$
$+6e^{30} + 2e^{32} - 2e^{34}) + b^6(2e^4 + 3e^6 - 2e^8 - 29e^{10} - 36e^{12} - 84e^{14}$
$-41e^{16} - 48e^{18} + 48e^{20} + 41e^{22} + 84e^{24} + 36e^{26} + 29e^{28} + 2e^{30} - 3e^{32}$
$-2e^{34}) + b^7(2e^4 - 2e^6 - 6e^8 - 17e^{10} - 32e^{12} - 44e^{14} - 20e^{16} - e^{18}$
$+72e^{20} + 76e^{22} + 82e^{24} + 38e^{26} + 21e^{28} - 10e^{30} - 3e^{32} - 5e^{34})$
$+b^8(-4e^6 + 3e^8 - 17e^{10} - 2e^{12} - 26e^{14} + 30e^{16} + 29e^{18} + 117e^{20} + 75e^{22}$
$+87e^{24} + 29e^{26} + 7e^{28} - 10e^{30} - 9e^{32} - 6e^{34} - e^{36}) + b^9(-e^8 - 3e^{10}$
$+10e^{12} + 10e^{14} + 58e^{16} + 55e^{18} + 107e^{20} + 49e^{22} + 64e^{24} - 7e^{26}$
$+5e^{28} - 24e^{30} - 10e^{32} - 10e^{34} - e^{36}) + b^{10}(-e^6 + e^8 + 2e^{10} + 4e^{12}$
$+11e^{14} + 22e^{16} + 31e^{18} + 29e^{20} + 11e^{22} - 12e^{24} - 20e^{26} - 27e^{28} - 26e^{30}$
$-16e^{32} - 8e^{34} - e^{36}) + b^{11}(2e^8 - 2e^{10} + 5e^{12} - e^{14} + 15e^{16} - 6e^{18} + 5e^{20}$
$-36e^{22} - 19e^{24} - 40e^{26} - 29e^{28} - 26e^{30} - 13e^{32} - 5e^{34} - e^{36}) + b^{12}(-e^{10}$
$+e^{12} - 4e^{14} - e^{16} - 18e^{18} - 11e^{20} - 35e^{22} - 13e^{24} - 35e^{26} - 11e^{28}$
$-18e^{30} - e^{32} - 4e^{34} + e^{36} - e^{38}),$
$R_1(b,e) = -2e + 4e^3 + e^5 + 15e^7 - 8e^9 + 21e^{11} - 18e^{13} + 26e^{15} - 27e^{17}$
$+6e^{19} - 19e^{21} + 7e^{23} - 6e^{25} + 2e^{27} - 2e^{29} + b(e + 5e^3 + 8e^5 + 5e^7 + e^9$
$+15e^{11} - 8e^{13} + 16e^{15} - 47e^{17} + 11e^{19} - 19e^{21} + 18e^{23} - 13e^{25} + 7e^{27}$
$-4e^{29} + 4e^{31}) + b^2(2e + 6e^3 + 2e^5 + 6e^7 + 9e^9 + 7e^{11} - 7e^{13} - 34e^{15}$
$-24e^{17} + 2e^{19} + 11e^{21} + 6e^{23} + 5e^{25} + 5e^{27} + 4e^{29} + 2e^{31} - 2e^{33}) + b^3(e$
$+4e^3 + 16e^7 - 11e^9 + 14e^{11} - 77e^{13} - e^{15} - 63e^{17} + 59e^{19} - 6e^{21}$
$+52e^{23} - 3e^{25} + 19e^{27} - 3e^{29} - e^{31}) + b^4(4e^3 + 5e^5 + 3e^7 - 9e^9$
$-26e^{11} - 60e^{13} + 4e^{15} - 35e^{17} + 88e^{19} - 11e^{21} + 51e^{23} - 24e^{25}$
$+30e^{27} - 20e^{29} + 8e^{31} - 8e^{33}) + b^5(4e^3 + 7e^7 - 23e^9 - 19e^{11} - 49e^{13}$
$+e^{15} + e^{17} + 73e^{19} + 12e^{21} + 30e^{23} - e^{25} - 9e^{27} - 16e^{29} - 10e^{31}$
$-5e^{33} + 4e^{35}) + b^6(e^3 + e^5 - 3e^7 - 16e^9 - 10e^{11} - 51e^{13} + 26e^{15}$
$-11e^{17} + 96e^{19} - 3e^{21} + 60e^{23} - 45e^{25} - 4e^{27} - 38e^{29} - 7e^{31} + e^{33}$

$$+3e^{35}) + b^7(-3e^5 + 2e^7 - 6e^9 - 21e^{11} - 23e^{13} + 18e^{15} + 26e^{17} + 73e^{19}$$
$$+7e^{21} + 11e^{23} - 39e^{25} - 15e^{27} - 39e^{29} + 7e^{31} - 4e^{33} + 6e^{35})$$
$$+b^8(6e^7 - 20e^9 - 24e^{13} + 61e^{15} + 9e^{17} + 88e^{19} - 49e^{21} + 20e^{23}$$
$$-56e^{25} - 18e^{27} - 25e^{29} - e^{31} + e^{33} + 6e^{35} + 2e^{37}) + b^9(-e^7 - e^9$$
$$+7e^{11} + 3e^{13} + 44e^{15} + 45e^{19} - 53e^{21} + 31e^{23} - 65e^{25} + 10e^{27} - 47e^{29}$$
$$+12e^{31} + 14e^{35} + e^{37}) + b^{10}(2e^7 - 2e^9 - e^{11} + 9e^{13} + 16e^{15} + 7e^{17}$$
$$-6e^{19} - 12e^{21} - 12e^{23} - 9e^{25} - 21e^{27} - 7e^{29} + 14e^{31} + 14e^{33} + 8e^{35})$$
$$+b^{11}(-4e^9 + 7e^{11} + e^{13} + 14e^{15} - 23e^{17} + 11e^{19} - 35e^{21} + 18e^{23}$$
$$-34e^{25} + 7e^{27} + 5e^{29} + 21e^{31} + 8e^{33} + 3e^{35} + e^{37}) + b^{12}(2e^{11} - 4e^{13}$$
$$-e^{15} - 15e^{17} + 8e^{19} - 21e^{21} + 18e^{23} - 26e^{25} + 27e^{27} - 6e^{29} + 19e^{31}$$
$$-7e^{33} + 6e^{35} - 2e^{37} + 2e^{39}),$$
$$R_2(b,e) = 4e^2 - 2e^4 + 7e^6 - 5e^8 + 27e^{10} - 13e^{12} + 20e^{14} - 29e^{16}$$
$$+14e^{18} - 17e^{20} + 5e^{22} - 14e^{24} + 3e^{26} - e^{28} + e^{30} + b(2e^2 - 3e^4 + 7e^6$$
$$+e^8 + 11e^{10} - 53e^{12} - 14e^{14} - 73e^{16} + 9e^{18} - 45e^{20} + 8e^{22} - 17e^{24}$$
$$+21e^{26} - 5e^{28} + 2e^{30} - 2e^{32}) + b^2(7e^6 - 12e^8 - 39e^{10} - 64e^{12} - 39e^{14}$$
$$-37e^{16} + 2e^{18} - 16e^{20} + 16e^{22} + 17e^{24} + 16e^{26} - 3e^{28} + e^{30} - e^{32} + e^{34})$$
$$+b^3(2e^2 - e^4 + e^6 - 42e^8 - 30e^{10} - 85e^{12} + 8e^{14} - 44e^{16} + 67e^{18} + e^{20}$$
$$+81e^{22} + 16e^{24} + 32e^{26} - 6e^{28} + e^{30} - e^{32}) + b^4(2e^2 - 5e^4 - 6e^6 - 30e^8$$
$$-18e^{10} - 53e^{12} + 65e^{14} - 3e^{16} + 159e^{18} + 44e^{20} + 143e^{22} + 9e^{24} + 40e^{26}$$
$$-47e^{28} + 5e^{30} - 7e^{32} + 4e^{34}) + b^5(e^2 - 5e^4 - 28e^8 - 11e^{10} - 14e^{12}$$
$$+84e^{14} + 57e^{16} + 172e^{18} + 51e^{20} + 89e^{22} - 23e^{24} - 25e^{26} - 43e^{28} + e^{30}$$
$$-4e^{32} + 2e^{34} - 2e^{36}) + b^6(-e^4 - e^6 - 26e^8 + e^{10} - 12e^{12} + 103e^{14} + 67e^{16}$$
$$+162e^{18} + 6e^{20} + 51e^{22} - 98e^{24} - 37e^{26} - 59e^{28} - 4e^{30} - 4e^{32} + 3e^{34})$$
$$+b^7(-e^6 - 13e^8 + 7e^{10} + 22e^{12} + 103e^{14} + 51e^{16} + 104e^{18} - 20e^{20} - 9e^{22}$$
$$-119e^{24} - 60e^{26} - 79e^{28} + 8e^{30} - 5e^{32} + 10e^{34} + e^{36}) + b^8(-2e^4$$
$$+2e^6 - 11e^8 + 26e^{10} + 23e^{12} + 87e^{14} + 79e^{18} - 92e^{20} - 24e^{22}$$
$$-141e^{24} - 68e^{26} - 61e^{28} + 12e^{30} + 3e^{32} + 14e^{34} + 3e^{36} - e^{38})$$
$$+b^9(3e^6 - 4e^8 + 22e^{10} + 5e^{12} + 55e^{14} - 30e^{16} + 30e^{18} - 137e^{20}$$
$$-41e^{22} - 183e^{24} - 39e^{26} - 53e^{28} + 43e^{30} + 10e^{32} + 18e^{34} - 2e^{36}$$
$$+e^{38}) + b^{10}(-e^8 + 7e^{10} - 2e^{12} - 4e^{14} - 53e^{16} - 42e^{18} - 104e^{20}$$
$$-79e^{22} - 99e^{24} - 10e^{26} + 11e^{28} + 45e^{30} + 12e^{32} + 10e^{34} + 4e^{36}$$
$$+3e^{38}) + b^{11}(e^6 - e^8 + 5e^{10} - 9e^{12} - 6e^{14} - 36e^{16} - 10e^{18} - 67e^{20}$$
$$-7e^{22} - 19e^{24} + 46e^{26} + 31e^{28} + 35e^{30} + 13e^{32} + 17e^{34} + 7e^{36})$$
$$+b^{12}(-2e^8 + 2e^{10} - 9e^{12} + 3e^{14} - 22e^{16} + 11e^{18} - 32e^{20} + 49e^{22}$$
$$-e^{24} + 69e^{26} + 15e^{28} + 43e^{30} + 8e^{32} + 19e^{34} - 2e^{36} + e^{38} - e^{40})$$
$$+b^{13}(e^{10} - e^{12} + e^{14} + e^{16} + 18e^{18} + 11e^{20} + 35e^{22} + 13e^{24} + 35e^{26}$$
$$+11e^{28} + 18e^{30} + e^{32} + 4e^{34} - e^{36} + e^{38}),$$
$$R_3(b,e) = -e + e^3 + e^5 - 3e^7 - 16e^9 - 38e^{11} - 45e^{13} - 46e^{15} - 56e^{17}$$
$$-50e^{19} - 32e^{21} - 14e^{23} - e^{27} - 2e^{29} + b(2e - e^3 - e^5 - 30e^7 - 35e^9$$
$$-87e^{11} - 44e^{13} - 95e^{15} - 44e^{17} - 33e^{19} + 27e^{21} + 15e^{23} + 22e^{25} - 5e^{27}$$
$$+3e^{29} + 4e^{31}) + b^2(e - 2e^3 - 14e^5 - 32e^7 - 48e^9 - 64e^{11} - 46e^{13}$$
$$-73e^{15} + 41e^{17} + 49e^{19} + 101e^{21} + 45e^{23} + 36e^{25} - e^{27} + 12e^{29} - 3e^{31}$$
$$-2e^{33}) + b^3(-3e^3 - 13e^5 - 31e^7 - 43e^9 - 45e^{11} - 10e^{13} + 77e^{15} + 156e^{17}$$
$$+197e^{19} + 174e^{21} + 105e^{23} + 42e^{25} + 7e^{27} - 5e^{29} - 5e^{31} + e^{33}) + b^4(-2e^3$$
$$-6e^5 - 31e^7 - 23e^9 - 46e^{11} + 110e^{13} + 113e^{15} + 271e^{17} + 189e^{19} + 148e^{21}$$
$$-16e^{23} - 19e^{25} - 61e^{27} - 10e^{29} - 6e^{31} - 7e^{33}) + b^5(-e^3 - 5e^5 - 20e^7$$

$$-23e^9 + 20e^{11} + 115e^{13} + 132e^{15} + 252e^{17} + 92e^{19} + 61e^{21} - 97e^{23} - 75e^{25}$$
$$-110e^{27} - 17e^{29} - 33e^{31} + 5e^{33} + 6e^{35}) + b^6(-4e^5 - 23e^7 + 14e^9 + 33e^{11}$$
$$+122e^{13} + 144e^{15} + 165e^{17} + 40e^{19} - 26e^{21} - 159e^{23} - 158e^{25} - 114e^{27}$$
$$-48e^{29} - 5e^{31} + 17e^{33} + 3e^{35} - e^{37}) + b^7(-4e^5 + e^7 + 14e^9 + 19e^{11}$$
$$+109e^{13} + 69e^{15} + 126e^{17} - 49e^{19} - 116e^{21} - 252e^{23} - 138e^{25} - 122e^{27}$$
$$+4e^{29} + 13e^{31} + 16e^{33} + 8e^{35}) + b^8(4e^5 - e^7 + e^9 + 31e^{11} + 52e^{13} + 31e^{15}$$
$$+17e^{17} - 147e^{19} - 199e^{21} - 230e^{23} - 154e^{25} - 84e^{27} + 24e^{29} + 6e^{31}$$
$$+33e^{33} + 11e^{35} + e^{37}) + b^9(-6e^7 + 21e^9 + 4e^{11} + 28e^{13} - 63e^{15} - 61e^{17}$$
$$-224e^{19} - 145e^{21} - 207e^{23} - 58e^{25} - 18e^{27} + 35e^{29} + 35e^{31} + 38e^{33}$$
$$+16e^{35} + 2e^{37} - e^{39}) + b^{10}(2e^7 - 8e^{11} + e^{13} - 51e^{15} - 36e^{17} - 122e^{19}$$
$$-33e^{21} - 59e^{23} + 84e^{25} + 23e^{27} + 97e^{29} + 44e^{31} + 48e^{33} + 11e^{35} + e^{37}$$
$$-2e^{39}) + b^{11}(-2e^7 + 3e^{11} - 8e^{13} - 24e^{15} - 26e^{17} - 21e^{19} + 20e^{21}$$
$$+58e^{23} + 56e^{25} + 84e^{27} + 82e^{29} + 50e^{31} + 28e^{33} + 2e^{35} - e^{37} + e^{39})$$
$$+b^{12}(4e^9 - 6e^{11} - 2e^{13} - 15e^{15} + 26e^{17} + 5e^{19} + 73e^{21} + 27e^{23} + 80e^{25}$$
$$+49e^{27} + 45e^{29} + 11e^{31} + 6e^{33} - 3e^{35} + 2e^{39}) + b^{13}(-2e^{11} + 4e^{13} + e^{15}$$
$$+15e^{17} - 8e^{19} + 21e^{21} - 18e^{23} + 26e^{25} - 27e^{27} + 6e^{29} - 19e^{31} + 7e^{33}$$
$$-6e^{35} + 2e^{37} - 2e^{39}),$$
$$R_4(b,e) = 3e^2 - 2e^4 - 2e^6 - 9e^8 + 8e^{10} - 17e^{12} + 11e^{14} - 43e^{16}$$
$$+26e^{18} - e^{20} + 25e^{22} - 6e^{24} + 6e^{26} - 3e^{28} + 4e^{30} + b(-2e^2 - e^4$$
$$-12e^6 + 7e^8 - 32e^{10} - 6e^{12} - 36e^{14} + 3e^{16} + 54e^{18} + 14e^{20} + 12e^{22}$$
$$-7e^{24} + 10e^{26} - 2e^{28} + 6e^{30} - 8e^{32}) + b^2(-2e^2 - 4e^4 - 3e^6 - 22e^8$$
$$-39e^{10} - 13e^{12} - 11e^{14} + 89e^{16} + 26e^{18} + 25e^{20} - 20e^{22} + 9e^{24}$$
$$-22e^{26} - 14e^{30} - 3e^{32} + 4e^{34}) + b^3(-4e^4 - 16e^6 - 30e^8 - 10e^{10}$$
$$-12e^{12} + 92e^{14} + 27e^{16} + 98e^{18} - 37e^{20} + 10e^{22} - 75e^{24} - 12e^{26}$$
$$-33e^{28} + 2e^{32}) + b^4(e^2 - 8e^4 - 22e^6 - 2e^8 - 20e^{10} + 92e^{12} + 44e^{14}$$
$$+88e^{16} + 32e^{18} - 72e^{20} - 49e^{22} - 54e^{24} - 19e^{26} - 32e^{28} + 17e^{30}$$
$$-12e^{32} + 16e^{34}) + b^5(-5e^4 - 10e^6 - 16e^8 + 27e^{10} + 61e^{12} + 59e^{14}$$
$$+96e^{16} - 31e^{18} - 71e^{20} - 89e^{22} - 54e^{24} - 54e^{26} + 46e^{28} + 7e^{30} + 32e^{32}$$
$$+10e^{34} - 8e^{36}) + b^6(-2e^4 - 13e^6 + 14e^8 + 10e^{10} + 49e^{12} + 73e^{14} + 35e^{16}$$
$$-14e^{18} - 113e^{20} - 84e^{22} - 105e^{24} + 48e^{26} + 22e^{28} + 62e^{30} + 23e^{32} - e^{34}$$
$$-4e^{36}) + b^7(-2e^4 + 2e^6 - e^8 + 8e^{10} + 81e^{12} + 25e^{14} + 32e^{16} - 99e^{18}$$
$$-108e^{20} - 101e^{22} + 18e^{24} + 17e^{26} + 67e^{28} + 58e^{30} + 4e^{32} + 11e^{34}$$
$$-10e^{36} - 2e^{38}) + b^8(-5e^6 + 39e^{10} + 28e^{12} + 20e^{14} - 39e^{16} - 77e^{18}$$
$$-133e^{20} + 10e^{22} - 29e^{24} + 68e^{26} + 77e^{28} + 33e^{30} + 30e^{32} - 2e^{34} - 15e^{36}$$
$$-5e^{38}) + b^9(17e^8 - 8e^{10} + 13e^{12} - 34e^{14} - 13e^{16} - 90e^{18} - 33e^{20} - 22e^{22}$$
$$+2e^{24} + 88e^{26} + 38e^{28} + 78e^{30} + 4e^{32} - 10e^{34} - 28e^{36} - 2e^{38}) + b^{10}(e^6$$
$$-2e^8 + 6e^{12} - 39e^{14} - 20e^{16} - 69e^{18} + 22e^{20} - 23e^{22} + 88e^{24} + 7e^{26}$$
$$+92e^{28} - 2e^{30} - 10e^{32} - 34e^{34} - 14e^{36} - 3e^{38}) + b^{11}(2e^8 + 2e^{10} - 14e^{12}$$
$$-13e^{14} - 14e^{16} + 5e^{18} - 14e^{20} + 55e^{22} - e^{24} + 73e^{26} - 20e^{30} - 29e^{32}$$
$$-17e^{34} - 10e^{36} - 4e^{38} - e^{40}) + b^{12}(-7e^{10} + 2e^{12} - 3e^{14} + 16e^{16} - 12e^{18}$$
$$+30e^{20} + 13e^{22} + 20e^{24} + 17e^{26} - 30e^{28} - 8e^{30} - 22e^{32} - 3e^{34} - 12e^{36}$$
$$+3e^{38} - 4e^{40}) + b^{13}(4e^{12} - 2e^{14} + 7e^{16} - 5e^{18} + 27e^{20} - 13e^{22} + 20e^{24}$$
$$-29e^{26} + 14e^{28} - 17e^{30} + 5e^{32} - 14e^{34} + 3e^{36} - e^{38} + e^{40}),$$
$$R_5(b,e) = -3e^3 - 2e^5 - 13e^9 - 38e^{11} - 30e^{13} - 40e^{15} - 3e^{17} - 26e^{19}$$
$$-5e^{21} - 6e^{23} + 16e^{25} - e^{27} + 2e^{29} - 2e^{31} + b(e - 3e^3 - 8e^7 - 35e^9$$
$$-20e^{11} + 8e^{13} + 26e^{15} + 64e^{17} + 26e^{19} + 53e^{21} + 36e^{23} + 19e^{25} - 18e^{27}$$

$$+2e^{29} - 4e^{31} + 4e^{33}) + b^2(-4e^5 - 22e^7 - 8e^9 + 37e^{11} + 60e^{13} + 81e^{15}$$
$$+78e^{17} + 56e^{19} + 57e^{21} + 15e^{23} - 29e^{25} - 13e^{27} - 5e^{29} - e^{31} + 2e^{33}$$
$$-2e^{35}) + b^3(-3e^3 - 8e^5 - 3e^7 + 19e^9 + 49e^{11} + 103e^{13} + 101e^{15} + 118e^{17}$$
$$+36e^{19} + 31e^{21} - 60e^{23} - 29e^{25} - 47e^{27} - 5e^{29} - 3e^{31} + 3e^{33}) + b^4(-5e^3$$
$$+4e^5 + 4e^7 + 8e^9 + 63e^{11} + 76e^{13} + 54e^{15} + 29e^{17} - 109e^{19} - 120e^{21}$$
$$-169e^{23} - 115e^{25} - 64e^{27} + 35e^{29} + 13e^{33} - 6e^{35}) + b^5(-e^3 + e^5 - 6e^7$$
$$+26e^9 + 59e^{11} + 38e^{13} + 3e^{15} - 82e^{17} - 204e^{19} - 143e^{21} - 159e^{23}$$
$$-68e^{25} + 28e^{27} + 28e^{29} + 16e^{31} + 11e^{33} - 3e^{35} + 3e^{37}) + b^6(-3e^5 + 2e^7$$
$$+26e^9 + 42e^{11} + 42e^{13} - 33e^{15} - 130e^{17} - 192e^{19} - 134e^{21} - 117e^{23}$$
$$+42e^{25} + 40e^{27} + 75e^{29} + 32e^{31} + 12e^{33} - 6e^{35}) + b^7(17e^9 + 30e^{11}$$
$$-16e^{13} - 90e^{15} - 123e^{17} - 165e^{19} - 110e^{21} - 20e^{23} + 70e^{25} + 120e^{27}$$
$$+119e^{29} + 22e^{31} + 11e^{33} - 13e^{35} - 3e^{37}) + b^8(2e^5 + 15e^9 + 2e^{11} - 49e^{13}$$
$$-78e^{15} - 109e^{17} - 145e^{19} + 3e^{21} - 4e^{23} + 131e^{25} + 143e^{27} + 92e^{29}$$
$$+23e^{31} + e^{33} - 23e^{35} - 5e^{37} + e^{39}) + b^9(-2e^7 + 5e^9 - 7e^{11} - 33e^{13}$$
$$-67e^{15} - 94e^{17} - 43e^{19} + 62e^{21} + 92e^{23} + 222e^{25} + 140e^{27} + 88e^{29} - 14e^{31}$$
$$-22e^{33} - 25e^{35} + 4e^{37} - 4e^{39}) + b^{10}(2e^9 - 6e^{11} - 17e^{13} - 7e^{15} + 32e^{17}$$
$$+77e^{19} + 139e^{21} + 187e^{23} + 172e^{25} + 104e^{27} + 9e^{29} - 34e^{31} - 25e^{33}$$
$$-14e^{35} - 10e^{37} - 5e^{39}) + b^{11}(-e^7 + e^9 - 5e^{11} - 6e^{13} + 12e^{15} + 25e^{17}$$
$$+52e^{19} + 110e^{21} + 61e^{23} + 55e^{25} - 13e^{27} - 30e^{29} - 37e^{31} - 28e^{33} - 34e^{35}$$
$$-11e^{37}) + b^{12}(2e^9 - 2e^{11} + 2e^{13} + 10e^{15} + 19e^{17} + 40e^{19} + 30e^{21} - 16e^{23}$$
$$-7e^{25} - 60e^{27} - 49e^{29} - 59e^{31} - 36e^{33} - 26e^{35} + 2e^{37} - 3e^{39} + 2e^{41})$$
$$+b^{13}(-e^{11} + e^{13} + e^{15} - 3e^{17} - 16e^{19} - 38e^{21} - 45e^{23} - 46e^{25} - 56e^{27}$$
$$-50e^{29} - 32e^{31} - 14e^{33} - e^{37} - 2e^{39}),$$
$$R_6(b,e) = 2e^2 - 2e^4 + e^6 - 16e^8 + 6e^{10} + 5e^{12} + 26e^{14} + 3e^{16}$$
$$+40e^{18} + 30e^{20} + 38e^{22} + 13e^{24} + 2e^{28} + 3e^{30} + b(-3e^2 - 7e^6 + 12e^{10}$$
$$+36e^{12} + 9e^{14} + 63e^{16} + 48e^{18} + 34e^{20} - e^{22} - 19e^{24} - 17e^{26} + 8e^{28}$$
$$-6e^{30} - 6e^{32}) + b^2(-2e^2 - 3e^4 - e^6 + 6e^8 + 19e^{10} + 19e^{12} + 35e^{14}$$
$$+94e^{16} - 10e^{18} + 3e^{20} - 72e^{22} - 38e^{24} - 33e^{26} - 6e^{28} - 20e^{30} + 6e^{32}$$
$$+3e^{34}) + b^3(-3e^4 - 4e^8 + 29e^{10} + 31e^{12} + 82e^{14} - 15e^{16} - 22e^{18} - 120e^{20}$$
$$-114e^{22} - 110e^{24} - 50e^{26} - 19e^{28} + 7e^{30} + 8e^{32} - 2e^{34}) + b^4(-e^4 - 11e^6$$
$$+4e^8 + 39e^{10} + 70e^{12} - 13e^{14} - 12e^{16} - 152e^{18} - 154e^{20} - 122e^{22} - 36e^{24}$$
$$+10e^{26} + 50e^{28} + 5e^{30} + 9e^{32} + 12e^{34}) + b^5(-3e^4 - 7e^6 + 12e^8 + 50e^{10}$$
$$-5e^{14} - 48e^{16} - 179e^{18} - 67e^{20} - 114e^{22} + 17e^{24} + 46e^{26} + 91e^{28} + 20e^{30}$$
$$+52e^{32} - 8e^{34} - 8e^{36}) + b^6(-3e^4 + 22e^8 + 8e^{10} + 7e^{12} - 19e^{14} - 93e^{16}$$
$$-93e^{18} - 96e^{20} - 65e^{22} + 58e^{24} + 120e^{26} + 98e^{28} + 69e^{30} + 8e^{32} - 20e^{34}$$
$$-e^{36}) + b^7(2e^6 - e^8 + 14e^{10} + 14e^{12} - 48e^{14} - 36e^{16} - 144e^{18} - 47e^{20}$$
$$+29e^{22} + 161e^{24} + 112e^{26} + 120e^{28} - 3e^{32} - 10e^{34} - 11e^{36} - e^{38})$$
$$+b^8(-6e^6 + 7e^8 + 24e^{10} - 32e^{12} - 2e^{14} - 73e^{16} - 65e^{18} + 33e^{20} + 85e^{22}$$
$$+136e^{24} + 159e^{26} + 78e^{28} - 3e^{30} + 13e^{32} - 40e^{34} - 11e^{36} - e^{38}) + b^9(e^6$$
$$+11e^8 - 16e^{10} - 14e^{12} - 26e^{14} - 10e^{16} + 9e^{18} + 98e^{20} + 61e^{22} + 170e^{24}$$
$$+78e^{26} + 23e^{28} + 4e^{30} - 32e^{32} - 36e^{34} - 16e^{36} - 5e^{38} + 2e^{40}) + b^{10}(-2e^8$$
$$+e^{10} - e^{12} - 13e^{14} + 17e^{16} - e^{18} + 90e^{20} + 35e^{22} + 86e^{24} - 51e^{26} + 14e^{28}$$
$$-82e^{30} - 33e^{32} - 50e^{34} - 12e^{36} - e^{38} + 3e^{40}) + b^{11}(3e^8 - e^{10} - 13e^{12}$$
$$+10e^{14} + e^{16} + 29e^{18} + 20e^{20} + 10e^{22} - 16e^{24} + 11e^{26} - 66e^{28} - 68e^{30}$$

$$-49e^{32} - 29e^{34} + 6e^{36} + 3e^{38} - 2e^{40}) + b^{12}(-6e^{10} + 4e^{12} + 9e^{14} + 8e^{16}$$
$$-2e^{18} + 5e^{20} - 30e^{22} + 10e^{24} - 59e^{26} - 56e^{28} - 31e^{30} - 11e^{32} + 4e^{34}$$
$$+7e^{36} - e^{38} - 2e^{40}) + b^{13}(3e^{12} - 2e^{14} - 2e^{16} - 9e^{18} + 8e^{20} - 17e^{22}$$
$$+11e^{24} - 43e^{26} + 26e^{28} - e^{30} + 25e^{32} - 6e^{34} + 6e^{36} - 3e^{38} + 4e^{40}),$$
$$R_{13-k}(b,e) = -b^{13}e^{45}R_k(b^{-1}, e^{-1}), \quad (k = \overline{0,6}).$$

Appendix 2

The Polynomials $R_{2k}(b, f)$, that Define $N_{1,5}(u, b, f)$

$R_0(b, f) = 1 + f - f^3 + f^4 + 4f^5 + 11f^6 + 16f^7 + 17f^8 + 13f^9$
$+ 13f^{10} + 13f^{11} + 17f^{12} + 16f^{13} + 11f^{14} + 4f^{15} + f^{16} - f^{17} + f^{19}$
$+ f^{20} + b(-2f - f^2 + 5f^3 + 12f^4 + 15f^5 + 15f^6 + 11f^7 + 10f^8$
$+ 17f^9 + 27f^{10} + 25f^{11} + 16f^{12} - f^{13} - 9f^{14} - 5f^{15} + 2f^{16} + 3f^{17}$
$+ 3f^{18} - 2f^{20} - 2f^{21}) + b^2(-f + 3f^2 + 6f^3 + 5f^4 - 3f^6 - 10f^7$
$- 13f^8 - 15f^9 - 18f^{10} - 32f^{11} - 41f^{12} - 51f^{13} - 43f^{14} - 29f^{15}$
$- 18f^{16} - 11f^{17} - 3f^{18} - 2f^{19} - 2f^{20} - f^{21} + f^{22}) + b^3(3f^2 + f^3$
$- 5f^4 - 11f^5 - 11f^6 - 15f^7 - 20f^8 - 35f^9 - 58f^{10} - 85f^{11} - 81f^{12}$
$- 61f^{13} - 21f^{14} - f^{15} - 7f^{17} - 7f^{18} - 8f^{19} - 2f^{20} + 3f^{21} + 4f^{22})$
$+ b^4(2f^2 - f^3 - 4f^4 - 6f^5 - 9f^6 - 18f^7 - 21f^8 - 21f^9 - 23f^{10}$
$- 22f^{11} - 6f^{12} + 6f^{13} + 22f^{14} + 23f^{15} + 21f^{16} + 21f^{17} + 18f^{18}$
$+ 9f^{19} + 6f^{20} + 4f^{21} + f^{22} - 2f^{23}) + b^5(-4f^3 - 3f^4 + 2f^5 + 8f^6$
$+ 7f^7 + 7f^8 + f^{10} + 21f^{11} + 61f^{12} + 81f^{13} + 85f^{14} + 58f^{15} + 35f^{16}$
$+ 20f^{17} + 15f^{18} + 11f^{19} + 11f^{20} + 5f^{21} - f^{22} - 3f^{23}) + b^6(-f^3 + f^4$
$+ 2f^5 + 2f^6 + 3f^7 + 11f^8 + 18f^9 + 29f^{10} + 43f^{11} + 51f^{12} + 41f^{13}$
$+ 32f^{14} + 18f^{15} + 15f^{16} + 13f^{17} + 10f^{18} + 3f^{19} - 5f^{21} - 6f^{22} - 3f^{23}$
$+ f^{24}) + b^7(2f^4 + 2f^5 - 3f^7 - 3f^8 - 2f^9 + 5f^{10} + 9f^{11} + f^{12} - 16f^{13}$
$- 25f^{14} - 27f^{15} - 17f^{16} - 10f^{17} - 11f^{18} - 15f^{19} - 15f^{20} - 12f^{21}$
$- 5f^{22} + f^{23} + 2f^{24}) + b^8(-f^5 - f^6 + f^8 - f^9 - 4f^{10} - 11f^{11} - 16f^{12}$
$- 17f^{13} - 13f^{14} - 13f^{15} - 13f^{16} - 17f^{17} - 16f^{18} - 11f^{19} - 4f^{20}$
$- f^{21} + f^{22} - f^{24} - f^{25}),$

$R_2(b, f) = -2f - f^2 + 5f^3 + 12f^4 + 15f^5 + 15f^6 + 11f^7 + 10f^8$
$+ 17f^9 + 27f^{10} + 25f^{11} + 16f^{12} - f^{13} - 9f^{14} - 5f^{15} + 2f^{16} + 3f^{17}$
$+ 3f^{18} - 2f^{20} - 2f^{21} + b(f + 9f^2 + 12f^3 + 3f^4 - 4f^5 - 3f^6 + 11f^7$
$+ 28f^8 + 34f^9 + 10f^{10} - 22f^{11} - 36f^{12} - 29f^{13} - 3f^{14} + 12f^{15} - f^{16}$
$- 12f^{17} - 11f^{18} - 7f^{19} + 4f^{21} + 4f^{22}) + b^2(2f + 6f^2 - 4f^3 - 15f^4$
$- 13f^5 - 7f^6 - 6f^7 - 20f^8 - 53f^9 - 92f^{10} - 107f^{11} - 84f^{12} - 44f^{13}$
$+ f^{14} + 9f^{15} - 3f^{16} + f^{17} + 4f^{18} + f^{19} + 3f^{20} + 4f^{21} + 2f^{22} - 2f^{23})$
$+ b^3(f - f^2 - 11f^3 - 10f^4 + f^5 - 3f^6 - 26f^7 - 64f^8 - 89f^9 - 85f^{10}$
$- 35f^{11} + 31f^{12} + 59f^{13} + 46f^{14} - 3f^{15} - 18f^{16} + 9f^{17} + 25f^{18}$
$+ 24f^{19} + 19f^{20} + 5f^{21} - 6f^{22} - 8f^{23}) + b^4(-3f^2 - 9f^3 + 9f^5$
$+ f^6 - 8f^7 - 13f^8 - 4f^9 + 15f^{10} + 65f^{11} + 99f^{12} + 99f^{13} + 75f^{14}$
$+ 42f^{15} + 35f^{16} + 34f^{17} + 6f^{18} - 6f^{19} - 6f^{20} - 8f^{21} - 8f^{22}$
$- 2f^{23} + 4f^{24}) + b^5(-2f^2 + 12f^4 + 11f^5 - 3f^6 - 6f^7 + 9f^8 + 45f^9$
$+ 97f^{10} + 139f^{11} + 118f^{12} + 52f^{13} - f^{14} - 20f^{15} + 7f^{16} + 16f^{17}$

171

$$-11f^{19} - 19f^{20} - 23f^{21} - 12f^{22} + 2f^{23} + 6f^{24}) + b^6(4f^3 + 7f^4 - f^5$$
$$-7f^6 + 4f^7 + 19f^8 + 37f^9 + 45f^{10} + 31f^{11} - 20f^{12} - 51f^{13} - 49f^{14}$$
$$-27f^{15} - 9f^{16} - 20f^{17} - 40f^{18} - 39f^{19} - 29f^{20} - 14f^{21} + 4f^{22}$$
$$+12f^{23} + 6f^{24} - 2f^{25}) + b^7(f^3 - f^4 - 7f^5 - 6f^6 + 2f^7 + 4f^8 - 2f^9$$
$$-23f^{10} - 59f^{11} - 88f^{12} - 74f^{13} - 45f^{14} - 25f^{15} - 28f^{16} - 46f^{17}$$
$$-47f^{18} - 19f^{19} + 6f^{20} + 21f^{21} + 21f^{22} + 8f^{23} - 5f^{24} - 5f^{25})$$
$$+b^8(-2f^4 - 2f^5 + 2f^6 + 4f^7 - 2f^8 - 10f^9 - 20f^{10} - 24f^{11} - 12f^{12}$$
$$+6f^{13} + 8f^{14} - 8f^{16} - 6f^{17} + 12f^{18} + 24f^{19} + 20f^{20} + 10f^{21} + 2f^{22}$$
$$-4f^{23} - 2f^{24} + 2f^{25} + 2f^{26}) + b^9(f^5 + f^6 - f^8 + f^9 + 4f^{10} + 11f^{11}$$
$$+16f^{12} + 17f^{13} + 13f^{14} + 13f^{15} + 13f^{16} + 17f^{17} + 16f^{18} + 11f^{19}$$
$$+4f^{20} + f^{21} - f^{22} + f^{24} + f^{25}),$$
$$R_4(b, f) = -f + 3f^2 + 6f^3 + 5f^4 - 3f^6 - 10f^7 - 13f^8 - 15f^9 - 18f^{10}$$
$$-32f^{11} - 41f^{12} - 51f^{13} - 43f^{14} - 29f^{15} - 18f^{16} - 11f^{17} - 3f^{18}$$
$$-2f^{19} - 2f^{20} - f^{21} + f^{22} + b(2f + 6f^2 - 4f^3 - 15f^4 - 13f^5 - 7f^6$$
$$-6f^7 - 20f^8 - 53f^9 - 92f^{10} - 107f^{11} - 84f^{12} - 44f^{13} + f^{14} + 9f^{15}$$
$$-3f^{16} + f^{17} + 4f^{18} + f^{19} + 3f^{20} + 4f^{21} + 2f^{22} - 2f^{23}) + b^2(f - 3f^2$$
$$-17f^3 - 11f^4 - 2f^5 - 13f^6 - 44f^7 - 69f^8 - 72f^9 - 37f^{10} + 18f^{11}$$
$$+81f^{12} + 114f^{13} + 126f^{14} + 96f^{15} + 85f^{16} + 83f^{17} + 50f^{18} + 19f^{19}$$
$$+9f^{20} + 5f^{21} - 3f^{23} + f^{24}) + b^3(-5f^2 - 8f^3 + 7f^4 + 5f^5 - 19f^6$$
$$-41f^7 - 28f^8 + 24f^9 + 104f^{10} + 190f^{11} + 251f^{12} + 233f^{13} + 175f^{14}$$
$$+95f^{15} + 74f^{16} + 52f^{17} + 13f^{18} + f^{19} + 4f^{20} - 4f^{21} - 10f^{22} - 5f^{23}$$
$$+4f^{24}) + b^4(-2f^2 + 2f^3 + 13f^4 - 4f^5 - 18f^6 + 9f^7 + 72f^8 + 131f^9$$
$$+165f^{10} + 175f^{11} + 133f^{12} + 64f^{13} + 14f^{14} - 17f^{15} - 23f^{16} - 65f^{17}$$
$$-98f^{18} - 67f^{19} - 37f^{20} - 23f^{21} - 12f^{22} + 2f^{23} + 5f^{24} - 2f^{25})$$
$$+b^5(8f^3 + 11f^4 - 8f^5 - 6f^6 + 23f^7 + 62f^8 + 84f^9 + 94f^{10}$$
$$+52f^{11} - 48f^{12} - 152f^{13} - 196f^{14} - 197f^{15} - 161f^{16} - 154f^{17}$$
$$-115f^{18} - 66f^{19} - 45f^{20} - 29f^{21} - 4f^{22} + 11f^{23} + 5f^{24} - 3f^{25})$$
$$+b^6(f^3 - 7f^4 - 11f^5 + 15f^6 + 41f^7 + 33f^8 - 13f^9 - 74f^{10} - 140f^{11}$$
$$-176f^{12} - 151f^{13} - 114f^{14} - 107f^{15} - 121f^{16} - 130f^{17} - 82f^{18}$$
$$-21f^{19} + 13f^{20} + 30f^{21} + 32f^{22} + 17f^{23} - 3f^{24} - 6f^{25} + f^{26})$$
$$+b^7(-4f^4 - 3f^5 + 5f^6 - 2f^7 - 23f^8 - 46f^9 - 64f^{10} - 80f^{11} - 66f^{12}$$
$$-30f^{13} - 10f^{14} + 10f^{16} + 30f^{17} + 66f^{18} + 80f^{19} + 64f^{20} + 46f^{21}$$
$$+23f^{22} + 2f^{23} - 5f^{24} + 3f^{25} + 4f^{26}) + b^8(5f^5 + 5f^6 - 8f^7 - 21f^8$$
$$-21f^9 - 6f^{10} + 19f^{11} + 47f^{12} + 46f^{13} + 28f^{14} + 25f^{15} + 45f^{16}$$
$$+74f^{17} + 88f^{18} + 59f^{19} + 23f^{20} + 2f^{21} - 4f^{22} - 2f^{23} + 6f^{24} + 7f^{25}$$
$$+f^{26} - f^{27}) + b^9(-2f^6 - f^7 + 5f^8 + 12f^9 + 15f^{10} + 15f^{11} + 11f^{12}$$
$$+10f^{13} + 17f^{14} + 27f^{15} + 25f^{16} + 16f^{17} - f^{18} - 9f^{19} - 5f^{20} + 2f^{21}$$
$$+3f^{22} + 3f^{23} - 2f^{25} - 2f^{26}),$$
$$R_6(b, f) = 3f^2 + f^3 - 5f^4 - 11f^5 - 11f^6 - 15f^7 - 20f^8 - 35f^9$$
$$-58f^{10} - 85f^{11} - 81f^{12} - 61f^{13} - 21f^{14} - f^{15} - 7f^{17} - 7f^{18} - 8f^{19}$$
$$-2f^{20} + 3f^{21} + 4f^{22} + b(f - f^2 - 11f^3 - 10f^4 + f^5 - 3f^6 - 26f^7$$
$$-64f^8 - 89f^9 - 85f^{10} - 35f^{11} + 31f^{12} + 59f^{13} + 46f^{14} - 3f^{15}$$
$$-18f^{16} + 9f^{17} + 25f^{18} + 24f^{19} + 19f^{20} + 5f^{21} - 6f^{22} - 8f^{23})$$
$$+b^2(-5f^2 - 8f^3 + 7f^4 + 5f^5 - 19f^6 - 41f^7 - 28f^8 + 24f^9$$
$$+104f^{10} + 190f^{11} + 251f^{12} + 233f^{13} + 175f^{14} + 95f^{15} + 74f^{16}$$

$$+52f^{17} + 13f^{18} + f^{19} + 4f^{20} - 4f^{21} - 10f^{22} - 5f^{23} + 4f^{24})$$
$$+b^3(-3f^2 + 3f^3 + 8f^4 - 15f^5 - 33f^6 - 6f^7 + 73f^8 + 167f^9 + 226f^{10}$$
$$+235f^{11} + 162f^{12} + 29f^{13} - 36f^{14} - 27f^{15} + 45f^{16} + 29f^{17} - 20f^{18}$$
$$-43f^{19} - 51f^{20} - 51f^{21} - 22f^{22} + 9f^{23} + 16f^{24}) + b^4(7f^3 + 4f^4$$
$$-16f^5 - 3f^6 + 45f^7 + 86f^8 + 92f^9 + 67f^{10} + 26f^{11} - 82f^{12} - 174f^{13}$$
$$-199f^{14} - 183f^{15} - 160f^{16} - 172f^{17} - 128f^{18} - 50f^{19} - 22f^{20} - 7f^{21}$$
$$+14f^{22} + 23f^{23} + 6f^{24} - 8f^{25}) + b^5(3f^3 - 8f^4 - 12f^5 + 23f^6 + 56f^7$$
$$+53f^8 - 80f^{10} - 194f^{11} - 290f^{12} - 276f^{13} - 176f^{14} - 110f^{15} - 100f^{16}$$
$$-137f^{17} - 91f^{18} - 41f^{19} + 2f^{20} + 46f^{21} + 65f^{22} + 35f^{23} - 6f^{24}$$
$$-13f^{25}) + b^6(-6f^4 + 7f^5 + 25f^6 + 9f^7 - 37f^8 - 79f^9 - 104f^{10}$$
$$-112f^{11} - 86f^{12} - 13f^{13} + 16f^{14} - 16f^{16} + 13f^{17} + 86f^{18}$$
$$+112f^{19} + 104f^{20} + 79f^{21} + 37f^{22} - 9f^{23} - 25f^{24} - 7f^{25} + 6f^{26})$$
$$+b^7(-f^4 + 6f^5 + 3f^6 - 17f^7 - 32f^8 - 30f^9 - 13f^{10} + 21f^{11}$$
$$+82f^{12} + 130f^{13} + 121f^{14} + 107f^{15} + 114f^{16} + 151f^{17} + 176f^{18}$$
$$+140f^{19} + 74f^{20} + 13f^{21} - 33f^{22} - 41f^{23} - 15f^{24} + 11f^{25} + 7f^{26}$$
$$-f^{27}) + b^8(2f^5 - 6f^6 - 12f^7 - 4f^8 + 14f^9 + 29f^{10} + 39f^{11} + 40f^{12}$$
$$+20f^{13} + 9f^{14} + 27f^{15} + 49f^{16} + 51f^{17} + 20f^{18} - 31f^{19} - 45f^{20}$$
$$-37f^{21} - 19f^{22} - 4f^{23} + 7f^{24} + f^{25} - 7f^{26} - 4f^{27}) + b^9(-f^6$$
$$+3f^7 + 6f^8 + 5f^9 - +3f^{11} - 10f^{12} - 13f^{13} - 15f^{14} - 18f^{15}$$
$$-32f^{16} - 41f^{17} - 51f^{18} - 43f^{19} - 29f^{20} - 18f^{21} - 11f^{22} - 3f^{23}$$
$$-2f^{24} - 2f^{25} - f^{26} + f^{27}),$$
$$R_8(b, f) = 2f^2 - f^3 - 4f^4 - 6f^5 - 9f^6 - 18f^7 - 21f^8 - 21f^9 - 23f^{10}$$
$$-22f^{11} - 6f^{12} + 6f^{13} + 22f^{14} + 23f^{15} + 21f^{16} + 21f^{17} + 18f^{18}$$
$$+9f^{19} + 6f^{20} + 4f^{21} + f^{22} - 2f^{23} + b(-3f^2 - 9f^3 + 9f^5 + f^6 - 8f^7$$
$$-13f^8 - 4f^9 + 15f^{10} + 65f^{11} + 99f^{12} + 99f^{13} + 75f^{14} + 42f^{15} + 35f^{16}$$
$$+34f^{17} + 6f^{18} - 6f^{19} - 6f^{20} - 8f^{21} - 8f^{22} - 2f^{23} + 4f^{24}) + b^2(-2f^2$$
$$+2f^3 + 13f^4 - 4f^5 - 18f^6 + 9f^7 + 72f^8 + 131f^9 + 165f^{10} + 175f^{11}$$
$$+133f^{12} + 64f^{13} + 14f^{14} - 17f^{15} - 23f^{16} - 65f^{17} - 98f^{18} - 67f^{19}$$
$$-37f^{20} - 23f^{21} - 12f^{22} + 2f^{23} + 5f^{24} - 2f^{25}) + b^3(7f^3 + 4f^4 - 16f^5$$
$$-3f^6 + 45f^7 + 86f^8 + 92f^9 + 67f^{10} + 26f^{11} - 82f^{12} - 174f^{13} - 199f^{14}$$
$$-183f^{15} - 160f^{16} - 172f^{17} - 128f^{18} - 50f^{19} - 22f^{20} - 7f^{21} + 14f^{22}$$
$$+23f^{23} + 6f^{24} - 8f^{25}) + b^4(f^3 - 8f^4 - 9f^5 + 33f^6 + 61f^7 + 20f^8$$
$$-67f^9 - 145f^{10} - 201f^{11} - 272f^{12} - 243f^{13} - 200f^{14} - 178f^{15}$$
$$-181f^{16} - 141f^{17} - 16f^{18} + 81f^{19} + 77f^{20} + 72f^{21} + 55f^{22} + 25f^{23}$$
$$-10f^{24} - 9f^{25} + 4f^{26}) + b^5(-8f^4 + +f^5 + 24f^6 + 7f^7 - 42f^8 - 89f^9$$
$$-123f^{10} - 182f^{11} - 209f^{12} - 130f^{13} - 55f^{14} + 55f^{16} + 130f^{17}$$
$$+209f^{18} + 182f^{19} + 123f^{20} + 89f^{21} + 42f^{22} - 7f^{23} - 24f^{24} - f^{25}$$
$$+8f^{26}) + b^6(13f^5 + 6f^6 - 35f^7 - 65f^8 - 46f^9 - 2f^{10} + 41f^{11}$$
$$+91f^{12} + 137f^{13} + 100f^{14} + 110f^{15} + 176f^{16} + 276f^{17} + 290f^{18}$$
$$+194f^{19} + 80f^{20} - 53f^{22} - 56f^{23} - 23f^{24} + 12f^{25} + 8f^{26} - 3f^{27})$$
$$+b^7(3f^5 - 5f^6 - 11f^7 + 4f^8 + 29f^9 + 45f^{10} + 66f^{11} + 115f^{12}$$
$$+154f^{13} + 161f^{14} + 197f^{15} + 196f^{16} + 152f^{17} + 48f^{18} - 52f^{19}$$
$$-94f^{20} - 84f^{21} - 62f^{22} - 23f^{23} + 6f^{24} + 8f^{25} - 11f^{26} - 8f^{27})$$
$$+b^8(-6f^6 - 2f^7 + 12f^8 + 23f^9 + 19f^{10} + 11f^{11} - 16f^{13} - 7f^{14}$$

$$+20f^{15} + f^{16} - 52f^{17} - 118f^{18} - 139f^{19} - 97f^{20} - 45f^{21} - 9f^{22}$$
$$+6f^{23} + 3f^{24} - 11f^{25} - 12f^{26} + 2f^{28}) + b^9(3f^7 + f^8 - 5f^9 - 11f^{10}$$
$$-11f^{11} - 15f^{12} - 20f^{13} - 35f^{14} - 58f^{15} - 85f^{16} - 81f^{17} - 61f^{18}$$
$$-21f^{19} - f^{20} - 7f^{22} - 7f^{23} - 8f^{24} - 2f^{25} + 3f^{26} + 4f^{27}),$$
$$R_{16-2k}(b, f) = -b^9 f^{30} R_{2k}(b^{-1}, f^{-1}), \quad (k = \overline{0, 3}).$$

Appendix 3

Polynomials that Define the Quantity G_1
for the System $s(1,2)$.
Expressions of Focus Pseudo-Quantities $G_{1,i}$
and $\sigma_{1,i}$, $B_{1,i}$ $(i = 0,1,2,3,4)$

$$G_{1,0} = -17c^3d^2e^3g^2 + 11cd^3e^4g^2 + 34c^4de^2fg^2 - 118c^2d^2e^3fg^2$$
$$+15d^3e^4fg^2 - 17c^5ef^2g^2 + 215c^3de^2f^2g^2 - 200cd^2e^3f^2g^2$$
$$-108c^4ef^3g^2 + 403c^2de^2f^3g^2 - 91d^2e^3f^3g^2 - 224c^3ef^4g^2$$
$$+247cde^2f^4g^2 - 174c^2ef^5g^2 + 29de^2f^5g^2 - 47cef^6g^2 - 6ef^7g^2$$
$$+34c^2d^2e^4gh - 4d^3e^5gh - 92c^3de^3fgh + 152cd^2e^4fgh + 58c^4e^2f^2gh$$
$$-400c^2de^3f^2gh + 118d^2e^4f^2gh + 276c^3e^2f^3gh - 400cde^3f^3gh$$
$$+354c^2e^2f^4gh - 80de^3f^4gh + 152ce^2f^5gh + 24e^2f^6gh - 24cd^2e^5h^2$$
$$+72c^2de^4fh^2 - 32d^2e^5fh^2 - 72c^3e^3f^2h^2 + 132cde^4f^2h^2$$
$$-172c^2e^3f^3h^2 + 44de^4f^3h^2 - 116ce^3f^4h^2 - 24e^3f^5h^2 + 12c^3de^4gk$$
$$-12cd^2e^5gk - 12c^4e^3fgk + 72c^2de^4fgk - 16d^2e^5fgk - 72c^3e^3f^2gk$$
$$+102cde^4f^2gk - 122c^2e^3f^3gk + 34de^4f^3gk - 70ce^3f^4gk - 12e^3f^5gk$$
$$-12c^2de^5hk + 4d^2e^6hk + 36c^3e^4fhk - 52cde^5fhk + 120c^2e^4f^2hk$$
$$-32de^5f^2hk + 104ce^4f^3hk + 24e^4f^4hk - 6c^3e^5k^2 + 5cde^6k^2$$
$$-23c^2e^5fk^2 + 5de^6fk^2 - 23ce^5f^2k^2 - 6e^5f^3k^2 + 9c^4d^2e^2gl$$
$$-13c^2d^3e^3gl - 4d^4e^4gl - 18c^5defgl + 65c^3d^2e^2fgl - 16cd^3e^3fgl$$
$$+9c^6f^2gl - 103c^4def^2gl + 90c^2d^2e^2f^2gl - 11d^3e^3f^2gl + 51c^5f^3gl$$
$$-140c^3def^3gl + 7cd^2e^2f^3gl + 76c^4f^4gl + 26c^2def^4gl - 31d^2e^2f^4gl$$
$$-10c^3f^5gl + 110cdef^5gl - 79c^2f^6gl + 29def^6gl - 41cf^7gl - 6f^8gl$$
$$-12c^3d^2e^3hl + 26cd^3e^4hl + 48c^4de^2fhl - 122c^2d^2e^3fhl + 22d^3e^4fhl$$
$$-36c^5ef^2hl + 250c^3de^2f^2hl - 104cd^2e^3f^2hl - 166c^4ef^3hl$$
$$+248c^2de^2f^3hl - 2d^2e^3f^3hl - 190c^3ef^4hl + 40cde^2f^4hl - 38c^2ef^5hl$$
$$-10de^2f^5hl + 34cef^6hl + 12ef^7hl - 12c^4de^3kl + 11c^2d^2e^4kl$$
$$+12c^5e^2fkl - 70c^3de^3fkl + 16cd^2e^4fkl + 65c^4e^2f^2kl - 109c^2de^3f^2kl$$
$$+5d^2e^4f^2kl + 105c^3e^2f^3kl - 58cde^3f^3kl + 53c^2e^2f^4kl - 7de^3f^4kl$$
$$-5ce^2f^5kl - 6e^2f^6kl + 3c^3d^3e^2l^2 - 5cd^4e^3l^2 - 6c^4d^2efl^2 + 20c^2d^3e^2fl^2$$
$$-5d^4e^3fl^2 + 3c^5df^2l^2 - 37c^3d^2ef^2l^2 + 26cd^3e^2f^2l^2 + 22c^4df^3l^2$$
$$-73c^2d^2ef^3l^2 + 9d^3e^2f^3l^2 + 58c^3df^4l^2 - 59cd^2ef^4l^2 + 68c^2df^5l^2$$
$$-17d^2ef^5l^2 + 35cdf^6l^2 + 6df^7l^2 - 12c^3d^2e^3gm + 2cd^3e^4gm$$
$$+24c^4de^2fgm - 90c^2d^2e^3fgm - 2d^3e^4fgm - 12c^5ef^2gm$$
$$+138c^3de^2f^2gm - 152cd^2e^3f^2gm - 62c^4ef^3gm + 200c^2de^2f^3gm$$
$$-82d^2e^3f^3gm - 70c^3ef^4gm + 32cde^2f^4gm + 50c^2ef^5gm$$

$$-58de^2f^5gm + 82cef^6gm + 12ef^7gm + 20c^2d^2e^4hm - 40c^3de^3fhm$$
$$+96cd^2e^4fhm + 44c^4e^2f^2hm - 212c^2de^3f^2hm + 76d^2e^4f^2hm$$
$$+164c^3e^2f^3hm - 152cde^3f^3hm + 76c^2e^2f^4hm + 20de^3f^4hm$$
$$-68ce^2f^5hm - 24e^2f^6hm + 12c^3de^4km - 6cd^2e^5km - 24c^4e^3fkm$$
$$+70c^2de^4fkm - 10d^2e^5fkm - 88c^3e^3f^2km + 80cde^4f^2km$$
$$-70c^2e^3f^3km + 14de^4f^3km + 10ce^3f^4km + 12e^3f^5km + 6c^4d^2e^2lm$$
$$-4c^2d^3e^3lm - 4d^4e^4lm - 12c^5deflm + 64c^3d^2e^2flm - 28cd^3e^3flm$$
$$+6c^6f^2lm - 92c^4def^2lm + 180c^2d^2e^2f^2lm - 32d^3e^3f^2lm + 32c^5f^3lm$$
$$-212c^3def^3lm + 148cd^2e^2f^3lm + 40c^4f^4lm - 168c^2def^4lm + 22d^2e^2f^4lm$$
$$-20c^3f^5lm - 48cdef^5lm - 46c^2f^6lm - 12def^6lm - 12cf^7lm - 4c^3d^2e^3m^2$$
$$+8c^4de^2fm^2 - 20c^2d^2e^3fm^2 - 8d^3e^4fm^2 - 4c^5ef^2m^2 + 64c^3de^2f^2m^2$$
$$-48cd^2e^3f^2m^2 - 20c^4ef^3m^2 + 148c^2de^2f^3m^2 - 48d^2e^3f^3m^2 - 20c^3ef^4m^2$$
$$+100cde^2f^4m^2 + 20c^2ef^5m^2 + 24cef^6m^2 + 31c^2d^2e^4gn - 38c^3de^3fgn$$
$$+96cd^2e^4fgn + 13c^4e^2f^2gn - 125c^2de^3f^2gn + 65d^2e^4f^2gn + 41c^3e^2f^3gn$$
$$-58cde^3f^3gn - 7c^2e^2f^4gn + 29de^3f^4gn - 41ce^2f^5gn - 6e^2f^6gn$$
$$-12c^3de^4hn - 46cd^2e^5hn + 78c^2de^4fhn - 50d^2e^5fhn - 56c^3e^3f^2hn$$
$$+88cde^4f^2hn - 38c^2e^3f^3hn - 10de^4f^3hn + 34ce^3f^4hn + 12e^3f^5hn$$
$$+6c^4e^4kn - 17c^2de^5kn + 4d^2e^6kn + 29c^3e^4fkn - 32cde^5fkn + 28c^2e^4f^2kn$$
$$-7de^5f^2kn - 5ce^4f^3kn - 6e^4f^4kn - 22c^3d^2e^3ln + 32cd^3e^4ln + 32c^4de^2fln$$
$$-134c^2d^2e^3fln + 28d^3e^4fln - 10c^5ef^2ln + 148c^3de^2f^2ln - 116cd^2e^3f^2ln$$
$$-34c^4ef^3ln + 126c^2de^2f^3ln - 12d^2e^3f^3ln - 2c^3ef^4ln + 14cde^2f^4ln$$
$$+34c^2ef^5ln + 12cef^6ln + 14c^2d^2e^4mn + 4d^3e^5mn - 52c^3de^3fmn$$
$$+88cd^2e^4fmn + 14c^4e^2f^2mn - 204c^2de^3f^2mn + 74d^2e^4f^2mn + 40c^3e^2f^3mn$$
$$-128cde^3f^3mn - 18c^2e^2f^4mn + 12de^3f^4mn - 36ce^2f^5mn + 6c^3de^4n^2$$
$$-31cd^2e^5n^2 + 63c^2de^4fn^2 - 27d^2e^5fn^2 - 14c^3e^3f^2n^2 + 43cde^4f^2n^2$$
$$+2c^2e^3f^3n^2 - 6de^4f^3n^2 + 12ce^3f^4n^2;$$
$$G_{1,1} = 2c^4d^2e^2g^2 + 9c^2d^3e^3g^2 - 4c^5defg^2 - 14c^3d^2e^2fg^2 + 41cd^3e^3fg^2$$
$$+2c^6f^2g^2 - 5c^4def^2g^2 - 102c^2d^2e^2f^2g^2 + 36d^3e^3f^2g^2 + 10c^5f^3g^2$$
$$+57c^3def^3g^2 - 129cd^2e^2f^3g^2 + 10c^4f^4g^2 + 121c^2def^4g^2 - 35d^2e^2f^4y^2$$
$$-10c^3f^5g^2 + 65cdef^5g^2 - 12c^2f^6g^2 + 6def^6g^2 - 22c^3d^2e^3gh - 12cd^3e^4gh$$
$$+56c^4de^2fgh - 52c^2d^2e^3fgh - 16d^3e^4fgh - 34c^5ef^2gh + 200c^3de^2f^2gh$$
$$-2cd^2e^3f^2gh - 160c^4ef^3gh + 144c^2de^2f^3gh + 28d^2e^3f^3gh - 198c^3ef^4gh$$
$$-24cde^2f^4gh - 76c^2ef^5gh - 12de^2f^5gh - 12cef^6gh + 24c^2d^2e^4h^2$$
$$-72c^3de^3fh^2 + 32cd^2e^4fh^2 + 72c^4e^2f^2h^2 - 132c^2de^3f^2h^2 + 172c^3e^2f^3h^2$$
$$-44cde^3f^3h^2 + 116c^2e^2f^4h^2 + 24ce^2f^5h^2 - 6c^4de^3gk + 4c^2d^2e^4gk$$
$$+6c^5e^2fgk - 32c^3de^3fgk + 40c^4e^2f^2gk - 34c^2de^3f^2gk - 8d^2e^4f^2gk$$
$$+70c^3e^2f^3gk + 6cde^3f^3gk + 38c^2e^2f^4gk + 6de^3f^4gk + 6ce^2f^5gk$$
$$+12c^3de^4hk - 4cd^2e^5hk - 36c^4e^3fhk + 52c^2de^4fhk - 120c^3e^3f^2hk$$
$$+32cde^4f^2hk - 104c^2e^3f^3hk - 24ce^3f^4hk + 6c^4e^4k^2 - 5c^2de^5k^2$$
$$+23c^3e^4fk^2 - 5cde^5fk^2 + 23c^2e^4f^2k^2 + 6ce^4f^3k^2 + c^3d^3e^2gl + 4cd^4e^3gl$$
$$+4c^4d^2efgl - 4c^2d^3e^2fgl - 5c^5df^2gl + 8c^3d^2ef^2gl + 23cd^3e^2f^2gl$$
$$-14c^4df^3gl - 53c^2d^2ef^3gl + 20d^3e^2f^3gl + 18c^3df^4gl - 76cd^2ef^4gl$$
$$+56c^2df^5gl - 23d^2ef^5gl + 35cdf^6gl + 6df^7gl - 10c^2d^3e^3hl - 12c^5defhl$$
$$+22c^3d^2e^2fhl - 6cd^3e^3fhl + 12c^6f^2hl - 50c^4def^2hl - 12c^2d^2e^2f^2hl$$
$$+50c^5f^3hl + 8c^3def^3hl - 26cd^2e^2f^3hl + 34c^4f^4hl + 64c^2def^4hl - 38c^3f^5hl$$

$$+22cdef^5hl - 46c^2f^6hl - 12cf^7hl + 6c^5de^2kl - 3c^3d^2e^3kl - 6c^6efkl$$
$$+30c^4de^2fkl - 33c^5ef^2kl + 41c^3de^2f^2kl + 3cd^2e^3f^2kl - 53c^4ef^3kl$$
$$+18c^2de^2f^3kl - 21c^3ef^4kl + cde^2f^4kl + 11c^2ef^5kl + 6cef^6kl + c^2d^4e^2l^2$$
$$+4c^3d^3efl^2 - 3cd^4e^2fl^2 - 5c^4d^2f^2l^2 + 24c^2d^3ef^2l^2 - 4d^4e^2f^2l^2 - 27c^3d^2f^3l^2$$
$$+31cd^3ef^3l^2 - 45c^2d^2f^4l^2 + 11d^3ef^4l^2 - 29cd^2f^5l^2 - 6d^2f^6l^2 + 6c^4d^2e^2gm$$
$$+6c^2d^3e^3gm - 12c^5defgm + 30c^3d^2e^2fgm + 34cd^3e^3fgm + 6c^6f^2gm$$
$$-54c^4def^2gm + 24d^3e^3f^2gm + 30c^5f^3gm - 40c^3def^3gm - 18cd^2e^2f^3gm$$
$$+30c^4f^4gm + 24c^2def^4gm - 2d^2e^2f^4gm - 30c^3f^5gm + 14cdef^5gm$$
$$-36c^2f^6gm - 12def^6gm - 20c^3d^2e^3hm + 40c^4de^2fhm - 96c^2d^2e^3fhm$$
$$-44c^5ef^2hm + 212c^3de^2f^2hm - 76cd^2e^3f^2hm - 164c^4ef^3hm$$
$$+152c^2de^2f^3hm - 76c^3ef^4hm - 20cde^2f^4hm + 68c^2ef^5hm + 24cef^6hm$$
$$-12c^4de^3km + 6c^2d^2e^4km + 24c^5e^2fkm - 70c^3de^3fkm + 10cd^2e^4fkm$$
$$+88c^4e^2f^2km - 80c^2de^3f^2km + 70c^3e^2f^3km - 14cde^3f^3km - 10c^2e^2f^4km$$
$$-12ce^2f^5km - 4c^3d^3e^2lm + 4cd^4e^3lm - 4c^4d^2eflm - 4c^2d^3e^2flm$$
$$+8c^5df^2lm - 28c^3d^2ef^2lm + 8cd^3e^2f^2lm + 52c^4df^3lm - 48c^2d^2ef^3lm$$
$$+112c^3df^4lm - 20cd^2ef^4lm + 92c^2df^5lm + 24cdf^6lm + 4c^4d^2e^2m^2$$
$$-8c^5defm^2 + 20c^3d^2e^2fm^2 + 8cd^3e^3fm^2 + 4c^6f^2m^2 - 64c^4def^2m^2$$
$$+48c^2d^2e^2f^2m^2 + 20c^5f^3m^2 - 148c^3def^3m^2 + 48cd^2e^2f^3m^2 + 20c^4f^4m^2$$
$$-100c^2def^4m^2 - 20c^3f^5m^2 - 24c^2f^6m^2 - 19c^3d^2e^3gn - 16cd^3e^4gn$$
$$+20c^4de^2fgn - 20c^2d^2e^3fgn - 16d^3e^4fgn - 7c^5ef^2gn + 45c^3de^2f^2gn$$
$$+3cd^2e^3f^2gn - 21c^4ef^3gn + 6c^2de^2f^3gn + 4d^2e^3f^3gn + 7c^3ef^4gn$$
$$-13cde^2f^4gn + 21c^2ef^5gn + 6de^2f^5gn + 12c^4de^3hn + 46c^2d^2e^4hn$$
$$-78c^3de^3fhn + 50cd^2e^4fhn + 56c^4e^2f^2hn - 88c^2de^3f^2hn + 38c^3e^2f^3hn$$
$$+10cde^3f^3hn - 34c^2e^2f^4hn - 12ce^2f^5hn - 6c^5e^3kn + 17c^3de^4kn$$
$$-4cd^2e^5kn - 29c^4e^3fkn + 32c^2de^4fkn - 28c^3e^3f^2kn + 7cde^4f^2kn$$
$$+5c^2e^3f^3kn + 6ce^3f^4kn + 10c^4d^2e^2ln - 16c^2d^3e^3ln - 14c^5defln$$
$$+58c^3d^2e^2fln - 12cd^3e^3fln + 4c^6f^2ln - 68c^4def^2ln + 48c^2d^2e^2f^2ln$$
$$+14c^5f^3ln - 74c^3def^3ln + 8cd^2e^2f^3ln + 2c^4f^4ln - 30c^2def^4ln - 14c^3f^5ln$$
$$-6cdef^5ln - 6c^2f^6ln - 14c^3d^2e^3mn - 4cd^3e^4mn + 52c^4de^2fmn$$
$$-88c^2d^2e^3fmn - 14c^5ef^2mn + 204c^3de^2f^2mn - 74cd^2e^3f^2mn$$
$$-40c^4ef^3mn + 128c^2de^2f^3mn + 18c^3ef^4mn - 12cde^2f^4mn + 36c^2ef^5mn$$
$$-6c^4de^3n^2 + 31c^2d^2e^4n^2 - 63c^3de^3fn^2 + 27cd^2e^4fn^2 + 14c^4e^2f^2n^2$$
$$-43c^2de^3f^2n^2 - 2c^3e^2f^3n^2 + 6cde^3f^3n^2 - 12c^2e^2f^4n^2;$$
$$G_{1,2} = -6c^3d^3e^2g^2 - 27cd^4e^3g^2 + 18c^4d^2efg^2 + 70c^2d^3e^2fg^2 - 31d^4e^3fg^2$$
$$-12c^5df^2g^2 - 23c^3d^2ef^2g^2 + 175cd^3e^2f^2g^2 - 38c^4df^3g^2 - 206c^2d^2ef^3g^2$$
$$+99d^3e^2f^3g^2 + 8c^3df^4g^2 - 195cd^2ef^4g^2 + 42c^2df^5g^2 - 18d^2ef^5g^2$$
$$+12c^4d^2e^2gh + 74c^2d^3e^3gh + 4d^4e^4gh - 36c^5defgh - 194c^3d^2e^2fgh$$
$$+52cd^3e^3fgh + 24c^6f^2gh + 106c^4def^2gh - 378c^2d^2e^2f^2gh - 30d^3e^3f^2gh$$
$$+86c^5f^3gh + 524c^3def^3gh - 178cd^2e^2f^3gh + 14c^4f^4gh + 518c^2def^4gh$$
$$-18d^2e^2f^4gh - 94c^3f^5gh + 120cdef^5gh - 30c^2f^6gh - 48c^3d^2e^3h^2 - 8cd^3e^4h^2$$
$$+144c^4de^2fh^2 + 8c^2d^2e^3fh^2 - 168c^5ef^2h^2 + 140c^3de^2f^2h^2 + 68cd^2e^3f^2h^2$$
$$-356c^4ef^3h^2 - 76c^2de^2f^3h^2 + 12d^2e^3f^3h^2 - 208c^3ef^4h^2 - 48cde^2f^4h^2$$
$$-36c^2ef^5h^2 + 6c^5de^2gk - 8c^3d^2e^3gk + 12cd^3e^4gk - 6c^6efgk + 34c^4de^2fgk$$
$$-44c^2d^2e^3fgk + 16d^3e^4fgk - 62c^5ef^2gk + 70c^3de^2f^2gk - 62cd^2e^3f^2gk$$
$$-146c^4ef^3gk + 58c^2de^2f^3gk - 26d^2e^3f^3gk - 120c^3ef^4gk + 34cde^2f^4gk$$

$$-44c^2ef^5gk + 6de^2f^5gk - 6cef^6gk - 24c^4de^3hk + 8c^2d^2e^4hk - 4d^3e^5hk$$
$$+96c^5e^2fhk - 124c^3de^3fhk + 24cd^2e^4fhk + 332c^4e^2f^2hk - 132c^2de^3f^2hk$$
$$+16d^2e^4f^2hk + 328c^3e^2f^3hk - 56cde^3f^3hk + 112c^2e^2f^4hk - 12de^3f^4hk$$
$$+12ce^2f^5hk - 18c^5e^3k^2 + 21c^3de^4k^2 - 5cd^2e^5k^2 - 75c^4e^3fk^2 + 43c^2de^4fk^2$$
$$-5d^2e^5fk^2 - 92c^3e^3f^2k^2 + 28cde^4f^2k^2 - 41c^2e^3f^3k^2 + 6de^4f^3k^2 - 6ce^3f^4k^2$$
$$-6c^4d^3egl - 7c^2d^4e^2gl + 4d^5e^3gl + 6c^5d^2fgl + 2c^3d^3efgl - 36cd^4e^2fgl$$
$$+23c^4d^2f^2gl + 81c^2d^3ef^2gl - 29d^4e^2f^2gl - 13c^3d^2f^3gl + 142cd^3ef^3gl$$
$$-113c^2d^2f^4gl + 57d^3ef^4gl - 93cd^2f^5gl - 18d^2f^6gl + 12c^5d^2ehl + 14c^3d^3e^2hl$$
$$-10cd^4e^3hl - 12c^6dfhl + 8c^4d^2efhl + 74c^2d^3e^2fhl - 6d^4e^3fhl - 58c^5df^2hl$$
$$-112c^3d^2ef^2hl + 74cd^3e^2f^2hl - 24c^4df^3hl - 262c^2d^2ef^3hl + 14d^3e^2f^3hl$$
$$+194c^3df^4hl - 166cd^2ef^4hl + 224c^2df^5hl - 24d^2ef^5hl + 60cdf^6hl - 6c^6dekl$$
$$-c^4d^2e^2kl - 3c^2d^3e^3kl + 6c^7fkl - 16c^5defkl - 7c^3d^2e^2fkl + 35c^6f^2kl$$
$$-10c^4def^2kl + 3d^3e^3f^2kl + 56c^5f^3kl + 7cd^2e^2f^3kl + 10c^2def^4kl + d^2e^2f^4kl$$
$$-56c^3f^5kl + 16cdef^5kl - 35c^2f^6kl + 6def^6kl - 6cf^7kl - 6c^3d^4el^2 + 5cd^5e^2l^2$$
$$+6c^4d^3fl^2 - 28c^2d^4efl^2 + 5d^5e^2fl^2 + 41c^3d^3f^2l^2 - 43cd^4ef^2l^2 + 92c^2d^3f^3l^2$$
$$-21d^4ef^3l^2 + 75cd^3f^4l^2 + 18d^3f^5l^2 - 18c^3d^3e^2gm - 18cd^4e^3gm+$$
$$+12c^4d^2efgm + 2c^2d^3e^2fgm - 14d^4e^3fgm - 30c^5df^2gm + 8c^3d^2ef^2gm$$
$$+122cd^3e^2f^2gm - 64c^4df^3gm - 238c^2d^2ef^3gm + 102d^3e^2f^3gm$$
$$+130c^3df^4gm - 186cd^2ef^4gm + 156c^2df^5gm + 36d^2ef^5gm + 24c^4d^2e^2hm$$
$$+28c^2d^3e^3hm - 24c^5defhm + 28c^3d^2e^2fhm + 72c^6f^2hm - 152c^4def^2hm$$
$$-28d^3e^3f^2hm + 240c^5f^3hm - 28cd^2e^2f^3hm + 152c^2def^4hm - 24d^2e^2f^4hm$$
$$-240c^3f^5hm + 24cdef^5hm - 72c^2f^6hm + 24c^5de^2km - 14c^3d^2e^3km$$
$$+6cd^3e^4km - 60c^6efkm + 166c^4de^2fkm - 74c^2d^2e^3fkm + 10d^3e^4fkm$$
$$-224c^5ef^2km + 262c^3de^2f^2km - 74cd^2e^3f^2km - 194c^4ef^3km$$
$$+112c^2de^2f^3km - 14d^2e^3f^3km + 24c^3ef^4km - 8cde^2f^4km + 58c^2ef^5km$$
$$-12de^2f^5km + 12cef^6km + 12c^4d^3elm - 16c^2d^4e^2lm + 4d^5e^3lm$$
$$-12c^5d^2flm + 56c^3d^3eflm - 24cd^4e^2flm - 112c^4d^2f^2lm + 132c^2d^3ef^2lm$$
$$-8d^4e^2f^2lm - 328c^3d^2f^3lm + 124cd^3ef^3lm - 332c^2d^2f^4lm + 24d^3ef^4lm$$
$$-96cd^2f^5lm - 12c^3d^3e^2m^2 + 48c^4d^2efm^2 - 68c^2d^3e^2fm^2 + 8d^4e^3fm^2$$
$$+36c^5df^2m^2 + 76c^3d^2ef^2m^2 - 8cd^3e^2f^2m^2 + 208c^4df^3m^2 - 140c^2d^2ef^3m^2$$
$$+48d^3e^2f^3m^2 + 356c^3df^4m^2 - 144cd^2ef^4m^2 + 168c^2df^5m^2 + 33c^4d^2e^2gn$$
$$+49c^2d^3e^3gn - 24c^5defgn - 35c^3d^2e^2fgn + 9c^6f^2gn - 8c^4def^2gn$$
$$-49d^3e^3f^2gn + 30c^5f^3gn + 35cd^2e^2f^3gn + 8c^2def^4gn - 33d^2e^2f^4gn$$
$$-30c^3f^5gn + 24cdef^5gn - 9c^2f^6gn - 36c^5de^2hn - 102c^3d^2e^3hn+$$
$$+14cd^3e^4hn + 186c^4de^2fhn - 122c^2d^2e^3fhn + 18d^3e^4fhn - 156c^5ef^2hn$$
$$+238c^3de^2f^2hn - 2cd^2e^3f^2hn - 130c^4ef^3hn - 8c^2de^2f^3hn + 18d^2e^3f^3hn$$
$$+64c^3ef^4hn - 12cde^2f^4hn + 30c^2ef^5hn + 18c^6e^2kn - 57c^4de^3kn$$
$$+29c^2d^2e^4kn - 4d^3e^5kn + 93c^5e^2fkn - 142c^3de^3fkn + 36cd^2e^4fkn$$
$$+113c^4e^2f^2kn - 81c^2de^3f^2kn + 7d^2e^4f^2kn + 13c^3e^2f^3kn - 2cde^3f^3kn$$
$$-23c^2e^2f^4kn + 6de^3f^4kn - 6ce^2f^5kn - 6c^5d^2eln + 26c^3d^3e^2ln - 16cd^4e^3ln$$
$$+6c^6dfln - 34c^4d^2efln + 62c^2d^3e^2fln - 12d^4e^3fln + 44c^5df^2ln$$
$$-58c^3d^2ef^2ln + 44cd^3e^2f^2ln + 120c^4df^3ln - 70c^2d^2ef^3ln + 8d^3e^2f^3ln$$
$$+146c^3df^4ln - 34cd^2ef^4ln + 62c^2df^5ln - 6d^2ef^5ln + 6cdf^6ln + 18c^4d^2e^2mn$$
$$+30c^2d^3e^3mn - 4d^4e^4mn - 120c^5defmn + 178c^3d^2e^2fmn - 52cd^3e^3fmn$$
$$+30c^6f^2mn - 518c^4def^2mn + 378c^2d^2e^2f^2mn - 74d^3e^3f^2mn + 94c^5f^3mn$$

$$-524c^3def^3mn + 194cd^2e^2f^3mn - 14c^4f^4mn - 106c^2def^4mn$$
$$-12d^2e^2f^4mn - 86c^3f^5mn + 36cdef^5mn - 24c^2f^6mn + 18c^5de^2n^2$$
$$-99c^3d^2e^3n^2 + 31cd^3e^4n^2 + 195c^4de^2fn^2 - 175c^2d^2e^3fn^2 + 27d^3e^4fn^2$$
$$-42c^5ef^2n^2 + 206c^3de^2f^2n^2 - 70cd^2e^3f^2n^2 - 8c^4ef^3n^2 + 23c^2de^2f^3n^2$$
$$+6d^2e^3f^3n^2 + 38c^3ef^4n^2 - 18cde^2f^4n^2 + 12c^2ef^5n^2;$$
$$G_{1,3} = -6c^3d^3efg^2 - 27cd^4e^2fg^2 + 12c^4d^2f^2g^2 + 43c^2d^3ef^2g^2$$
$$-31d^4e^2f^2g^2 + 2c^3d^2f^3g^2 + 63cd^3ef^3g^2 - 14c^2d^2f^4g^2 + 6d^3ef^4g^2$$
$$+12c^4d^2efgh + 74c^2d^3e^2fgh + 4d^4e^3fgh - 36c^5df^2gh - 128c^3d^2ef^2gh$$
$$+88cd^3e^2f^2gh - 18c^4df^3gh - 204c^2d^2ef^3gh + 14d^3e^2f^3gh + 40c^3df^4gh$$
$$-52cd^2ef^4gh + 14c^2df^5gh - 48c^3d^2e^2fh^2 - 8cd^3e^3fh^2 + 24c^6f^2h^2$$
$$+100c^4def^2h^2 - 48c^2d^2e^2f^2h^2 + 20c^5f^3h^2 + 148c^3def^3h^2 - 20cd^2e^2f^3h^2$$
$$-20c^4f^4h^2 + 64c^2def^4h^2 - 4d^2e^2f^4h^2 - 20c^3f^5h^2 + 8cdef^5h^2 - 4c^2f^6h^2$$
$$+6c^5defgk - 8c^3d^2e^2fgk + 12cd^3e^3fgk + 6c^6f^2gk + 30c^4def^2gk$$
$$-48c^2d^2e^2f^2gk + 16d^3e^3f^2gk + 14c^5f^3gk + 74c^3def^3gk - 58cd^2e^2f^3gk$$
$$-2c^4f^4gk + 68c^2def^4gk - 10d^2e^2f^4gk - 14c^3f^5gk + 14cdef^5gk$$
$$-4c^2f^6gk - 24c^6efhk + 20c^4de^2fhk - 8c^2d^2e^3fhk - 4d^3e^4fhk$$
$$-92c^5ef^2hk + 48c^3de^2f^2hk + 4cd^2e^3f^2hk - 112c^4ef^3hk + 28c^2de^2f^3hk$$
$$+4d^2e^3f^3hk - 52c^3ef^4hk + 4cde^2f^4hk - 8c^2ef^5hk + 6c^6e^2k^2$$
$$-11c^4de^3k^2 + 4c^2d^2e^4k^2 + 29c^5e^2fk^2 - 31c^3de^3fk^2 + 3cd^2e^4fk^2$$
$$+45c^4e^2f^2k^2 - 24c^2de^3f^2k^2 - d^2e^4f^2k^2 + 27c^3e^2f^3k^2 - 4cde^3f^3k^2$$
$$+5c^2e^2f^4k^2 - 6c^4d^3fgl - 7c^2d^4efgl + 4d^5e^2fgl - 5c^3d^3f^2gl$$
$$-32cd^4ef^2gl + 28c^2d^3f^3gl - 17d^4ef^3gl + 29cd^3f^4gl + 6d^3f^5gl$$
$$+12c^5d^2fhl + 14c^3d^3efhl - 10cd^4e^2fhl + 10c^4d^2f^2hl + 80c^2d^3ef^2hl$$
$$-6d^4e^2f^2hl - 70c^3d^2f^3hl + 70cd^3ef^3hl - 88c^2d^2f^4hl + 12d^3ef^4hl$$
$$-24cd^2f^5hl - 6c^6dfkl - c^4d^2efkl - 3c^2d^3e^2fkl - 11c^5df^2kl$$
$$-18c^3d^2ef^2kl + 21c^4df^3kl - 41c^2d^2ef^3kl + 3d^3e^2f^3kl + 53c^3df^4kl$$
$$-30cd^2ef^4kl + 33c^2df^5kl - 6d^2ef^5kl + 6cdf^6kl - 6c^3d^4fl^2 + 5cd^5efl^2$$
$$-23c^2d^4f^2l^2 + 5d^5ef^2l^2 - 23cd^4f^3l^2 - 6d^4f^4l^2 + 12c^5d^2fgm$$
$$-10c^3d^3efgm - 50cd^4e^2fgm + 34c^4d^2f^2gm + 88c^2d^3ef^2gm$$
$$-46d^4e^2f^2gm - 38c^3d^2f^3gm + 78cd^3ef^3gm - 56c^2d^2f^4gm - 12d^3ef^4gm$$
$$-24c^6dfhm + 20c^4d^2efhm + 76c^2d^3e^2fhm - 68c^5df^2hm - 152c^3d^2ef^2hm$$
$$+96cd^3e^2f^2hm + 76c^4df^3hm - 212c^2d^2ef^3hm + 20d^3e^2f^3hm$$
$$+164c^3df^4hm - 40cd^2ef^4hm + 44c^2df^5hm + 12c^7fkm - 22c^5defkm$$
$$+26c^3d^2e^2fkm + 6cd^3e^3fkm + 46c^6f^2km - 64c^4def^2km + 12c^2d^2e^2f^2km$$
$$+10d^3e^3f^2km + 38c^5f^3km - 8c^3def^3km - 22cd^2e^2f^3km - 34c^4f^4km$$
$$+50c^2def^4km - 50c^3f^5km + 12cdef^5km - 12c^2f^6km + 24c^4d^3flm$$
$$-32c^2d^4eflm + 4d^5e^2flm + 104c^3d^3f^2lm - 52cd^4ef^2lm + 120c^2d^3f^3lm$$
$$-12d^4ef^3lm + 36cd^3f^4lm - 24c^5d^2fm^2 + 44c^3d^3efm^2 - 32cd^4e^2fm^2$$
$$-116c^4d^2f^2m^2 + 132c^2d^3ef^2m^2 - 24d^4e^2f^2m^2 - 172c^3d^2f^3m^2 +$$
$$+72cd^3ef^3m^2 - 72c^2d^2f^4m^2 - 6c^5d^2egn - 4c^3d^3e^2gn + 16cd^4e^3gn$$
$$+13c^4d^2efgn - 3c^2d^3e^2fgn + 16d^4e^3fgn - 21c^5df^2gn - 6c^3d^2ef^2gn$$
$$+20cd^3e^2f^2gn - 7c^4df^3gn - 45c^2d^2ef^3gn + 19d^3e^2f^3gn + 21c^3df^4gn$$
$$-20cd^2ef^4gn + 7c^2df^5gn + 12c^6dehn + 2c^4d^2e^2hn - 24c^2d^3e^3hn$$
$$-14c^5defhn + 18c^3d^2e^2fhn - 34cd^3e^3fhn + 36c^6f^2hn - 24c^4def^2hn$$
$$-6d^3e^3f^2hn + 30c^5f^3hn + 40c^3def^3hn - 30cd^2e^2f^3hn - 30c^4f^4hn$$

$$+54c^2def^4hn - 6d^2e^2f^4hn - 30c^3f^5hn + 12cdef^5hn - 6c^2f^6hn - 6c^7ekn$$
$$+23c^5de^2kn - 20c^3d^2e^3kn - 35c^6efkn + 76c^4de^2fkn - 23c^2d^2e^3fkn$$
$$-4d^3e^4fkn - 56c^5ef^2kn + 53c^3de^2f^2kn + 4cd^2e^3f^2kn - 18c^4ef^3kn$$
$$-8c^2de^2f^3kn - d^2e^3f^3kn + 14c^3ef^4kn - 4cde^2f^4kn + 5c^2ef^5kn$$
$$-6c^4d^3eln + 8c^2d^4e^2ln - 6c^5d^2fln - 6c^3d^3efln - 38c^4d^2f^2ln$$
$$+34c^2d^3ef^2ln - 4d^4e^2f^2ln - 70c^3d^2f^3ln + 32cd^3ef^3ln - 40c^2d^2f^4ln$$
$$+6d^3ef^4ln - 6cd^2f^5ln + 12c^5d^2emn - 28c^3d^3e^2mn + 16cd^4e^3mn$$
$$+12c^6dfmn + 24c^4d^2efmn + 2c^2d^3e^2fmn + 12d^4e^3fmn + 76c^5df^2mn$$
$$-144c^3d^2ef^2mn + 52cd^3e^2f^2mn + 198c^4df^3mn - 200c^2d^2ef^3mn$$
$$+22d^3e^2f^3mn + 160c^3df^4mn - 56cd^2ef^4mn + 34c^2df^5mn - 6c^6den^2$$
$$+35c^4d^2e^2n^2 - 36c^2d^3e^3n^2 - 65c^5defn^2 + 129c^3d^2e^2fn^2 - 41cd^3e^3fn^2$$
$$+12c^6f^2n^2 - 121c^4def^2n^2 + 102c^2d^2e^2f^2n^2 - 9d^3e^3f^2n^2 + 10c^5f^3n^2$$
$$-57c^3def^3n^2 + 14cd^2e^2f^3n^2 - 10c^4f^4n^2 + 5c^2def^4n^2 - 2d^2e^2f^4n^2$$
$$-10c^3f^5n^2 + 4cdef^5n^2 - 2c^2f^6n^2;$$
$$G_{1,4} = 6c^3d^4eg^2 + 27cd^5e^2g^2 - 12c^4d^3fg^2 - 43c^2d^4efg^2 + 31d^5e^2fg^2$$
$$-2c^3d^3f^2g^2 - 63cd^4ef^2g^2 + 14c^2d^3f^3g^2 - 6d^4ef^3g^2 - 12c^4d^3egh$$
$$-74c^2d^4e^2gh - 4d^5e^3gh + 36c^5d^2fgh + 128c^3d^3efgh - 88cd^4e^2fgh$$
$$+18c^4d^2f^2gh + 204c^2d^3ef^2gh - 14d^4e^2f^2gh - 40c^3d^2f^3gh + 52cd^3ef^3gh$$
$$-14c^2d^2f^4gh + 48c^3d^3e^2h^2 + 8cd^4e^3h^2 - 24c^6dfh^2 - 100c^4d^2efh^2$$
$$+48c^2d^3e^2fh^2 - 20c^5df^2h^2 - 148c^3d^2ef^2h^2 + 20cd^3e^2f^2h^2 + 20c^4df^3h^2$$
$$-64c^2d^2ef^3h^2 + 4d^3e^2f^3h^2 + 20c^3df^4h^2 - 8cd^2ef^4h^2 + 4c^2df^5h^2$$
$$+12c^3d^3e^2gk - 28cd^4e^3gk - 12c^6dfgk - 14c^4d^2efgk + 116c^2d^3e^2fgk$$
$$-32d^4e^3fgk - 34c^5df^2gk - 126c^3d^2ef^2gk + 134cd^3e^2f^2gk + 2c^4df^3gk$$
$$-148c^2d^2ef^3gk + 22d^3e^2f^3gk + 34c^3df^4gk - 32cd^2ef^4gk + 10c^2df^5gk$$
$$+12c^6dehk - 22c^4d^2e^2hk + 32c^2d^3e^3hk + 4d^4e^4hk + 12c^7fhk$$
$$+48c^5defhk - 148c^3d^2e^2fhk + 28cd^3e^3fhk + 46c^6f^2hk + 168c^4def^2hk$$
$$-180c^2d^2e^2f^2hk + 4d^3e^3f^2hk + 20c^5f^3hk + 212c^3def^3hk - 64cd^2e^2f^3hk$$
$$-40c^4f^4hk + 92c^2def^4hk - 6d^2e^2f^4hk - 32c^3f^5hk + 12cdef^5hk$$
$$-6c^2f^6hk - 6c^7ek^2 + 17c^5de^2k^2 - 9c^3d^2e^3k^2 + 5cd^3e^4k^2 - 35c^6efk^2$$
$$+59c^4de^2fk^2 - 26c^2d^2e^3fk^2 + 5d^3e^4fk^2 - 68c^5ef^2k^2 + 73c^3de^2f^2k^2$$
$$-20cd^2e^3f^2k^2 - 58c^4ef^3k^2 + 37c^2de^2f^3k^2 - 3d^2e^3f^3k^2 - 22c^3ef^4k^2$$
$$+6cde^2f^4k^2 - 3c^2ef^5k^2 + 6c^4d^4gl + 7c^2d^5egl - 4d^6e^2gl + 5c^3d^4fgl$$
$$+32cd^5efgl - 28c^2d^4f^2gl + 17d^5ef^2gl - 29cd^4f^3gl - 6d^4f^4gl - 12c^5d^3hl$$
$$-14c^3d^4ehl + 10cd^5e^2hl - 10c^4d^3fhl - 80c^2d^4efhl + 6d^5e^2fhl$$
$$+70c^3d^3f^2hl - 70cd^4ef^2hl + 88c^2d^3f^3hl - 12d^4ef^3hl + 24cd^3f^4hl$$
$$+6c^6d^2kl + 7c^4d^3ekl - 5c^2d^4e^2kl + 5c^5d^2fkl + 58c^3d^3efkl - 16cd^4e^2fkl$$
$$-53c^4d^2f^2kl + 109c^2d^3ef^2kl - 11d^4e^2f^2kl - 105c^3d^2f^3kl + 70cd^3ef^3kl$$
$$-65c^2d^2f^4kl + 12d^3ef^4kl - 12cd^2f^5kl + 6c^3d^5l^2 - 5cd^6el^2 + 23c^2d^5fl^2$$
$$-5d^6efl^2 + 23cd^5f^2l^2 + 6d^5f^3l^2 - 12c^5d^3gm + 10c^3d^4egm + 50cd^5e^2gm$$
$$-34c^4d^3fgm - 88c^2d^4efgm + 46d^5e^2fgm + 38c^3d^3f^2gm - 78cd^4ef^2gm$$
$$+56c^2d^3f^3gm + 12d^4ef^3gm + 24c^6d^2hm - 20c^4d^3ehm - 76c^2d^4e^2hm$$
$$+68c^5d^2fhm + 152c^3d^3efhm - 96cd^4e^2fhm - 76c^4d^2f^2hm$$
$$+212c^2d^3ef^2hm - 20d^4e^2f^2hm - 164c^3d^2f^3hm + 40cd^3ef^3hm$$
$$-44c^2d^2f^4hm - 12c^7dkm + 10c^5d^2ekm + 2c^3d^3e^2km - 22cd^4e^3km$$
$$-34c^6dfkm - 40c^4d^2efkm + 104c^2d^3e^2fkm - 26d^4e^3fkm + 38c^5df^2km$$

$$-248c^3d^2ef^2km + 122cd^3e^2f^2km + 190c^4df^3km - 250c^2d^2ef^3km$$
$$+12d^3e^2f^3km + 166c^3df^4km - 48cd^2ef^4km + 36c^2df^5km - 24c^4d^4lm$$
$$+32c^2d^5elm - 4d^6e^2lm - 104c^3d^4flm + 52cd^5eflm - 120c^2d^4f^2lm$$
$$+12d^5ef^2lm - 36cd^4f^3lm + 24c^5d^3m^2 - 44c^3d^4em^2 + 32cd^5e^2m^2$$
$$+116c^4d^3fm^2 - 132c^2d^4efm^2 + 24d^5e^2fm^2 + 172c^3d^3f^2m^2$$
$$-72cd^4ef^2m^2 + 72c^2d^3f^3m^2 + 6c^6d^2gn - 29c^4d^3egn - 65c^2d^4e^2gn$$
$$+41c^5d^2fgn + 58c^3d^3efgn - 96cd^4e^2fgn + 7c^4d^2f^2gn + 125c^2d^3ef^2gn$$
$$-31d^4e^2f^2gn - 41c^3d^2f^3gn + 38cd^3ef^3gn - 13c^2d^2f^4gn - 12c^7dhn$$
$$+58c^5d^2ehn + 82c^3d^3e^2hn + 2cd^4e^3hn - 82c^6dfhn - 32c^4d^2efhn$$
$$+152c^2d^3e^2fhn - 2d^4e^3fhn - 50c^5df^2hn - 200c^3d^2ef^2hn$$
$$+90cd^3e^2f^2hn + 70c^4df^3hn - 138c^2d^2ef^3hn + 12d^3e^2f^3hn + 62c^3df^4hn$$
$$-24cd^2ef^4hn + 12c^2df^5hn + 6c^8kn - 29c^6dekn + 31c^4d^2e^2kn$$
$$+11c^2d^3e^3kn + 4d^4e^4kn + 41c^7fkn - 110c^5defkn - 7c^3d^2e^2fkn$$
$$+16cd^3e^3fkn + 79c^6f^2kn - 26c^4def^2kn - 90c^2d^2e^2f^2kn + 13d^3e^3f^2kn$$
$$+10c^5f^3kn + 140c^3def^3kn - 65cd^2e^2f^3kn - 76c^4f^4kn + 103c^2def^4kn$$
$$-9d^2e^2f^4kn - 51c^3f^5kn + 18cdef^5kn - 9c^2f^6kn + 12c^5d^3ln$$
$$-34c^3d^4eln + 16cd^5e^2ln + 70c^4d^3fln - 102c^2d^4efln + 12d^5e^2fln$$
$$+122c^3d^3f^2ln - 72cd^4ef^2ln + 72c^2d^3f^3ln - 12d^4ef^3ln + 12cd^3f^4ln$$
$$-24c^6d^2mn + 80c^4d^3emn - 118c^2d^4e^2mn + 4d^5e^3mn - 152c^5d^2fmn$$
$$+400c^3d^3efmn - 152cd^4e^2fmn - 354c^4d^2f^2mn + 400c^2d^3ef^2mn$$
$$-34d^4e^2f^2mn - 276c^3d^2f^3mn + 92cd^3ef^3mn - 58c^2d^2f^4mn + 6c^7dn^2$$
$$-29c^5d^2en^2 + 91c^3d^3e^2n^2 - 15cd^4e^3n^2 + 47c^6dfn^2 - 247c^4d^2efn^2$$
$$+200c^2d^3e^2fn^2 - 11d^4e^3fn^2 + 174c^5df^2n^2 - 403c^3d^2ef^2n^2$$
$$+118cd^3e^2f^2n^2 + 224c^4df^3n^2 - 215c^2d^2ef^3n^2 + 17d^3e^2f^3n^2$$
$$+108c^3df^4n^2 - 34cd^2ef^4n^2 + 17c^2df^5n^2;$$

$$\sigma_{1,0} = 12c^4d^2e^4 - 22c^2d^3e^5 + 8d^4e^6 - 24c^5de^3f + 114c^3d^2e^4f - 76cd^3e^5f$$
$$+12c^6e^2f^2 - 162c^4de^3f^2 + 252c^2d^2e^4f^2 - 22d^3e^5f^2 + 70c^5e^2f^3$$
$$-308c^3de^3f^3 + 114cd^2e^4f^3 + 124c^4e^2f^4 - 162c^2de^3f^4 + 12d^2e^4f^4$$
$$+70c^3e^2f^5 - 24cde^3f^5 + 12c^2e^2f^6;$$

$$\sigma_{1,1} = 6c^5d^2e^3 + 11c^3d^3e^4 - 4cd^4e^5 + 12c^6de^2f - 51c^4d^2e^3f + 27c^2d^3e^4f$$
$$+4d^4e^5f - 6c^7ef^2 + 69c^5de^2f^2 - 69c^3d^2e^3f^2 - 27cd^3e^4f^2 - 29c^6ef^3$$
$$+73c^4de^2f^3 + 69c^2d^2e^3f^3 - 11d^3e^4f^3 - 27c^5ef^4 - 73c^3de^2f^4$$
$$+51cd^2e^3f^4 + 27c^4ef^5 - 69c^2de^2f^5 + 6d^2e^3f^5 + 29c^3ef^6$$
$$-12cde^2f^6 + 6c^2ef^7;$$

$$\sigma_{1,2} = 6c^6d^2e^2 - 23c^4d^3e^3 + 26c^2d^4e^4 - 8d^5e^5 - 12c^7def + 69c^5d^2e^2f$$
$$-130c^3d^3e^3f + 68cd^4e^4f + 6c^8f^2 - 69c^6def^2 + 180c^4d^2e^2f^2$$
$$-198c^2d^3e^3f^2 + 26d^4e^4f^2 + 23c^7f^3 - 74c^5def^3 + 170c^3d^2e^2f^3$$
$$-130cd^3e^3f^3 - 2c^6f^4 + 22c^4def^4 + 180c^2d^2e^2f^4 - 23d^3e^3f^4 - 54c^5f^5$$
$$-74c^3def^5 + 69cd^2e^2f^5 - 2c^4f^6 - 69c^2def^6 + 6d^2e^2f^6 + 23c^3f^7$$
$$-12cdef^7 + 6c^2f^8;$$

$$\sigma_{1,3} = 6c^5d^3e^2 - 11c^3d^4e^3 + 4cd^5e^4 - 12c^6d^2ef + 51c^4d^3e^2f - 27c^2d^4e^3f$$
$$-4d^5e^4f + 6c^7df^2 - 69c^5d^2ef^2 + 69c^3d^3e^2f^2 + 27cd^4e^3f^2 + 29c^6df^3$$
$$-73c^4d^2ef^3 - 69c^2d^3e^2f^3 + 11d^4e^3f^3 + 27c^5df^4 + 73c^3d^2ef^4$$
$$-51cd^3e^2f^4 - 27c^4df^5 + 69c^2d^2ef^5 - 6d^3e^2f^5 - 29c^3df^6 + 12cd^2ef^6$$
$$-6c^2df^7;$$

$$\sigma_{1,4} = 12c^4d^4e^2 - 22c^2d^5e^3 + 8d^6e^4 - 24c^5d^3ef + 114c^3d^4e^2f$$
$$-76cd^5e^3f + 12c^6d^2f^2 - 162c^4d^3ef^2 + 252c^2d^4e^2f^2 - 22d^5e^3f^2$$
$$+70c^5d^2f^3 - 308c^3d^3ef^3 + 114cd^4e^2f^3 + 124c^4d^2f^4 - 162c^2d^3ef^4$$
$$+12d^4e^2f^4 + 70c^3d^2f^5 - 24cd^3ef^5 + 12c^2d^2f^6;$$

$$B_{1,i} = 4k_i(c+f)(cf-de)^2(2c^2-de+5cf+2f^2)(3c^2-4de+10cf+3f^2),$$
$$(k_0 = k_1 = k_3 = k_4 = 1, \ k_2 = 3).$$

$$(.38)$$

Appendix 4

Matrices that Define a Linear System of Equations $A_2 B_2 = C_2$ for the Quantity G_2 in Case of the Differential Systems$(1, 2)$

$$A_2 = [A_2' | A_2'' | A_2''' | A_2''''],$$

$$A_2' = \begin{pmatrix}
3c & 3e & 0 & 0 & 0 & 0 & 0 \\
3d & 3(2c+f) & 6e & 0 & 0 & 0 & 0 \\
0 & 6d & 3(2c+f) & 3e & 0 & 0 & 0 \\
0 & 0 & 3d & 3f & 0 & 0 & 0 \\
3g & 3l & 0 & 0 & 4c & 4e & 0 \\
6h & 6(g+m) & 6l & 0 & 4d & 4(f+3c) & 12e \\
3k & 3(4h+n) & 3(g+4m) & 3l & 0 & 12d & 12(c+f) \\
0 & 6k & 6(h+n) & 6m & 0 & 0 & 12d \\
0 & 0 & 3k & 3n & 0 & 0 & 0 \\
0 & 0 & 0 & 0 & 4g & 4l & 0 \\
0 & 0 & 0 & 0 & 8h & 4(3g+2m) & 12l \\
0 & 0 & 0 & 0 & 4k & 4(6h+n) & 12(g+2m) \\
0 & 0 & 0 & 0 & 0 & 12k & 12(2h+n) \\
0 & 0 & 0 & 0 & 0 & 0 & 12k \\
0 & 0 & 0 & 0 & 0 & 0 & 0 \\
0 & 0 & 0 & 0 & 0 & 0 & 0 \\
0 & 0 & 0 & 0 & 0 & 0 & 0 \\
0 & 0 & 0 & 0 & 0 & 0 & 0 \\
0 & 0 & 0 & 0 & 0 & 0 & 0 \\
0 & 0 & 0 & 0 & 0 & 0 & 0 \\
0 & 0 & 0 & 0 & 0 & 0 & 0
\end{pmatrix},$$

$$A_2'' = \begin{pmatrix}
0 & 0 & 0 & 0 & 0 & 0 \\
0 & 0 & 0 & 0 & 0 & 0 \\
0 & 0 & 0 & 0 & 0 & 0 \\
0 & 0 & 0 & 0 & 0 & 0 \\
0 & 0 & -e^2 & 0 & 0 & 0 \\
0 & 0 & 2e(c-f) & 0 & 0 & 0 \\
12e & 0 & 2de-(c-f)^2 & 0 & 0 & 0 \\
4(3f+c) & 4l & 2d(f-c) & 0 & 0 & 0 \\
4d & 4f & -d^2 & 0 & 0 & 0 \\
0 & 0 & 0 & 5c & 5e & 0 \\
0 & 0 & 0 & 5d & 5(4c+f) & 20e \\
12l & 0 & 0 & 0 & 20d & 10(3c+2f) \\
4(g+6m) & 4l & 0 & 0 & 0 & 30d \\
4(2h+3n) & 8m & 0 & 0 & 0 & 0 \\
4k & 4n & 0 & 0 & 0 & 0 \\
0 & 0 & 0 & 5g & 5l & 0 \\
0 & 0 & 0 & 10h & 10(2g+m) & 20l \\
0 & 0 & 0 & 5k & 5(8h+n) & 10(3g+4m) \\
0 & 0 & 0 & 0 & 20k & 20(3h+n) \\
0 & 0 & 0 & 0 & 0 & 30k \\
0 & 0 & 0 & 0 & 0 & 0 \\
0 & 0 & 0 & 0 & 0 & 0
\end{pmatrix},$$

$$A_2''' = \begin{pmatrix}
0 & 0 & 0 & 0 & 0 & 0 \\
0 & 0 & 0 & 0 & 0 & 0 \\
0 & 0 & 0 & 0 & 0 & 0 \\
0 & 0 & 0 & 0 & 0 & 0 \\
0 & 0 & 0 & 0 & 0 & 0 \\
0 & 0 & 0 & 0 & 0 & 0 \\
0 & 0 & 0 & 0 & 0 & 0 \\
0 & 0 & 0 & 0 & 0 & 0 \\
0 & 0 & 0 & 0 & 0 & 0 \\
0 & 0 & 0 & 0 & 0 & 0 \\
0 & 0 & 0 & 0 & 0 & 0 \\
30e & 0 & 0 & 0 & 0 & 0 \\
10(2c+3f) & 20e & 0 & 0 & 0 & 0 \\
20d & 5(c+4f) & 5e & 0 & 0 & 0 \\
0 & 5d & 5f & 0 & 0 & 0 \\
0 & 0 & 0 & 6c & 6e & 0 \\
0 & 0 & 0 & 6d & 6(5c+f) & 30e \\
30l & 0 & 0 & 0 & 30d & 30(2c+f) \\
20(g+3m) & 20l & 0 & 0 & 0 & 60d \\
10(4h+3n) & 5(g+8m) & 5l & 0 & 0 & 0 \\
20k & 10(h+2n) & 10m & 0 & 0 & 0 \\
0 & 5k & 5n & 0 & 0 & 0
\end{pmatrix},$$

$$
A_2'''' = \begin{pmatrix}
0 & 0 & 0 & 0 & 0 \\
0 & 0 & 0 & 0 & 0 \\
0 & 0 & 0 & 0 & 0 \\
0 & 0 & 0 & 0 & 0 \\
0 & 0 & 0 & 0 & 0 \\
0 & 0 & 0 & 0 & 0 \\
0 & 0 & 0 & 0 & 0 \\
0 & 0 & 0 & 0 & 0 \\
0 & 0 & 0 & 0 & 0 \\
0 & 0 & 0 & 0 & 0 \\
0 & 0 & 0 & 0 & 0 \\
0 & 0 & 0 & 0 & 0 \\
0 & 0 & 0 & 0 & 0 \\
0 & 0 & 0 & 0 & 0 \\
0 & 0 & 0 & 0 & 0 \\
0 & 0 & 0 & 0 & e^3 \\
0 & 0 & 0 & 0 & 3e^2(f-c) \\
60e & 0 & 0 & 0 & 3e[(c-f)^2 - de] \\
60(c+f) & 60e & 0 & 0 & (f-c)[(c-f)^2 - 6de] \\
60d & 30(c+2f) & 30e & 0 & 3d[de - (c-f)^2] \\
0 & 30d & 6(c+5f) & 6e & 3d^2(f-c) \\
0 & 0 & 6d & 6f & -d^3
\end{pmatrix},
$$

$$B_2 = \begin{pmatrix} a_0 \\ a_1 \\ a_2 \\ a_3 \\ b_0 \\ b_1 \\ b_2 \\ b_3 \\ b_4 \\ G_1 \\ c_0 \\ c_1 \\ c_2 \\ c_3 \\ c_4 \\ c_5 \\ d_0 \\ d_1 \\ d_2 \\ d_3 \\ d_4 \\ d_5 \\ d_6 \\ G_2 \end{pmatrix} , \; C_2 = \begin{pmatrix} 2eg + (f-c)l \\ (f-c)(g+2m) - 2dl + 4eh \\ (f-c)(2h+n) + 2ek - 4dm \\ (f-c)k - 2dn \\ 0 \\ 0 \\ 0 \\ 0 \\ 0 \\ 0 \\ 0 \\ 0 \\ 0 \\ 0 \\ 0 \\ 0 \\ 0 \\ 0 \\ 0 \\ 0 \\ 0 \\ 0 \\ 0 \end{pmatrix} .$$

Appendix 5

Polynomials that Define the Quantity
G_1 for the System $s(1,3)$
Expressions of Focus Pseudo-Quantities $G_{1,i}$, and $\sigma_{1,i}$,
$B_{1,i}$ $(i = 0, 1, 2, 3, 4)$

$G_{1,0} = 7c^2de^2p - 6d^2e^3p - 7c^3efp + 35cde^2fp - 29c^2ef^2p + 22de^2f^2p$
$-25cef^3p - 3ef^4p - 9cde^3q + 9c^2e^2fq - 21de^3fq + 30ce^2f^2q + 9e^2f^3q$
$+6de^4r - 15ce^3fr - 9e^3f^2r + 3ce^4s + 3e^4fs - 3c^3det + 5cd^2e^2t + 3c^4ft$
$-15c^2deft + 5d^2e^2ft + 10c^3f^2t - 5cdef^2t + 7def^3t - 10cf^4t - 3f^5t$
$+3c^2de^2u - 6d^2e^3u - 3c^3efu + 15cde^2fu - 9c^2ef^2u + 6de^2f^2u + 3cef^3u$
$+9ef^4u - 3cde^3v + 3c^2e^2fv - 15de^3fv + 6ce^2f^2v - 9e^2f^3v + 6de^4w$
$-3ce^3fw + 3e^3f^2w;$

$G_{1,1} = -c^3dep - 2cd^2e^2p + c^4fp - c^2defp - 8d^2e^2fp + 3c^3f^2p$
$+12cdef^2p - c^2f^3p + 6def^3p - 3cf^4p + 9c^2de^2q - 9c^3efq + 21cde^2fq$
$-30c^2ef^2q - 9cef^3q - 6cde^3r + 15c^2e^2fr + 9ce^2f^2r - 3c^2e^3s - 3ce^3fs$
$-c^2d^2et + c^3dft - 5cd^2eft + 5c^2df^2t - 4d^2ef^2t + 7cdf^3t + 3df^4t$
$-3c^3deu + 6cd^2e^2u + 3c^4fu - 15c^2defu + 9c^3f^2u - 6cdef^2u - 3c^2f^3u$
$-9cf^4u + 3c^2de^2v - 3c^3efv + 15cde^2fv - 6c^2ef^2v + 9cef^3v - 6cde^3w$
$+3c^2e^2fw - 3ce^2f^2w;$

$G_{1,2} = 3(c^2d^2ep + 2d^3e^2p - c^3dfp - 3cd^2efp - 2c^2df^2p - 6d^2ef^2p$
$+3cdf^3p - 3c^3deq - 5cd^2e^2q + 3c^4fq + 7c^2defq - d^2e^2fq + 7c^3f^2q$
$+17cdef^2q - 7c^2f^3q + 3def^3q - 3cf^4q + 6c^2de^2r - 2d^2e^3r - 15c^3efr$
$+7cde^2fr - 14c^2ef^2r + 3de^2f^2r - 3cef^3r + 3c^3e^2s - cde^3s + 4c^2e^2fs$
$-de^3fs + ce^2f^2s + cd^3et - c^2d^2ft + d^3eft - 4cd^2f^2t - 3d^2f^3t$
$-3c^2d^2eu + 2d^3e^2u + 3c^3dfu - 7cd^2efu + 14c^2df^2u - 6d^2ef^2u$
$+15cdf^3u - 3c^3dev + cd^2e^2v + 3c^4fv - 17c^2defv + 5d^2e^2fv + 7c^3f^2v$
$-7cdef^2v - 7c^2f^3v + 3def^3v - 3cf^4v + 6c^2de^2w - 2d^2e^3w - 3c^3efw$
$+3cde^2fw + 2c^2ef^2w - de^2f^2w + cef^3w);$

$G_{1,3} = 3c^2d^2fp + 6d^3efp - 3cd^2f^2p - 9c^3dfq - 15cd^2efq + 6c^2df^2q$
$-3d^2ef^2q + 3cdf^3q + 9c^4fr + 6c^2defr - 6d^2e^2fr + 3c^3f^2r + 15cdef^2r$
$-9c^2f^3r + 3def^3r - 3cf^4r - 3c^4es + 4c^2de^2s - 7c^3efs + 5cde^2fs$
$-5c^2ef^2s + de^2f^2s - cef^3s + 3cd^3ft + 3d^3f^2t - 9c^2d^2fu + 6d^3efu$
$-15cd^2f^2u + 9c^3dfv - 21cd^2efv + 30c^2df^2v - 9d^2ef^2v + 9cdf^3v$
$-6c^3dew + 8cd^2e^2w + 3c^4fw - 12c^2defw + 2d^2e^2fw + c^3f^2w$
$+cdef^2w - 3c^2f^3w + def^3w - cf^4w;$

$G_{1,4} = -(3c^2d^3p + 6d^4ep - 3cd^3fp - 9c^3d^2q - 15cd^3eq + 6c^2d^2fq$
$-3d^3efq + 3cd^2f^2q + 9c^4dr + 6c^2d^2er - 6d^3e^2r + 3c^3dfr + 15cd^2efr$

$$-9c^2df^2r + 3d^2ef^2r - 3cdf^3r - 3c^5s + 7c^3des + 5cd^2e^2s - 10c^4fs$$
$$-5c^2defs + 5d^2e^2fs - 15cdef^2s + 10c^2f^3s - 3def^3s + 3cf^4s + 3cd^4t$$
$$+3d^4ft - 9c^2d^3u + 6d^4eu - 15cd^3fu + 9c^3d^2v - 21cd^3ev + 30c^2d^2fv$$
$$-9d^3efv + 9cd^2f^2v - 3c^4dw + 22c^2d^2ew - 6d^3e^2w - 25c^3dfw + 35cd^2efw$$
$$-29c^2df^2w + 7d^2ef^2w - 7cdf^3w);$$
$$\sigma_{1,0} = -6c^2de^3 + 8d^2e^4 + 6c^3e^2f - 28cde^3f + 20c^2e^2f^2 - 6de^3f^2 + 6ce^2f^3;$$
$$\sigma_{1,1} = 3c^3de^2 - 4cd^2e^3 - 3c^4ef + 11c^2de^2f + 4d^2e^3f - 7c^3ef^2 - 11cde^2f^2$$
$$+7c^2ef^3 - 3de^2f^3 + 3cef^4;$$
$$\sigma_{1,2} = -3c^4de + 10c^2d^2e^2 - 8d^3e^3 + 3c^5f - 14c^3def + 20cd^2e^2f + 4c^4f^2$$
$$+2c^2def^2 + 10d^2e^2f^2 - 14c^3f^3 - 14cdef^3 + 4c^2f^4 - 3def^4 + 3cf^5;$$
$$\sigma_{1,3} = -3c^3d^2e + 4cd^3e^2 + 3c^4df - 11c^2d^2ef - 4d^3e^2f + 7c^3df^2 + 11cd^2ef^2$$
$$-7c^2df^3 + 3d^2ef^3 - 3cdf^4;$$
$$\sigma_{1,4} = -6c^2d^3e + 8d^4e^2 + 6c^3d^2f - 28cd^3ef + 20c^2d^2f^2 - 6d^3ef^2 + 6cd^2f^3;$$
$$B_{1,i} = 4k_i(c+f)(-de+cf)(3c^2 - 4de + 10cf + 3f^2),$$
$$(k_0 = k_1 = k_3 = k_4 = 1, \; k_2 = 3).$$

Appendix 6

Matrices that Define a Linear System of Equations
$A_2 B_2 = C_2$ for the Quantity G_2 in Case of the Differential
System $s(1, 3)$

$$A_2 = [A_2'|A_2''|A_2'''],$$

$$A_2' = \begin{pmatrix}
4c & 4e & 0 & 0 & 0 \\
4d & 12c + 4f & 12e & 0 & 0 \\
0 & 12d & 12c + 12f & 12e & 0 \\
0 & 0 & 12d & 4c + 12f & 4e \\
0 & 0 & 0 & 4d & 4f \\
4p & 4t & 0 & 0 & 0 \\
12q & 12p + 12u & 12t & 0 & 0 \\
12r & 36q + 12v & 12p + 36u & 12t & 0 \\
4s & 36r + 4w & 36q + 36v & 4p + 36u & 4t \\
0 & 12s & 36r + 12w & 12q + 36v & 12u \\
0 & 0 & 12s & 12r + 12w & 12v \\
0 & 0 & 0 & 4s & 4w
\end{pmatrix},$$

$$A_2'' = \begin{pmatrix}
-e^2 & 0 & 0 & 0 & 0 \\
2ce - 2ef & 0 & 0 & 0 & 0 \\
-c^2 + 2de + 2cf - f^2 & 0 & 0 & 0 & 0 \\
-2cd + 2df & 0 & 0 & 0 & 0 \\
-d^2 & 0 & 0 & 0 & 0 \\
0 & 6c & 6e & 0 & 0 \\
0 & 6d & 30c + 6f & 30e & 0 \\
0 & 0 & 30d & 60c + 30f & 60e \\
0 & 0 & 0 & 60d & 0 \\
0 & 0 & 0 & 0 & 60d \\
0 & 0 & 0 & 0 & 0 \\
0 & 0 & 0 & 0 & 0
\end{pmatrix},$$

$$A_2''' = \begin{pmatrix} 0 & 0 & 0 & 0 \\ 0 & 0 & 0 & 0 \\ 0 & 0 & 0 & 0 \\ 0 & 0 & 0 & 0 \\ 0 & 0 & 0 & 0 \\ 0 & 0 & 0 & e^3 \\ 0 & 0 & 0 & -3e^2(c-f) \\ 0 & 0 & 0 & 3e[(c-f)^2 - de] \\ 60e & 0 & -(c-f)[(c-f)^2 - 6de] & \\ 30c + 60f & 30e & -3d[(c-f)^2 - de] & \\ 30d & 6c + 30f & 6e & -3d^2(c-f) \\ 0 & 6d & 6f & -d^3 \end{pmatrix},$$

$$B_2 = \begin{pmatrix} b_0 \\ b_1 \\ b_2 \\ b_3 \\ b_4 \\ G_1 \\ d_0 \\ d_1 \\ d_2 \\ d_3 \\ d_4 \\ d_5 \\ d_6 \\ G_2 \end{pmatrix}, \quad C_2 = \begin{pmatrix} 2ep - ct + ft \\ -cp + fp + 6eq - 2dt - 3cu + 3fu \\ -3cq + 3fq + 6er - 6du - 3cv + 3fv \\ -3cr + 3fr + 2es - 6dv - cw + fw \\ -cs + fs - 2dw \\ 0 \\ 0 \\ 0 \\ 0 \\ 0 \\ 0 \\ 0 \end{pmatrix}.$$

Appendix 7

Matrices that Define a Linear System of Equations
$A_3 B_3 = C_3$ for the Quantity G_3 in Case of the Differential
System $s(1,4)$

$$A_3 = [A_3'|A_3''|A_3'''|A_3''''],$$

$$
A_3' = \begin{pmatrix}
5c & 5e & 0 & 0 & 0 & 0 \\
5d & 20c+5f & 20e & 0 & 0 & 0 \\
0 & 20d & 30c+20f & 30e & 0 & 0 \\
0 & 0 & 30d & 20c+30f & 20e & 0 \\
0 & 0 & 0 & 20d & 5c+20f & 5e \\
0 & 0 & 0 & 0 & 5d & 5f \\
5g & 5l & 0 & 0 & 0 & 0 \\
20h & 20g+20m & 20l & 0 & 0 & 0 \\
30i & 80h+30n & 30g+80m & 30l & 0 & 0 \\
20j & 120i+20o & 120h+120n & 20g+120m & 20l & 0 \\
5k & 80j+5p & 180i+80o & 80h+180n & 5g+80m & 5l \\
0 & 20k & 120j+20p & 120i+120o & 20h+120n & 20m \\
0 & 0 & 30k & 80j+30p & 30i+80o & 30n \\
0 & 0 & 0 & 20k & 20j+20p & 20o \\
0 & 0 & 0 & 0 & 5k & 5p
\end{pmatrix},
$$

$$
A_3'' = \begin{pmatrix}
0 & 0 & 0 & 0 & 0 \\
0 & 0 & 0 & 0 & 0 \\
0 & 0 & 0 & 0 & 0 \\
0 & 0 & 0 & 0 & 0 \\
0 & 0 & 0 & 0 & 0 \\
0 & 0 & 0 & 0 & 0 \\
8c & 8e & 0 & 0 & 0 \\
8d & 56c+8f & 56e & 0 & 0 \\
0 & 56d & 168c+56f & 168e & 0 \\
0 & 0 & 168d & 280c+168f & 280e \\
0 & 0 & 0 & 280d & 280c+280f \\
0 & 0 & 0 & 0 & 280d \\
0 & 0 & 0 & 0 & 0 \\
0 & 0 & 0 & 0 & 0 \\
0 & 0 & 0 & 0 & 0
\end{pmatrix},
$$

$$A_3''' = \begin{pmatrix} 0 & 0 & 0 & 0 \\ 0 & 0 & 0 & 0 \\ 0 & 0 & 0 & 0 \\ 0 & 0 & 0 & 0 \\ 0 & 0 & 0 & 0 \\ 0 & 0 & 0 & 0 \\ 0 & 0 & 0 & 0 \\ 0 & 0 & 0 & 0 \\ 0 & 0 & 0 & 0 \\ 0 & 0 & 0 & 0 \\ 280e & 0 & 0 & 0 \\ 168c + 280f & 168e & 0 & 0 \\ 168d & 56c + 168f & 56e & 0 \\ 0 & 56d & 8c + 56f & 8e \\ 0 & 0 & 8d & 8f \end{pmatrix},$$

$$A_3'''' = \begin{pmatrix} 0 \\ 0 \\ 0 \\ 0 \\ 0 \\ 0 \\ -e^4 \\ 4ce^3 - 4e^3 f \\ -6c^2 e^2 + 4de^3 + 12ce^2 f - 6e^2 f^2 \\ 4c^3 e - 12cde^2 - 12c^2 ef + 12de^2 f + 12cef^2 - 4ef^3 \\ -c^4 + 12c^2 de - 6d^2 e^2 + 4c^3 f - 24cdef - 6c^2 f^2 + 12def^2 \\ +4cf^3 - f^4 \\ -4c^3 d + 12cd^2 e + 12c^2 df - 12d^2 ef - 12cdf^2 + 4df^3 \\ -6c^2 d^2 + 4d^3 e + 12cd^2 f - 6d^2 f^2 \\ -4cd^3 + 4d^3 f \\ -d^4 \end{pmatrix},$$

$$B_3 = \begin{pmatrix} c_0 \\ c_1 \\ c_2 \\ c_3 \\ c_4 \\ c_5 \\ f_0 \\ f_1 \\ f_2 \\ f_3 \\ f_4 \\ f_5 \\ f_6 \\ f_7 \\ f_8 \\ G_3 \end{pmatrix}, \quad C_3 = \begin{pmatrix} 2eg - cl + fl \\ -cg + fg + 8eh - 2dl - 4cm + 4fm \\ -4ch + 4fh + 12ei - 8dm - 6cn + 6fn \\ -6ci + 6fi + 8ej - 12dn - 4co + 4fo \\ -4cj + 4fj + 2ek - 8do - cp + fp \\ -ck + fk - 2dp \\ 0 \\ 0 \\ 0 \\ 0 \\ 0 \\ 0 \\ 0 \\ 0 \end{pmatrix}.$$

Appendix 8

Matrices that Define a Linear System of Equations $A_2 B_2 = C_2$ for the Quantity G_2 in Case of Differential System $s(1,5)$

$$A_2 = [A_2' | A_2''],$$

$$A_2' = \begin{pmatrix} 6c & 6e & 0 & 0 & 0 & 0 \\ 6d & 30c + 6f & 30e & 0 & 0 & 0 \\ 0 & 30d & 60c + 30f & 60e & 0 & 0 \\ 0 & 0 & 60d & 60c + 60f & 60e & 0 \\ 0 & 0 & 0 & 60d & 30c + 60f & 30e \\ 0 & 0 & 0 & 0 & 30d & 6c + 30f \\ 0 & 0 & 0 & 0 & 0 & 6d \end{pmatrix},$$

$$A_2'' = \begin{pmatrix} 0 & e^3 \\ 0 & -3ce^2 + 3e^2 f \\ 0 & 3c^2 e - 3de^2 - 6cef + 3ef^2 \\ 0 & -c^3 + 6cde + 3c^2 f - 6def - 3cf^2 + f^3 \\ 0 & -3c^2 d + 3d^2 e + 6cdf - 3df^2 \\ 6e & -3cd^2 + 3d^2 f \\ 6f & -d^3 \end{pmatrix},$$

$$B_2 = \begin{pmatrix} d_0 \\ d_1 \\ d_2 \\ d_3 \\ d_4 \\ d_5 \\ d_6 \\ G_2 \end{pmatrix}, \quad C_2 = \begin{pmatrix} 2eg - cp + fp \\ -cg + fg + 10eh - 2dp - 5cq + 5fq \\ -5ch + 5fh + 20ek - 10dq - 10cr + 10fr \\ -10ck + 10fk + 20el - 20dr - 10cs + 10fs \\ -10cl + 10fl + 10em - 20ds - 5cu + 5fu \\ -5cm + 5fm + 2en - 10du - cv + fv \\ -cn + fn - 2dv \end{pmatrix}.$$

Appendix 9

Polynomials that Define the Quantity G_2 for
the Differential System $s(1,5)$
Expressions of Focus Pseudo-Quantities $G_{2,i}$ and $\sigma_{2,i}$,
$D_{2,i}$ ($i = 0, 1, 2, 3, 4, 5, 6$).

$G_{20} = 22c^4de^2g - 76c^2d^2e^3g + 20d^3e^4g - 22c^5efg + 265c^3de^2fg$

$\quad - 282cd^2e^3fg - 189c^4ef^2g + 741c^2de^2f^2g - 186d^2e^3f^2g - 479c^3ef^3g$

$\quad + 635cde^2f^3g - 449c^2ef^4g + 137de^2f^4g - 147cef^5g - 10ef^6g - 25c^3de^3h$

$\quad + 70cd^2e^4h + 25c^4e^2fh - 275c^2de^3fh + 150d^2e^4fh + 205c^3e^2f^2h$

$\quad - 615cde^3f^2h + 465c^2e^2f^3h - 285de^3f^3h + 335ce^2f^4h + 50e^2f^5h+$

$\quad + 30c^2de^4k - 40d^2e^5k - 30c^3e^3fk + 260cde^4fk - 220c^2e^3f^2k$

$\quad + 270de^4f^2k - 370ce^3f^3k - 100e^3f^4k - 40cde^5l + 40c^2e^4fl - 120de^5fl$

$\quad + 220ce^4f^2l + 100e^4f^3l + 20de^6m - 70ce^5fm - 50e^5f^2m + 10ce^6n$

$\quad + 10e^6fn - 10c^5dep + 39c^3d^2e^2p - 22cd^3e^3p + 10c^6fp - 116c^4defp$

$\quad + 147c^2d^2e^2fp - 22d^3e^3fp + 77c^5f^2p - 265c^3def^2p + 81cd^2e^2f^2p$

$\quad + 140c^4f^3p - 59c^2def^3p - 27d^2e^2f^3p + 167cdef^4p - 140c^2f^5p + 67def^5p$

$\quad - 77cf^6p - 10f^7p + 10c^4de^2q - 40c^2d^2e^3q + 20d^3e^4q - 10c^5efq$

$\quad + 115c^3de^2fq - 150cd^2e^3fq - 75c^4ef^2q + 255c^2de^2f^2q - 90d^2e^3f^2q$

$\quad - 125c^3ef^3q + 65cde^2f^3q + 25c^2ef^4q - 85de^2f^4q + 135cef^5q + 50ef^6q$

$\quad - 10c^3de^3r + 40cd^2e^4r + 10c^4e^2fr - 110c^2de^3fr + 120d^2e^4fr$

$\quad + 70c^3e^2f^2r - 210cde^3f^2r + 90c^2e^2f^3r - 30de^3f^3r - 70ce^2f^4r$

$\quad - 100e^2f^5r + 10c^2de^4s - 40d^2e^5s - 10c^3e^3fs + 100cde^4fs - 60c^2e^3f^2s$

$\quad + 130de^4f^2s - 30ce^3f^3s + 100e^3f^4s - 10cde^5u + 10c^2e^4fu - 90de^5fu$

$\quad + 40ce^4f^2u - 50e^4f^3u + 20de^6v - 10ce^5fv + 10e^5f^2v;$

$G_{21} = 2c^5deg - 2c^3d^2e^2g - 12cd^3e^3g - 2c^6fg + 17c^4defg + 36c^2d^2e^2fg$

$\quad - 32d^3e^3fg - 15c^5f^2g + c^3def^2g + 132cd^2e^2f^2g - 25c^4f^3g$

$\quad - 105c^2def^3g + 74d^2e^2f^3g + 5c^3f^4g - 111cdef^4g + 27c^2f^5g - 20def^5g$

$\quad + 10cf^6g - 25c^4de^2h + 70c^2d^2e^3h + 25c^5efh - 275c^3de^2fh$

$\quad + 150cd^2e^3fh + 205c^4ef^2h - 615c^2de^2f^2h + 465c^3ef^3h - 285cde^2f^3h$

$$+ 335c^2ef^4h + 50cef^5h + 30c^3de^3k - 40cd^2e^4k - 30c^4e^2fk$$
$$+ 260c^2de^3fk - 220c^3e^2f^2k + 270cde^3f^2k - 370c^2e^2f^3k - 100ce^2f^4k$$
$$- 40c^2de^4l + 40c^3e^3fl - 120cde^4fl + 220c^2e^3f^2l + 100ce^3f^3l$$
$$+ 20cde^5m - 70c^2e^4fm - 50ce^4f^2m + 10c^2e^5n + 10ce^5fn + 2c^4d^2ep$$
$$- 6c^2d^3e^2p - 2c^5dfp + 25c^3d^2efp - 22cd^3e^2fp - 19c^4df^2p$$
$$+ 81c^2d^2ef^2p - 16d^3e^2f^2p - 59c^3df^3p + 95cd^2ef^3p - 79c^2df^4p$$
$$+ 37d^2ef^4p - 47cdf^5p - 10df^6p + 10c^5deq - 40c^3d^2e^2q + 20cd^3e^3q$$
$$- 10c^6fq + 115c^4defq - 150c^2d^2e^2fq - 75c^5f^2q + 255c^3def^2q$$
$$- 90cd^2e^2f^2q - 125c^4f^3q + 65c^2def^3q + 25c^3f^4q - 85cdef^4q$$
$$+ 135c^2f^5q + 50cf^6q - 10c^4de^2r + 40c^2d^2e^3r + 10c^5efr - 110c^3de^2fr$$
$$+ 120cd^2e^3fr + 70c^4ef^2r - 210c^2de^2f^2r + 90c^3ef^3r - 30cde^2f^3r$$
$$- 70c^2ef^4r - 100cef^5r + 10c^3de^3s - 40cd^2e^4s - 10c^4e^2fs$$
$$+ 100c^2de^3fs - 60c^3e^2f^2s + 130cde^3f^2s - 30c^2e^2f^3s + 100ce^2f^4s$$
$$- 10c^2de^4u + 10c^3e^3fu - 90cde^4fu + 40c^2e^3f^2u - 50ce^3f^3u + 20cde^5v$$
$$- 10c^2e^4fv + 10ce^4f^2v;$$

$$G_{22} = 5(-c^4d^2eg + 4d^4e^3g + c^5dfg - 7c^3d^2efg - 22cd^3e^2fg + 7c^4df^2g$$
$$+ 9c^2d^2ef^2g - 34d^3e^2f^2g + 9c^3df^3g + 51cd^2ef^3g - 7c^2df^4g + 20d^2ef^4g$$
$$- 10cdf^5g + 5c^5deh - c^3d^2e^2h - 18cd^3e^3h - 5c^6fh + 37c^4defh$$
$$+ 99c^2d^2e^2fh - 2d^3e^3fh - 36c^5f^2h - 29c^3def^2h + 165cd^2e^2f^2h$$
$$- 52c^4f^3h - 239c^2def^3h + 17d^2e^2f^3h + 26c^3f^4h - 124cdef^4h$$
$$+ 57c^2f^5h - 10def^5h + 10cf^6h - 30c^4de^2k + 46c^2d^2e^3k - 8d^3e^4k$$
$$+ 30c^5efk - 272c^3de^2fk + 60cd^2e^3fk + 226c^4ef^2k - 366c^2de^2f^2k$$
$$+ 54d^2e^3f^2k + 414c^3ef^3k - 128cde^2f^3k + 174c^2ef^4k - 20de^2f^4k$$
$$+ 20cef^5k + 40c^3de^3l - 8cd^2e^4l - 40c^4e^2fl + 136c^2de^3fl - 24d^2e^4fl$$
$$- 228c^3e^2f^2l + 68cde^3f^2l - 144c^2e^2f^3l + 20de^3f^3l - 20ce^2f^4l$$
$$- 20c^2de^4m + 4d^2e^5m + 70c^3e^3fm - 18cde^4fm + 64c^2e^3f^2m$$
$$- 10de^4f^2m + 10ce^3f^3m - 10c^3e^4n + 2cde^5n - 12c^2e^4fn + 2de^5fn$$
$$- 2ce^4f^2n - c^3d^3ep + 2cd^4e^2p + c^4d^2fp - 11c^2d^3efp + 2d^4e^2fp$$
$$+ 9c^3d^2f^2p - 27cd^3ef^2p + 25c^2d^2f^3p - 17d^3ef^3p + 27cd^2f^4p$$
$$+ 10d^2f^5p + 5c^4d^2eq - 12c^2d^3e^2q + 4d^4e^3q - 5c^5dfq + 59c^3d^2efq$$
$$- 34cd^3e^2fq - 47c^4df^2q + 171c^2d^2ef^2q - 34d^3e^2f^2q - 141c^3df^3q$$
$$+ 153cd^2ef^3q - 169c^2df^4q + 20d^2ef^4q - 70cdf^5q + 10c^5der$$
$$- 42c^3d^2e^2r + 8cd^3e^3r - 10c^6fr + 114c^4defr - 150c^2d^2e^2fr$$

$$+ 24d^3e^3fr - 72c^5f^2r + 246c^3def^2r - 66cd^2e^2f^2r - 104c^4f^3r$$
$$+ 90c^2def^3r - 6d^2e^2f^3r + 52c^3f^4r - 8cdef^4r + 114c^2f^5r - 20def^5r$$
$$+ 20cf^6r - 10c^4de^2s + 42c^2d^2e^3s - 8d^3e^4s + 10c^5efs - 104c^3de^2fs$$
$$+ 28cd^2e^3fs + 62c^4ef^2s - 162c^2de^2f^2s + 26d^2e^3f^2s + 42c^3ef^3s$$
$$- 32cde^2f^3s - 94c^2ef^4s + 20de^2f^4s - 20cef^5s + 10c^3de^3u - 2cd^2e^4u$$
$$- 10c^4e^2fu + 94c^2de^3fu - 18d^2e^4fu - 42c^3e^2f^2u + 26cde^3f^2u$$
$$+ 42c^2e^2f^3u - 10de^3f^3u + 10ce^2f^4u - 20c^2de^4v + 4d^2e^5v$$
$$+ 10c^3e^3fv - 6cde^4fv - 8c^2e^3f^2v + 2de^4f^2v - 2ce^3f^3v);$$

$$G_{23} = 10(-c^3d^3eg - 2cd^4e^2g + c^4d^2fg - 4c^2d^3efg - 14d^4e^2fg$$
$$+ 6c^3d^2f^2g + 21cd^3ef^2g + 3c^2d^2f^3g + 20d^3ef^3g - 10cd^2f^4g + 5c^4d^2eh$$
$$+ 9c^2d^3e^2h - 5c^5dfh + 22c^3d^2efh + 64cd^3e^2fh - 31c^4df^2h$$
$$- 93c^2d^2ef^2h + 7d^3e^2f^2h - 21c^3df^3h - 104cd^2ef^3h + 47c^2df^4h$$
$$- 10d^2ef^4h + 10cdf^5h - 10c^5dek - 13c^3d^2e^2k + 4cd^3e^3k + 10c^6fk$$
$$- 54c^4defk - 105c^2d^2e^2fk + 28d^3e^3fk + 67c^5f^2k + 128c^3def^2k$$
$$- 99cd^2e^2f^2k + 73c^4f^3k + 244c^2def^3k - 47d^2e^2f^3k - 73c^3f^4k$$
$$+ 114cdef^4k - 67c^2f^5k + 10def^5k - 10cf^6k + 40c^4de^2l - 28c^2d^2e^3l$$
$$- 40c^5efl + 176c^3de^2fl - 88cd^2e^3fl - 248c^4ef^2l + 246c^2de^2f^2l$$
$$- 12d^2e^3f^2l - 258c^3ef^3l + 104cde^2f^3l - 92c^2ef^4l + 10de^2f^4l$$
$$- 10cef^5l - 20c^3de^3m + 14cd^2e^4m + 70c^4e^2fm - 63c^2de^3fm$$
$$+ 2d^2e^4fm + 99c^3e^2f^2m - 44cde^3f^2m + 42c^2e^2f^3m - 5de^3f^3m$$
$$+ 5ce^2f^4m - 10c^4e^3n + 7c^2de^4n - 17c^3e^3fn + 8cde^4fn - 8c^2e^3f^2n$$
$$+ de^4f^2n - ce^3f^3n - c^2d^4ep + c^3d^3fp - 8cd^4efp + 8c^2d^3f^2p - 7d^4ef^2p$$
$$+ 17cd^3f^3p + 10d^3f^4p + 5c^3d^3eq - 2cd^4e^2q - 5c^4d^2fq + 44c^2d^3efq$$
$$- 14d^4e^2fq - 42c^3d^2f^2q + 63cd^3ef^2q - 99c^2d^2f^3q + 20d^3ef^3q$$
$$- 70cd^2f^4q - 10c^4d^2er + 12c^2d^3e^2r + 10c^5dfr - 104c^3d^2efr$$
$$+ 88cd^3e^2fr + 92c^4df^2r - 246c^2d^2ef^2r + 28d^3e^2f^2r + 258c^3df^3r$$
$$- 176cd^2ef^3r + 248c^2df^4r - 40d^2ef^4r + 40cdf^5r - 10c^5des + 47c^3d^2e^2s$$
$$- 28cd^3e^3s + 10c^6fs - 114c^4defs + 99c^2d^2e^2fs - 4d^3e^3fs + 67c^5f^2s$$
$$- 244c^3def^2s + 105cd^2e^2f^2s + 73c^4f^3s - 128c^2def^3s + 13d^2e^2f^3s$$
$$- 73c^3f^4s + 54cdef^4s - 67c^2f^5s + 10def^5s - 10cf^6s + 10c^4de^2u$$
$$- 7c^2d^2e^3u - 10c^5efu + 104c^3de^2fu - 64cd^2e^3fu - 47c^4ef^2u$$
$$+ 93c^2de^2f^2u - 9d^2e^3f^2u + 21c^3ef^3u - 22cde^2f^3u + 31c^2ef^4u$$
$$- 5de^2f^4u + 5cef^5u - 20c^3de^3v + 14cd^2e^4v + 10c^4e^2fv - 21c^2de^3fv$$

$$+ 2d^2e^4fv - 3c^3e^2f^2v + 4cde^3f^2v - 6c^2e^2f^3v + de^3f^3v - ce^2f^4v);$$

$$G_{24} = 5(2c^2d^4eg + 4d^5e^2g - 2c^3d^3fg - 6cd^4efg - 8c^2d^3f^2g$$

$$- 20d^4ef^2g + 10cd^3f^3g - 10c^3d^3eh - 18cd^4e^2h + 10c^4d^2fh$$

$$+ 26c^2d^3efh - 2d^4e^2fh + 42c^3d^2f^2h + 94cd^3ef^2h - 42c^2d^2f^3h$$

$$+ 10d^3ef^3h - 10cd^2f^4h + 20c^4d^2ek + 26c^2d^3e^2k - 8d^4e^3k - 20c^5dfk$$

$$- 32c^3d^2efk + 28cd^3e^2fk - 94c^4df^2k - 162c^2d^2ef^2k + 42d^3e^2f^2k$$

$$+ 42c^3df^3k - 104cd^2ef^3k + 62c^2df^4k - 10d^2ef^4k + 10cdf^5k - 20c^5del$$

$$- 6c^3d^2e^2l + 24cd^3e^3l + 20c^6fl - 8c^4defl - 66c^2d^2e^2fl + 8d^3e^3fl$$

$$+ 114c^5f^2l + 90c^3def^2l - 150cd^2e^2f^2l + 52c^4f^3l + 246c^2def^3l$$

$$- 42d^2e^2f^3l - 104c^3f^4l + 114cdef^4l - 72c^2f^5l + 10def^5l - 10cf^6l$$

$$+ 20c^4de^2m - 34c^2d^2e^3m + 4d^3e^4m - 70c^5efm + 153c^3de^2fm$$

$$- 34cd^2e^3fm - 169c^4ef^2m + 171c^2de^2f^2m - 12d^2e^3f^2m - 141c^3ef^3m$$

$$+ 59cde^2f^3m - 47c^2ef^4m + 5de^2f^4m - 5cef^5m + 10c^5e^2n - 17c^3de^3n$$

$$+ 2cd^2e^4n + 27c^4e^2fn - 27c^2de^3fn + 2d^2e^4fn + 25c^3e^2f^2n$$

$$- 11cde^3f^2n + 9c^2e^2f^3n - de^3f^3n + ce^2f^4n + 2cd^5ep - 2c^2d^4fp$$

$$+ 2d^5efp - 12cd^4f^2p - 10d^4f^3p - 10c^2d^4eq + 4d^5e^2q + 10c^3d^3fq$$

$$- 18cd^4efq + 64c^2d^3f^2q - 20d^4ef^2q + 70cd^3f^3q + 20c^3d^3er$$

$$- 24cd^4e^2r - 20c^4d^2fr + 68c^2d^3efr - 8d^4e^2fr - 144c^3d^2f^2r$$

$$+ 136cd^3ef^2r - 228c^2d^2f^3r + 40d^3ef^3r - 40cd^2f^4r - 20c^4d^2es$$

$$+ 54c^2d^3e^2s - 8d^4e^3s + 20c^5dfs - 128c^3d^2efs + 60cd^3e^2fs$$

$$+ 174c^4df^2s - 366c^2d^2ef^2s + 46d^3e^2f^2s + 414c^3df^3s - 272cd^2ef^3s$$

$$+ 226c^2df^4s - 30d^2ef^4s + 30cdf^5s - 10c^5deu + 17c^3d^2e^2u - 2cd^3e^3u$$

$$+ 10c^6fu - 124c^4defu + 165c^2d^2e^2fu - 18d^3e^3fu + 57c^5f^2u$$

$$- 239c^3def^2u + 99cd^2e^2f^2u + 26c^4f^3u - 29c^2def^3u - d^2e^2f^3u$$

$$- 52c^3f^4u + 37cdef^4u - 36c^2f^5u + 5def^5u - 5cf^6u + 20c^4de^2v$$

$$- 34c^2d^2e^3v + 4d^3e^4v - 10c^5efv + 51c^3de^2fv - 22cd^2e^3fv - 7c^4ef^2v$$

$$+ 9c^2de^2f^2v + 9c^3ef^3v - 7cde^2f^3v + 7c^2ef^4v - de^2f^4v + cef^5v);$$

$$G_{25} = -10c^2d^4fg - 20d^5efg + 10cd^4f^2g + 50c^3d^3fh + 90cd^4efh$$

$$- 40c^2d^3f^2h + 10d^4ef^2h - 10cd^3f^3h - 100c^4d^2fk - 130c^2d^3efk$$

$$+ 40d^4e^2fk + 30c^3d^2f^2k - 100cd^3ef^2k + 60c^2d^2f^3k - 10d^3ef^3k$$

$$+ 10cd^2f^4k + 100c^5dfl + 30c^3d^2efl - 120cd^3e^2fl + 70c^4df^2l$$

$$+ 210c^2d^2ef^2l - 40d^3e^2f^2l - 90c^3df^3l + 110cd^2ef^3l - 70c^2df^4l$$

$$+ 10d^2ef^4l - 10cdf^5l - 50c^6fm + 85c^4defm + 90c^2d^2e^2fm - 20d^3e^3fm$$

$$- 135c^5f^2m - 65c^3def^2m + 150cd^2e^2f^2m - 25c^4f^3m - 255c^2def^3m$$
$$+ 40d^2e^2f^3m + 125c^3f^4m - 115cdef^4m + 75c^2f^5m - 10def^5m$$
$$+ 10cf^6m + 10c^6en - 37c^4de^2n + 16c^2d^2e^3n + 47c^5efn - 95c^3de^2fn$$
$$+ 22cd^2e^3fn + 79c^4ef^2n - 81c^2de^2f^2n + 6d^2e^3f^2n + 59c^3ef^3n$$
$$- 25cde^2f^3n + 19c^2ef^4n - 2de^2f^4n + 2cef^5n - 10cd^5fp - 10d^5f^2p$$
$$+ 50c^2d^4fq - 20d^5efq + 70cd^4f^2q - 100c^3d^3fr + 120cd^4efr$$
$$- 220c^2d^3f^2r + 40d^4ef^2r - 40cd^3f^3r + 100c^4d^2fs - 270c^2d^3efs$$
$$+ 40d^4e^2fs + 370c^3d^2f^2s - 260cd^3ef^2s + 220c^2d^2f^3s - 30d^3ef^3s$$
$$+ 30cd^2f^4s - 50c^5dfu + 285c^3d^2efu - 150cd^3e^2fu - 335c^4df^2u$$
$$+ 615c^2d^2ef^2u - 70d^3e^2f^2u - 465c^3df^3u + 275cd^2ef^3u - 205c^2df^4u$$
$$+ 25d^2ef^4u - 25cdf^5u + 20c^5dev - 74c^3d^2e^2v + 32cd^3e^3v - 10c^6fv$$
$$+ 111c^4defv - 132c^2d^2e^2fv + 12d^3e^3fv - 27c^5f^2v + 105c^3def^2v$$
$$- 36cd^2e^2f^2v - 5c^4f^3v - c^2def^3v + 2d^2e^2f^3v + 25c^3f^4v - 17cdef^4v$$
$$+ 15c^2f^5v - 2def^5v + 2cf^6v;$$

$$G_{26} = 10c^2d^5g + 20d^6eg - 10cd^5fg - 50c^3d^4h - 90cd^5eh + 40c^2d^4fh$$
$$- 10d^5efh + 10cd^4f^2h + 100c^4d^3k + 130c^2d^4ek - 40d^5e^2k - 30c^3d^3fk$$
$$+ 100cd^4efk - 60c^2d^3f^2k + 10d^4ef^2k - 10cd^3f^3k - 100c^5d^2l$$
$$- 30c^3d^3el + 120cd^4e^2l - 70c^4d^2fl - 210c^2d^3efl + 40d^4e^2fl$$
$$+ 90c^3d^2f^2l - 110cd^3ef^2l + 70c^2d^2f^3l - 10d^3ef^3l + 10cd^2f^4l$$
$$+ 50c^6dm - 85c^4d^2em - 90c^2d^3e^2m + 20d^4e^3m + 135c^5dfm$$
$$+ 65c^3d^2efm - 150cd^3e^2fm + 25c^4df^2m + 255c^2d^2ef^2m - 40d^3e^2f^2m$$
$$- 125c^3df^3m + 115cd^2ef^3m - 75c^2df^4m + 10d^2ef^4m - 10cdf^5m$$
$$- 10c^7n + 67c^5den - 27c^3d^2e^2n - 22cd^3e^3n - 77c^6fn + 167c^4defn$$
$$|\ 81c^2d^2e^2fn - 22d^3e^3fn - 140c^5f^2n - 59c^3def^2n + 147cd^2e^2f^2n$$
$$- 265c^2def^3n + 39d^2e^2f^3n + 140c^3f^4n - 116cdef^4n + 77c^2f^5n$$
$$- 10def^5n + 10cf^6n + 10cd^6p + 10d^6fp - 50c^2d^5q + 20d^6eq$$
$$- 70cd^5fq + 100c^3d^4r - 120cd^5er + 220c^2d^4fr - 40d^5efr + 40cd^4f^2r$$
$$- 100c^4d^3s + 270c^2d^4es - 40d^5e^2s - 370c^3d^3fs + 260cd^4efs$$
$$- 220c^2d^3f^2s + 30d^4ef^2s - 30cd^3f^3s + 50c^5d^2u - 285c^3d^3eu$$
$$+ 150cd^4e^2u + 335c^4d^2fu - 615c^2d^3efu + 70d^4e^2fu + 465c^3d^2f^2u$$
$$- 275cd^3ef^2u + 205c^2d^2f^3u - 25d^3ef^3u + 25cd^2f^4u - 10c^6dv$$
$$+ 137c^4d^2ev - 186c^2d^3e^2v + 20d^4e^3v - 147c^5dfv + 635c^3d^2efv$$
$$- 282cd^3e^2fv - 449c^4df^2v + 741c^2d^2ef^2v - 76d^3e^2f^2v - 479c^3df^3v+$$

$$265cd^2ef^3v - 189c^2df^4v + 22d^2ef^4v - 22cdf^5v;$$

$$\sigma_{20} = 2e^3(de - cf)(-5c^2 + 16de - 26cf - 5f^2)(-2c^2 + de - 5cf - 2f^2);$$

$$\sigma_{21} = e^2(c - f)(de - cf)(-5c^2 + 16de - 26cf - 5f^2)(-2c^2 + de - 5cf \\ - 2f^2);$$

$$\sigma_{22} = 2e(de - cf)(-5c^2 + 16de - 26cf - 5f^2)(-2c^2 + de - 5cf - 2f^2) \cdot \\ \cdot (-c^2 + de + 2cf - f^2);$$

$$\sigma_{23} = (c - f)(-de + cf)(c^2 - 6de - 2cf + f^2)(2c^2 - de + 5cf + 2f^2) \cdot \\ \cdot (5c^2 - 16de + 26cf + 5f^2);$$

$$\sigma_{24} = 2d(de - cf)(-5c^2 + 16de - 26cf - 5f^2)(-2c^2 + de - 5cf - 2f^2) \cdot \\ \cdot (-c^2 + de + 2cf - f^2);$$

$$\sigma_{25} = d^2(c - f)(de - cf)(-5c^2 + 16de - 26cf - 5f^2)(-2c^2 + de - 5cf \\ - 2f^2);$$

$$\sigma_{26} = 2d^3(de - cf)(-5c^2 + 16de - 26cf - 5f^2)(-2c^2 + de - 5cf - 2f^2);$$

$$\begin{aligned} D_{20} = {}&-60c^5de + 222c^3d^2e^2 - 96cd^3e^3 + 60c^6f - 744c^4def \\ &+ 954c^2d^2e^2f - 96d^3e^3f + 522c^5f^2 - 2220c^3def^2 + 954cd^2e^2f^2 \\ &+ 1362c^4f^3 - 2220c^2def^3 + 222d^2e^2f^3 + 1362c^3f^4 - 744cdef^4 \\ &+ 522c^2f^5 - 60def^5 + 60cf^6; \end{aligned}$$

$$\begin{aligned} D_{21} = {}&60c^5de - 222c^3d^2e^2 + 96cd^3e^3 - 60c^6f + 744c^4def - 954c^2d^2e^2f \\ &+ 96d^3e^3f - 522c^5f^2 + 2220c^3def^2 - 954cd^2e^2f^2 - 1362c^4f^3 \\ &+ 2220c^2def^3 - 222d^2e^2f^3 - 1362c^3f^4 + 744cdef^4 - 522c^2f^5 \\ &+ 60def^5 - 60cf^6; \end{aligned}$$

$$\begin{aligned} D_{22} = 5({}&60c^5de - 222c^3d^2e^2 + 96cd^3e^3 - 60c^6f + 744c^4def \\ &- 954c^2d^2e^2f + 96d^3e^3f - 522c^5f^2 + 2220c^3def^2 - 954cd^2e^2f^2 \\ &- 1362c^4f^3 + 2220c^2def^3 - 222d^2e^2f^3 - 1362c^3f^4 + 744cdef^4 \\ &- 522c^2f^5 + 60def^5 - 60cf^6); \end{aligned}$$

$$\begin{aligned} D_{23} = 10({}&-60c^5de + 222c^3d^2e^2 - 96cd^3e^3 + 60c^6f - 744c^4def \\ &+ 954c^2d^2e^2f - 96d^3e^3f + 522c^5f^2 - 2220c^3def^2 + 954cd^2e^2f^2 \\ &+ 1362c^4f^3 - 2220c^2def^3 + 222d^2e^2f^3 + 1362c^3f^4 - 744cdef^4 \\ &+ 522c^2f^5 - 60def^5 + 60cf^6); \end{aligned}$$

$$\begin{aligned} D_{24} = 5({}&-60c^5de + 222c^3d^2e^2 - 96cd^3e^3 + 60c^6f - 744c^4def \\ &+ 954c^2d^2e^2f - 96d^3e^3f + 522c^5f^2 - 2220c^3def^2 + 954cd^2e^2f^2 \\ &+ 1362c^4f^3 - 2220c^2def^3 + 222d^2e^2f^3 + 1362c^3f^4 - 744cdef^4 \\ &+ 522c^2f^5 - 60def^5 + 60cf^6); \end{aligned}$$

$$D_{25} = 60c^5de - 222c^3d^2e^2 + 96cd^3e^3 - 60c^6f + 744c^4def - 954c^2d^2e^2f$$
$$+ 96d^3e^3f - 522c^5f^2 + 2220c^3def^2 - 954cd^2e^2f^2 - 1362c^4f^3$$
$$+ 2220c^2def^3 - 222d^2e^2f^3 - 1362c^3f^4 + 744cdef^4 - 522c^2f^5$$
$$+ 60def^5 - 60cf^6;$$

$$D_{26} = 60c^5de - 222c^3d^2e^2 + 96cd^3e^3 - 60c^6f + 744c^4def - 954c^2d^2e^2f$$
$$+ 96d^3e^3f - 522c^5f^2 + 2220c^3def^2 - 954cd^2e^2f^2 - 1362c^4f^3$$
$$+ 2220c^2def^3 - 222d^2e^2f^3 - 1362c^3f^4 + 744cdef^4 - 522c^2f^5$$
$$+ 60def^5 - 60cf^6.$$

Appendix 10

Matrices that Define a System of Linear Equations $\tilde{A}_2 \tilde{B}_2 = \tilde{C}_2$ for the Quantity G_2 in Case of the Differential System $s(1,2,3)$

$$\tilde{A}_2 = [\tilde{A}_2'|\tilde{A}_2''|\tilde{A}_2'''|\tilde{A}_2''''|\tilde{A}_2'''''],$$

$$\tilde{A}_2' = \begin{pmatrix}
3c & 3e & 0 & 0 & 0 & 0 & 0 \\
3d & 6c+3f & 6e & 0 & 0 & 0 & 0 \\
0 & 6d & 3c+6f & 3e & 0 & 0 & 0 \\
0 & 0 & 3d & 3f & 0 & 0 & 0 \\
3g & 3l & 0 & 0 & 4c & 4e & 0 \\
6h & 6g+6m & 6l & 0 & 4d & 12c+4f & 12e \\
3k & 12h+3n & 3g+12m & 3l & 0 & 12d & 12c+12f \\
0 & 6k & 6h+6n & 6m & 0 & 0 & 12d \\
0 & 0 & 3k & 3n & 0 & 0 & 0 \\
3p & 3t & 0 & 0 & 4g & 4l & 0 \\
9q & 6p+9u & 6t & 0 & 8h & 12g+8m & 12l \\
9r & 18q+9v & 3p+18u & 3t & 4k & 24h+4n & 12g+24m \\
3s & 18r+3w & 9q+18v & 9u & 0 & 12k & 24h+12n \\
0 & 6s & 9r+6w & 9v & 0 & 0 & 12k \\
0 & 0 & 3s & 3w & 0 & 0 & 0 \\
0 & 0 & 0 & 0 & 4p & 4t & 0 \\
0 & 0 & 0 & 0 & 12q & 12p+12u & 12t \\
0 & 0 & 0 & 0 & 12r & 36q+12v & 12p+36u \\
0 & 0 & 0 & 0 & 4s & 36r+4w & 36q+36v \\
0 & 0 & 0 & 0 & 0 & 12s & 36r+12w \\
0 & 0 & 0 & 0 & 0 & 0 & 12s \\
0 & 0 & 0 & 0 & 0 & 0 & 0
\end{pmatrix},$$

$$
\tilde{A}_2'' =
\begin{pmatrix}
0 & 0 & 0 & 0 & 0 \\
0 & 0 & 0 & 0 & 0 \\
0 & 0 & 0 & 0 & 0 \\
0 & 0 & 0 & 0 & 0 \\
0 & 0 & -e^2 & 0 & 0 \\
0 & 0 & 2ce - 2ef & 0 & 0 \\
12e & 0 & -c^2 + 2de + 2cf - f^2 & 0 & 0 \\
4c + 12f & 4e & -2cd + 2df & 0 & 0 \\
4d & 4f & -d^2 & 0 & 0 \\
0 & 0 & 0 & 5c & 5e \\
0 & 0 & 0 & 5d & 20c + 5f \\
12l & 0 & 0 & 0 & 20d \\
4g + 24m & 4l & 0 & 0 & 0 \\
8h + 12n & 8m & 0 & 0 & 0 \\
4k & 4n & 0 & 0 & 0 \\
0 & 0 & 0 & 5g & 5l \\
0 & 0 & 0 & 10h & 20g + 10m \\
12t & 0 & 0 & 5k & 40h + 5n \\
4p + 36u & 4t & 0 & 0 & 20k \\
12q + 36v & 12u & 0 & 0 & 0 \\
12r + 12w & 12v & 0 & 0 & 0 \\
4s & 4w & 0 & 0 & 0
\end{pmatrix},
$$

$$
\tilde{A}_2''' =
\begin{pmatrix}
0 & 0 & 0 & 0 & 0 \\
0 & 0 & 0 & 0 & 0 \\
0 & 0 & 0 & 0 & 0 \\
0 & 0 & 0 & 0 & 0 \\
0 & 0 & 0 & 0 & 0 \\
0 & 0 & 0 & 0 & 0 \\
0 & 0 & 0 & 0 & 0 \\
0 & 0 & 0 & 0 & 0 \\
0 & 0 & 0 & 0 & 0 \\
0 & 0 & 0 & 0 & 0 \\
20e & 0 & 0 & 0 & 0 \\
30c + 20f & 30e & 0 & 0 & 0 \\
30d & 20c + 30f & 20e & 0 & 0 \\
0 & 20d & 5c + 20f & 5e & 0 \\
0 & 0 & 5d & 5f & 0 \\
0 & 0 & 0 & 0 & 6c \\
20l & 0 & 0 & 0 & 6d \\
30g + 40m & 30l & 0 & 0 & 0 \\
60h + 20n & 20g + 60m & 20l & 0 & 0 \\
30k & 40h + 30n & 5g + 40m & 5l & 0 \\
0 & 20k & 10h + 20n & 10m & 0 \\
0 & 0 & 5k & 5n & 0
\end{pmatrix},
$$

$$
\widetilde{A}_2'''' =
\begin{pmatrix}
0 & 0 & 0 & 0 & 0 \\
0 & 0 & 0 & 0 & 0 \\
0 & 0 & 0 & 0 & 0 \\
0 & 0 & 0 & 0 & 0 \\
0 & 0 & 0 & 0 & 0 \\
0 & 0 & 0 & 0 & 0 \\
0 & 0 & 0 & 0 & 0 \\
0 & 0 & 0 & 0 & 0 \\
0 & 0 & 0 & 0 & 0 \\
0 & 0 & 0 & 0 & 0 \\
0 & 0 & 0 & 0 & 0 \\
0 & 0 & 0 & 0 & 0 \\
0 & 0 & 0 & 0 & 0 \\
0 & 0 & 0 & 0 & 0 \\
0 & 0 & 0 & 0 & 0 \\
6e & 0 & 0 & 0 & 0 \\
30c + 6f & 30e & 0 & 0 & 0 \\
30d & 60c + 30f & 60e & 0 & 0 \\
0 & 60d & 60c + 60f & 60e & 0 \\
0 & 0 & 60d & 30c + 60f & 30e \\
0 & 0 & 0 & 30d & 6c + 30f \\
0 & 0 & 0 & 0 & 6d
\end{pmatrix},
$$

$$
\widetilde{A}_2''''' =
\begin{pmatrix}
0 & 0 \\
0 & 0 \\
0 & 0 \\
0 & 0 \\
0 & 0 \\
0 & 0 \\
0 & 0 \\
0 & 0 \\
0 & 0 \\
0 & 0 \\
0 & 0 \\
0 & 0 \\
0 & 0 \\
0 & 0 \\
0 & 0 \\
0 & e^3 \\
0 & 3e^2(f - c) \\
0 & 3e[(c + f)^2 - de] \\
0 & -(c - f)[(c - f)^2 - 6de] \\
0 & 3d[-(c - f)^2 + de] \\
6e & 3d^2(f - c) \\
6f & -d^3
\end{pmatrix},
$$

$$
\widetilde{B}_2 = \begin{pmatrix} a_0 \\ a_1 \\ a_2 \\ a_3 \\ b0 \\ b1 \\ b2 \\ b3 \\ b4 \\ G_1 \\ c_0 \\ c_1 \\ c_2 \\ c_3 \\ c_4 \\ c_5 \\ d_0 \\ d_1 \\ d_2 \\ d_3 \\ d_4 \\ d_5 \\ d_6 \\ G_2 \end{pmatrix}, \quad
\widetilde{C}_2 = \begin{pmatrix} 2eg - cl + fl \\ -cg + fg + 4eh - 2dl - 2cm + 2fm \\ -2ch + 2fh + 2ek - 4dm - cn + fn \\ -ck + fk - 2dn \\ 2ep - ct + ft \\ -cp + fp + 6eq - 2dt - 3cu + 3fu \\ -3cq + 3fq + 6er - 6du - 3cv + 3fv \\ -3cr + 3fr + 2es - 6dv - cw + fw \\ -cs + fs - 2dw \\ 0 \\ 0 \\ 0 \\ 0 \\ 0 \\ 0 \\ 0 \\ 0 \\ 0 \\ 0 \\ 0 \\ 0 \\ 0 \\ 0 \end{pmatrix}.
$$

Bibliography

[1] Alekseev V. G. *Theory of rational invariants of binary forms.* Yuryev, 1899, 232 p. (in Russian).

[2] Arnold V. I., Afraimovich V. S., Ilyashenko Yu. S., Shilnikov L. P. *Bifurcation Theory. Results of science and technology. Modern math problems. Fundamental directions.* V.5, Moscow, VINITI, 1986, 218 p. (in Russian).

[3] Arzhantsev I. V. *Graduated algebras and the 14th Hilbert problem.* Textbook, Moscow, Ed. MCCME, 2009, 62 p. (in Russian).

[4] Baltag V. A. *Topological classification of cubic zero-systems of differential equations.* Preprint, Institute of Mathematics with Computing Center., Chişinău, 1989, 59 p. (in Russian).

[5] Bautin N. N. On the number of limit cycles which appear with the variation of coefficients from equilibrium position of focus or center type. *Math. Sb.*, 1952, vol. 30(72), pp. 181–196 (Amer. Math. Soc. Transl., 1954, vol. 100, pp. 397–413).

[6] Calin Iu., Ciubotaru S. $GL(2, R)$-comitants and Lyapunov quantities for bidimensional polynomial systems of differential equations with nonlinearities of the fourth degree. *Conference Mathematics and Information Technologies: Research and Education (MITRE–2016)*, Abstracts, Chişinău, June 23–26, 2016, pp. 13–15.

[7] Calin Iu., Ciubotaru S. The center conditions for a class of bidimensional polynomial systems of differential equations with nonlinearities of the fourth degree. *Conference Mathematics and Information Technologies: Research and Education (MITRE–2016)*, Abstracts, Chişinău, June 23–26, 2016, pp. 22–23.

[8] Calin Iu., Baltag V. The invariant center conditions for a class of cubic systems of differential equations. *Conference Mathematics and Information Technologies: Research and Education (MITRE–2009)*, Abstracts, Chişinău, October 8-9, 2009, pp. 6–7.

[9] Ciobanu M., Rotaru T. 130 years of trouble to solve Poincare's problem about center and focus. *Akademos. Sci. J. Innov. Cult. Art, ASM*, 2013, vol. 3(30), pp. 13–21. (in Romanian).

[10] Cozma D. *Integrability of Cubic Systems with Invariant Straight Lines and Invariant Conics.* Chişinău, Shtiintsa, 2013, 240 p.

[11] Derksen H. *Constructive Invariant Theory.* In H. E. A. Eddy Campbell and D. L. Wehlau (Eds.) *Invariant Theory in All Characteristics.* CRM Proceedings and Lecture Notes, Vol. 35, American Mathematical Society, Providence, RI, 2004, pp. 11–36.

[12] Derksen Harm, Kemper Gregor. *Computational Invariant Theory.* Encyclopaedia of Mathematical Sciences, vol. 130, Springer-Verlag, Berlin, 2002, 268 p.

[13] Dubé Thomas. *Inductive Proof of Macaulay's Theorem.* Technical Report, №,455, Robotics Report №,202, June, 1989, New-York University. Dep. of Computer Science. Courant Institute of Mathematical Sciences.

[14] Franklin F. On the calculation of the generating function and tabs if ground-forms for binary quantities. *Am. J. Math.*, 1880, vol. 3(1), pp. 128–153.

[15] Graf V., Bothmer H. C., Kröker J. Focal Values of Plane Cubic Centers. *Qual. Theory Dyn. Syst.*, 2010, vol. 9, pp. 319–324.

[16] Gurevich G. B. *Foundations of the theory of algebraic invariants.* Groningen: Noordhoff 1964, 429 p.

[17] Gherştega N. N., Popa M. N., Pricop V. V. Generators of the algebras of invariants for differential systems with homogeneous nonlinearities of odd degree. *Bul. Acad. de Ştiinţe a Repub. Moldova, Matematica*, 2012, vol. 2(69), pp. 43–58.

[18] Gherştega N. N., Popa M. N., Pricop V. V. About characteristic of graded algebras $S_{1,4}$ and $SI_{1,4}$. *Bul. Acad. de Ştiinţe a Repub. Moldova, Matematica*, 2010, vol. 1(62), pp. 23–32.

[19] Ilyashenko Yu. S. The occurrence of limit cycles in perturbation of the equation $dw/dz = -Rz/Rw$, where $R(z,w)$ is a polynomial. *Mat. Sat*, 1969, vol. 78, pp. 360–373. (in Russian).

[20] Liapounoff A. Probléme général de la stabilité du mouvement. *Annales de la Faculté des Sciences de Touluose*, 1907, vol. 2(9), pp. 204–477. Reproduction in Annals of Mathematics Studies 17, Princenton: Princenton University Press, 1947, reprinted, Kraus Reprint Corporation, New York, 1965.

[21] Lunkevich V. A., Sibirsky K. S. Quadratic zero-systems of differential equations. *Diff. equations*, 1989, Minsk, vol. 25(6), pp. 1056-1058. (in Russian).

[22] Markushevich A. I. *Short course of the theory of analytic functions.* Moscow, Nauka, 1978, 415 p. (in Russian).

[23] *Mathematical encyclopedia. Center and focus problem.* https://www.encyclopediaofmath.org//index.php?title=Centre_and_ focus_problem&oldid=13304.

[24] Ovsyannikov L. V. *Group analysis of differential equations.* Academic Press, 1982, 432 p.

[25] Poincaré Henri. *Science and hypothesis. Included in the book ABOUT SCIENCE.* Moscow, Nauka, The main edition of the physical and mathematical literature, 1983, 560 p. (in Russian). http://ilib.mccme.ru/Poincare/O-nauke.htm

[26] Poincaré H. Mémoire sur les curbes définies par une équation différentielle. *J. Math. Pure et Appl.* (Sér 3), 1881, vol. 7, pp. 375–422; (Sér 3), 1882, vol. 8, pp. 251–296; (Sér 4), 1885, vol. 1, pp. 167–244; (Sér 4), 1886, vol. 2, pp. 151–217.

[27] Pontryagin L. S. *Ordinary differential equations.* Nauka, Moscow, 1974, 331 p. (in Russian).

[28] Popa M., Repeşco V. *Lie algebras and dynamical systems in plane.* Optional course. Tiraspol State University, Chişinău, Moldova, 2016, 237 p. (in Romanian).

[29] Popa M. N., Pricop V. V. Applications of algebraic methods in solving the center-focus problem. *Bul. Acad. de Ştiinţe a Repub. Moldova, Matematica,* 2013, vol. 1(71), pp. 45–71. http://arxiv.org/abs/1310.4343.pdf

[30] Popa M. N., Pricop V. V. About a solution of the center-focus problem. *XV International Scientific Conference on Differential Equations, Erugin Readings, 2013,* Yanka Kupala State University of Grodno, Abstracts, Part I, Grodno, May 13-16, 2013, pp. 69–70.

[31] Popa M. N. Applications of algebraic methods to the center-focus problem. *The 20th Conference on Applied and Industrial Mathematics Dedicated to Academician Mitrofan M. Ciobanu, CAIM*-2012, Communications, Chişinău, August 22-25, 2012, pp. 184–186.

[32] Popa M. N., Pricop V. V. Applications of generating functions and Hilbert series to the center-focus problem. *The 8th International Algebraic Conference in Ukraine Dedicated to the Memory of Professor Vitaliy Mikhaylovich Usenco,* Lugansk Taras Shevcenko National University, IACONU 2011, Book of abstracts, July 5-12, Lugansk, 2011, p. 10.

[33] Popa M. N. Algebraic methods for differential systems. *Piteshti Univ. Ser. Appl. Ind. Math.,* 2004, vol. 15, p. 340 (in Romanian).

[34] Popa M. N. *Applications of algebras to differential systems.* Chişinău, IMI, ASM, 2001, 224 p.(in Russian).

[35] Sadovsky A. P. *Polynomial ideals and varieties.* Minsk, BSU, 2008, 199 p. (in Russian).

[36] Saint Luke (Voyno-Yasenetsky). *Spirit, Soul, Body.* Ed. St. Leo, Pope of Rome, 2002 (With the blessing of His Beatitude Metropolitan of Kiev and All Ukraine), 149 p. (in Russian).

[37] Sibirsky K. S. *Introduction to the algebraic theory of invariants of differential equations.* Manchester-New York, Manchester University Press 1988. VII, 169 p., ISBN 0-7190-2069-5 (Nonlinear Science: Theory and Applications).

[38] Sibirsky K. S. *Algebraic invariants of differential equations and matrices.* Chişinău, Shtiintsa, 1976, 268 p. (in Russian).

[39] Sibirsky K. S. *The method of invariants in the qualitative theory of differential equations.* Chişinău, RIO AN MSSR, 1968, 184 p. (in Russian).

[40] Springer T. A. *Invariant Theory.* Springer-Verlag Berlin Heidelberg, 1977, 120 p.

[41] Ufnarovskij V. A. Combinatorial and Asymptotic Methods in Algebra. In Algebra VI, A. I. Kostrikin and I. R. Shafarevich (Eds.), *Encyclopaedia of Mathematical Sciences*, Vol. 57, Springer, Berlin, New York, 1995, 196 p.

[42] Vulpe N. I., Macari P. M., Popa M. N. An integral algebraic basis of centro-affine invariants of differential system with homogeneous right-hand sides of the fourth order. *Bul. Acad. de Ştiinţe a Repub. Moldova, Matematica*, 1996, vol. 1(20), pp. 48–55. (in Russian).

[43] Vulpe N. I. *Polynomial bases of comitants of differential systems and their applications in qualitative theory.* Chişinău, Shtiintsa, 1986, 171 p. (in Russian).

[44] Vulpe N. I. *An integral algebraic basis of centro-affine invariants of a homogeneous cubic differential system.* In the book: Algebraic invariants of dynamical systems. Mat. Issled., no. 55, Chişinău, Shtiintsa, 1980, pp. 37–45. (in Russian).

[45] Żołądek H. On certain generalization of the Bautin's theorem. *Nonlinearity*, 1994, vol. 7, pp. 273–279.

Index

Printed in the United States
by Baker & Taylor Publisher Services